Models, Simulations, and Representations

Routledge Studies in the Philosophy of Science

1 **Evolution, Rationality and Cognition**
 A Cognitive Science for the Twenty-First Century
 Edited by António Zilhão

2 **Conceptual Systems**
 Harold I. Brown

3 **Nancy Cartwright's Philosophy of Science**
 Edited by Stephan Hartmann, Carl Hoefer, and Luc Bovens

4 **Fictions in Science**
 Philosophical Essays on Modeling and Idealization
 Edited by Mauricio Suárez

5 **Karl Popper's Philosophy of Science**
 Rationality without Foundations
 Stefano Gattei

6 **Emergence in Science and Philosophy**
 Edited by Antonella Corradini and Timothy O'Connor

7 **Popper's Critical Rationalism**
 A Philosophical Investigation
 Darrell Rowbottom

8 **Conservative Reductionism**
 Michael Esfeld and Christian Sachse

9 **Models, Simulations, and Representations**
 Edited by Paul Humphreys and Cyrille Imbert

Models, Simulations, and Representations

Edited by Paul Humphreys and Cyrille Imbert

First published 2012
by Routledge
711 Third Avenue, New York, NY 10017

Simultaneously published in the UK
by Routledge
2 Park Square, Milton Park, Abingdon, Oxon OX14 4RN

*Routledge is an imprint of the Taylor & Francis Group,
an informa business*

© 2012 Taylor & Francis

The right of Paul Humphreys and Cyrille Imbert to be identified as the authors of the editorial material, and of the authors for their individual chapters, has been asserted by them in accordance with sections 77 and 78 of the Copyright, Designs and Patents Act 1988.

Typeset in Sabon by IBT Global.
Printed and bound in the United States of America on acid-free paper by IBT Global.

All rights reserved. No part of this book may be reprinted or reproduced or utilised in any form or by any electronic, mechanical, or other means, now known or hereafter invented, including photocopying and recording, or in any information storage or retrieval system, without permission in writing from the publishers.

Trademark Notice: Product or corporate names may be trademarks or registered trademarks, and are used only for identification and explanation without intent to infringe.

Library of Congress Cataloging-in-Publication Data
　Models, simulations, and representations / edited by Paul Humphreys and Cyrille Imbert.
　　p. cm. — (Routledge studies in the philosophy of science ; 9)
　Includes bibliographical references and index.
　1. Science—Philosophy.　I. Humphreys, Paul.　II. Imbert, Cyrille, 1977–
　Q175.3.M624 2011
　003—dc22
　　　　　　　　　　　　　　　　　2011003850

ISBN13: 978-0-415-89196-7 (hbk)
ISBN13: 978-0-203-80841-2 (ebk)

To Patrick Suppes, pioneer. [PH]

To Marianne, who has grown up over the months with this volume. [CI]

Contents

List of Figures   ix
List of Tables   xi
Preface   xiii

PART I
Models

1   The Productive Tension: Mechanisms vs. Templates in Modeling the Phenomena   3
TARJA KNUUTTILA AND ANDREA LOETTGERS

2   Theories or Models? The Case of Algebraic Quantum Field Theory   25
TRACY LUPHER

3   Modeling and Experimenting   42
ISABELLE PESCHARD

4   Model Dynamics: Epistemological Perspectives on Science and Its Educational Practices   62
MICHAEL STÖLTZNER

PART II
Simulations

5   Weak Emergence and Computer Simulation   91
MARK A. BEDAU

6	Agent-Based Modeling and the Fallacies of Individualism BRIAN EPSTEIN	115
7	Scientific Models, Simulation, and the Experimenter's Regress AXEL GELFERT	145
8	Simulation and the Sense of Understanding JAAKKO KUORIKOSKI	168
9	Models and Simulations in Brain Experiments PATRICK SUPPES	188

PART III
Representation

10	Representing with Physical Models RONALD N. GIERE	209
11	The Truth of False Idealizations in Modeling USKALI MÄKI	216
12	Idealized Models as Inferentially Veridical Representations: A Conceptual Framework JUHA SAATSI	234
13	Formats of Representation in Scientific Theorizing MARION VORMS	250
	Contributors	275
	Index	279

Figures

2.1	Isotony.	28
2.2	Microcausality.	28
3.1	Wake behind a cylinder, seen from above.	43
4.1	Scheme of a learning process from Clement (2000, 1042).	67
4.2	An epistemological supposition of the nature of our conceptions about the entities in the world.	79
4.3	A series of mental models of the Earth.	82
5.1	Population dynamics and mutation rate.	104
6.1	Macroentity represented as a hexagon.	116
6.2	Interaction among individual agents.	116
6.3	Multiscale organization with employee subagents.	118
6.4	Ontological dependence and causal interactions.	122
6.5	Radically heterogeneous set of other entities and properties.	125
9.1	The 10-20 system of EEG sensors.	190
9.2	The EEG recording of one sensor to hearing the spoken word 'circle'.	190
9.3	Miller-Nicely data on the left and brain data on the right.	198
9.4	Invariant partial order that is the intersection of the perceptual and brain-data semiorders.	198
9.5	Trees derived from the two semiorders, one perceptual and the other brain data, with the perceptual (Miller and Nicely 1955) results on the left and the brain results on the right.	199
9.6	Prototypes and test samples of brain waves of six names of capitals or countries.	199
12.1	A schematic representation of a model, comprising a model description and a model system (modified from Giere 1988).	235
13.1	By John Kulvicki, "Knowing with Images: Medium and Message," *Philosophy of Science* 77: 303 (2010), University of Chicago Press.	252
13.2	By John Kulvicki, "Knowing with Images: Medium and Message," *Philosophy of Science* 77: 304 (2010), University of Chicago Press.	252

13.3 By John Kulvicki, "Knowing with Images: Medium and Message," *Philosophy of Science* 77: 305 (2010), University of Chicago Press. 252
13.4 Diagram and its corresponding equation. 255
13.5 Evolution of temperature in Paris. 259

Tables

9.1 Results for the "Stop-Go" Experiment — 193
9.2 Data of Perceptual and Brain Confusion Matrices with Perceptual Data on the Left for Each Entry and Brain Data on the Right — 195
9.3 Conditional Probability Matrices for Miller-Nicely Perception Experiment and the Brain Experiment, with Perception Probabilities on the Left and Brain Data on the Right — 197

Preface

Over the course of several decades, scientific models have taken their place alongside scientific theories as the main vehicles for representing parts of the world that lend themselves to scientific investigation. Models differ from theories in being less general and in being tailored to specific applications. The application of models also requires that greater attention be paid to the role of approximations, idealizations, and simplifications than does the investigation of the logical structure of theories. More recently, philosophers of science have recognized that computer simulations have become one of the most widely used methods of investigation in many sciences, ranging from astrophysics to sociology, and that the scientific use of computers can also create epistemological problems of its own. Although many simulations are based on models, they are distinctively different in that they allow the exploration of phenomena that lie beyond the reach of traditional a priori methods of mathematics, in being dynamically implemented on concrete machines rather than directly by humans, in requiring different methods of justification and validation, and in encouraging different modes of representation, the most prominent of which are visualization techniques. Despite these differences, models and simulations raise similar questions, even if the answers may differ in each case, and the simultaneous study of models and simulations now constitutes an actively investigated and rapidly expanding philosophical field.

Models and simulations both pose essentially philosophical problems about the nature of representation. Because the central representational questions and answers in these areas tend to differ significantly from related questions and answers in the philosophy of mind, where representation and intentionality are much debated topics, and in philosophy of language, where the focus is rarely upon scientific languages and representations, new solutions to problems of representation must be developed within the philosophy of science. This volume brings together some of the best work by established senior philosophers and by younger researchers, with the organization of the chapters arranged so that the similarities and differences of these methods can be examined. All of the articles are previously unpublished.

xiv *Preface*

In Part I, Tarja Knuuttila and Andrea Loettgers explore the merits of two different styles of modeling: those that emphasize the mechanisms producing the data and those that abstract from the particularities of the mechanisms to emphasize formal similarities. This highlights the fact that, as far as modeling is concerned, the units of scientific reasoning are sometimes wider than representations of particular target systems, the focus of many philosophical studies about models. The authors use as a running example what is now called the Lotka-Volterra model, which is widely used in population biology to model fluctuations in the populations of predators and their prey and also in political science to model levels of defense expenditures. In addition to their philosophical claims, the authors make an important historical point in showing that Lotka and Volterra arrived at similar formal models using very different motivations corresponding to the two styles mentioned above.

In his contribution, Tracy Lupher considers various criteria for deciding whether two theories or two models are equivalent, using the framework of algebraic quantum theory to provide a sharp representational apparatus within which to consider the question. Because most of the discussions of physical theories and models have taken place within the context of classical mechanics, traditional quantum mechanics, or theories of space-time, this chapter is a valuable contribution to the growing but still under-developed area of theory and model identity.

Isabelle Peschard's chapter also explores the relations between theories and models, this time within the context of models of turbulence, which are among the most intensively studied models in science. One of the more interesting aspects of her article is the example of a feature occurring in a region between stable flow and turbulent flow that is only revealed when a particular type of visual representation is used. Peschard's analysis also raises the normative question of whether certain 'data' should be ignored on the grounds that they are artifacts of the apparatus, an issue that has been discussed in the context of experimentation by Ian Hacking and Peter Galison. Overall, her chapter demonstrates that selecting the relevant parameters for the model-based or experimental study of some phenomenon is a creative process of conceptual innovation.

Michael Stöltzner's chapter is concerned with the role of models in education and in particular with the dynamics of learning an increasingly sophisticated hierarchy of models as students move through the educational process. Among other topics, he discusses whether models in the sequence are incommensurable with one another and whether intermediate level models are merely heuristic devices on the way to the expert level model or are essential tools of learning that enable students to eventually construct their own models. He also displays evidence that historical models follow a Lakatosian research program and that textbooks would provide a clearer presentation of scientific materials if better use were made of historical models.

Turning to Part II, Mark Bedau's definition of weak emergence has attracted considerable attention within the philosophical literature on emergence and complexity theory. In his contribution, Bedau explores the connections between his original definition of weak emergence, which was formulated in terms of emergent states being those that are underivable except by simulation, and his more recent reformulation in terms of incompressible explanations. The original definition appeared to many to be epistemological, whereas the later definition was explicitly designed to be ontological. Bedau uses a running example of a simulation in artificial life to argue that simulations of systems exhibiting emergence allow us to attain a different kind of understanding than do the traditional methods of observation, experiment, and theoretical modeling.

Brian Epstein examines two assumptions that are built into most individualist models in social sciences. The first is that macroscopic social properties locally supervene on the microscopic properties of individuals in the society. Epstein argues that many social properties attached to individuals cannot be localized in the way required for local supervenience—this is not a trivial analytic consequence of the properties being social but a result of changes in other parts of society affecting spatially distant members. The second assumption is that a separation of social properties into levels based on mereological composition cannot be a general account of the supposed hierarchy of social properties. Epstein shows that these two assumptions are common in individually based simulations in the social sciences, but he also shows how satisfactory modeling practices can take place despite the failure of these assumptions.

The experimenter's regress has been much discussed by sociologists and philosophers of science. It hinges on the issue of whether or not there is evidence for the correct operating of an experiment other than the results from the experiment itself. Axel Gelfert explores whether there is a similar problem present in computer simulations. He investigates the extent to which it is appropriate to construe simulations as particular types of experiments and also draws on criteria developed by Allan Franklin as a response to the experimenter's regress to develop related but different criteria for simulations. Gelfert concludes that the simulationist's regress is conceptually different from the experimenter's regress and is not an insuperable difficulty but cannot be eliminated entirely.

Jaakko Kuorikoski's chapter emphasizes that the notion of understanding, usually studied independently of tractability issues, requires elaboration in the context of computational science, in which the complexity of simulations usually goes beyond what can be grasped by cognitively unaided human beings. He distinguishes between the psychological sense of understanding and a more objective sense of what he calls 'proper understanding,' a conflation of the two sometimes giving rise to an illusion of understanding. This illusion can, for example, easily be fostered by the common use of visualization techniques for the outputs of simulations. By employing a deflationary account of proper understanding, he connects the issues involved in understanding results from simulations with the well-established literature on scientific explanation. The arguments given by Kuorikoski should have wide application in philosophy of science.

Finally, Patrick Suppes presents some of his empirical research on the identification of linguistic items in brain wave patterns and connects this with results from simulations in computational neuroscience using weakly coupled phase oscillators. The interest of this work lies not only in its demonstration that very sophisticated phenomena (here, linguistic data) can be reproduced using remarkably simple mechanisms based on a commonly occurring biological process but also in Suppes's philosophical commentary that suggests that our understanding of neural phenomena may be essentially limited in ways that traditional foundationalist projects have not recognized.

In Part III, Ronald Giere's chapter presents the argument that physical models such as ball and stick models in chemistry represent in the same way as do theoretical models and computational models. Adopting the view that representation involves a model, a target, and an intentional agent, he uses similarity relations to provide a unified account of the three different types of representations just mentioned.

The next two chapters are about a topic—idealized representations—which, though much studied in the last decades, still resists a satisfactory analysis. In his contribution, Uskali Mäki builds on his earlier influential work to explore how false models can provide us with understanding. Previous approaches have tended to take false models as a whole and to analyze their representational content from that perspective. Mäki decomposes models into components with different functions and assesses each type of idealization with respect to its justification and candidacy for truth. Juha Saatsi's chapter takes on a similar problem of false models but uses an account of partial truth to do so. Distinguishing between approximate truth and partial truth, Saatsi focuses on the non-linguistic semantic account of theories as models and uses the concept of an inferentially veridical representation to capture both implicitly and explicitly attributed properties that feature in the model.

The last contribution studies how the materiality of representations is crucial to their scientific use, a philosophical question missed both by logical positivists, in spite of their concern for linguistic issues and followers of the model-theoretic approach of theories, who usually discard linguistic issues in their approach of philosophical questions. Marion Vorms uses the idea of a format for a representation, examples of which are diagrams and sentences and, more specifically, the use of Newtonian, Lagrangian, and Hamiltonian formulations of classical mechanics. She argues that a rule-based approach to syntax and semantics cannot account for pragmatic features of representations and that it is the mastery of the rules of construction, transformation, and interpretation that are key to using a format. The formats are applicable both to data and to theoretical representations and different formats play different roles in the agent's inferential processes.

We would like to thank the College of Arts and Sciences at the University of Virginia and the John Templeton Foundation for financial support of this project, the many referees who provided detailed comments on earlier versions of these chapters, and Ashley Kennedy and Charles Rathkopf for logistical help.

Paul Humphreys and Cyrille Imbert

Part I
Models

1 The Productive Tension
Mechanisms vs. Templates in Modeling the Phenomena

Tarja Knuuttila and Andrea Loettgers

> [T]he ability to support a tension that can occasionally become almost unbearable is one of the prime requisites for the best sort of scientific research.
>
> (Thomas Kuhn 1977, 226)

INTRODUCTION

One characteristic feature of scientific modeling is the way modelers recycle equations, algorithms, and other formalisms around different domains and in which process the formalisms obtain different interpretations, depending on the domains they are applied to. Often these domains are situated far away from each other, with seemingly nothing in common between them. Although this feature of modeling has been frequently noticed by the philosophers of science, as yet it has not really been targeted by philosophical analysis. There may be several reasons for this neglect: it seems partly to be due to the representational approach to models, which focuses on the relationship of a single model and its real-world target system. Such a narrow unit of analysis loses sight of the cross-disciplinary nature of modeling: theoretical and methodological dissemination between different disciplines happens frequently through modeling. On the other hand, although philosophers of science have discussed the importance of analogies, metaphors, and off-the-shelf models in theoretical transfer, these topics have remained somewhat disparate in the absence of a common concept that would draw them together. What is needed, then, is a new unit of analysis that would simultaneously extend the traditional perspective and offer a concrete key to the interdisciplinary exchange of concepts and representational means characteristic of modeling.

It seems to us that the concept of computational template introduced by Paul Humphreys (2002, 2004) meets the bill. Computational templates are genuinely cross-disciplinary computational devices, such as functions, sets of equations, and computational methods, which can be applied to different problems in various domains. Examples of computational templates are, for instance, the Poisson distribution, the Ising model (Hughes 1999), and the Lotka-Volterra model and different agent-based models. As purely syntactic objects computational templates are interestingly double faced. On the one hand, Humphreys claims that "syntax matters" meaning that the specific form of a mathematical representation is crucial for application (2004, 97, see also Vorms, this volume). On the other hand, he stresses that computational templates are results of complex processes of construction, which endow them with intended interpretation and initial justification that are not readable solely from the syntactic form.

Computational templates may have their origin in formal disciplines like the Poisson distribution in the probability theory or they may have been intended as theoretical models of a certain system and subsequently applied to different domains (like the Ising model and the Lotka-Volterra model). In this latter case, a theoretical template underlying a theoretical model becomes a genuine computational template first when it is separated from the original theoretical context and used to model other often different types of phenomena.[1] The main motivation for this transfer lies in the former successes and the tractability of the template: The distinguishing mark of computational templates is their tractability; theoretical templates need not be tractable. "The inability to draw out computational templates from theoretical templates," argues Humphreys, "underlies the enormous importance of a relatively small number of computational templates" (2004, 64; 68). Apart from tractability considerations Humphreys explains the versatility of computational templates by their generality, which he attributes to the fact that: "[. . .] templates [use] components that are highly general, such as conservation laws and mathematical approximations, and it would be a mistake to identify many of these components with a specific theory or subject matter" (2004, 90).

Computational templates usually occur embedded in *computational models*. Often theoretical or computational templates form a basis for a family of models taking into account different characteristics of the phenomenon of interest. A computational model, according to Humphreys, is a constellation on different components. Apart from the computational template, it consists of the construction assumptions that were used in arriving at the computational template, an initial justification of the template, a correction set, an interpretation, and an output representation. The construction assumptions of the computational template consist of an ontology, idealizations, abstractions, constraints, and approximations. An ontology specifies the kinds of objects referred to by a model. The correction set, in turn, is linked to the construction assumptions in that it relaxes some of the idealizations, abstractions, constraints, and approximations made and thus determines which parts of the model are intended to be interpreted realistically. Complemented with all these components, a computational template converts into a fully fledged model.

Humphreys's emphasis on the role cross-disciplinary computational templates play in contemporary modeling practice resonates interestingly with the historian of science Giorgio Israel's (1993) views on mathematical modeling. He claims that in the 20th century a new kind of idea on the relationship of mathematics to the reality was born. Whereas the classical reductionist approach relied on the "fundamental uniqueness of the mathematical representation" (472) the modeling approach makes use of the "*multiplicity* of representation" (473). The quest for general and unifying principles gave way to applying same abstract mathematical representations to multitude of domains. This modeling activity revolves around "formal structures capable of representing a large number of isomorphic phenomena" and used by way of mathematical analogy (478). Israel's formal structures, it seems to us, come close to Humphrey's templates.

The overall usability of computational templates is thus based on their tractability and generality that make them suitable for modeling different and heterogeneous phenomena showing some similarities under certain descriptions. But the question then arises whether the perceived similarities between different phenomena are really produced by the same kinds of mechanisms. We take it that a great deal of scientific modeling aims to study the mechanisms that produce natural and social phenomena. The goal is to learn about the "behavior of a system in terms of the functions performed by its parts and the interactions between these parts" as it has been put by Bechtel and Richardson (1993, 16). In the recent discussion on mechanisms several notions of mechanisms have been suggested, all of them geared towards capturing the characteristics of certain kinds of mechanisms (e.g., Machamer et al. 2000; Bechtel and Abrahamsen 2005; Glennan 2005). What is common to these accounts is their focus on the causal productivity of mechanisms that is due to the component parts, their various operations, and their interaction. Mechanisms are seen as real entities, and any model trying to depict them should thus aim for specifying their key components, operations, and organization. But how apt are general computational templates in accomplishing this task?

Computational templates are usually applied to modeling certain phenomena if they succeed to exhibit some overall features of it. Yet the success of a computational template in this task does not guarantee that it has captured the specific mechanisms underlying the phenomenon. Quite the contrary, it is doubtful that the processes of (re)interpreting and correcting a general computational template can deliver a mathematical description of a specific mechanism decomposed into its components and operations. Consequently, there seems to be a tension inherent in the modeling practice that is due to scientists' aim to depict the basic mechanisms underlying some specific phenomena in a certain domain and the general cross-disciplinary templates used in this task. This tension, we suggest, is a central driving force of modeling practice, being productive in different ways. On the one hand modelers borrow from other disciplines computational templates, which often do not easily lend themselves to the description of the phenomenon of interest and may thus require adjustments on the theoretical level in form of, for example, conceptual changes. In this process the templates themselves are also molded to better suit the domain of interest, which often gives rise to a family of models aiming at depicting their targets more realistically in detail. On the other hand, the interest in a certain phenomenon may lead to the construction of a new template as the available mathematical means turn out unsuitable or lacking. The Lotka-Volterra model is a case in point.

In the following we will illuminate the tension inherent in capturing the specific mechanisms by means of general cross-disciplinary templates through examining the Lotka-Volterra model both from the historical and philosophical perspective.[2] The Lotka-Volterra model, we suggest, displays this tension present in the modeling practice in two different levels. First, the tension is reflected by the different modeling strategies of Lotka and Volterra. Second, it is also felt internally in their respective modeling endeavors, especially in the work of Volterra, who eventually accomplished something else than what he set out to do. Although he aimed to isolate a mechanism, he created a template. We will also highlight how the tension arises partly from the conflict between the limited mathematical means available and the complexity of the real systems to be modeled. Finally, our analysis sheds new light on the notion of computational template by showing that its tractability may come in degrees and can be historically evolving and how complicated the process of transferring mathematical formalisms actually is. It typically includes the attempt to apply a certain *method of modeling* that has proven successful elsewhere with its characteristic *mathematical tools, problem solutions,* and *associated concepts*. They are introduced by drawing *analogies* between the two domains, a procedure that Volterra utilized repeatedly in his theorizing. Lotka's template-oriented approach, on the other hand, pointed towards the emerging systems theory with its study of complex forms of interaction.

TWO DIFFERENT APPROACHES TO THE DESIGN OF THE LOTKA-VOLTERRA MODEL

The Lotka-Volterra model of predator-prey dynamics is a central model in population biology, but it is also used outside of population biology in such areas as chemistry and social sciences (Epstein 1997). The model was independently introduced in the 1920s by Alfred Lotka and Vito Volterra. Few scientists at that time were aware of the different origins of the model because they mostly read Volterra's paper. Lotka's work was regarded as eclectic. Later on the model sparked a lot of discussion about the use of mathematical models in biology and had an impact on the development of alternative modeling approaches in population biology (see Levins 1966; Kingsland 1985). What makes the comparison of Volterra and Lotka especially interesting is that although they ended up in presenting a model that from the formal point of view looks the same—and was subsequently treated like that—they nevertheless followed different kinds of modeling strategies. Whereas Volterra attempted to isolate the essential or "sufficient" components of the predator-prey system and their interaction in "sea fisheries," Lotka started from a very general perspective and applied his model template both to the analysis of biological and chemical systems.

Volterra and the Idea of an Analytical Biology

Vito Volterra (1860–1940) was a world-renowned and very influential Italian mathematician and theoretical physicist. One of his main interests was to bring mathematics into the field of biology and social sciences. He aimed to transform these fields into analytical sciences that make use of quantitative methods. An insight into this program is provided by his inaugural address delivered at the opening of the academic year at the University of Rome in 1901. The address was entitled *On the Attempts to Apply Mathematics to the Biological and Social Sciences* (Volterra 1901), but it shows that Volterra's program went beyond using mathematics merely as a tool in these fields. He wanted to "translate natural phenomena into arithmetical or geometrical language" and by doing so "open a new avenue for mathematics" and to transfer social sciences and biology into analytical sciences. This goal was

> [. . .] to study the laws of the variation of measurable entities, to idealize these entities, to strip them of particular properties or attribute some property to them, to establish one or more elementary hypotheses that regulate their simultaneous and complex variation—all this marks the moment when we lay the foundation on which we can erect the entire analytical edifice. (Volterra 1901, 250)

Thus measurable entities and empirical data provided the basis on which analytical biology and social sciences should be built. A further important ingredient in this transformation process was mechanics: biology and social sciences should be modeled on mechanics. For Volterra mechanics constituted "together with geometry, if not the most brilliant then surely the most dependable and secure body of knowledge" (ibid). What Volterra especially appreciated in mechanics was the practice of idealization and abstraction, which meant for him the identification of the essential components and interactions contributing to the observed phenomena and separating them from mere perturbations.

However, there seemed to be also something anachronistic about Volterra's endeavor as physics at the beginning of the 20th century was marked by debates concerning the failure of the mechanistic worldview. Regarding this crisis Volterra wrote

[...] we have abandoned many illusions about giving a mechanical explanation of the universe. If we are no longer confident of explaining all physical phenomena by laws like that of universal gravitation, or by a single mechanism, we substitute mechanical models for those collapsed hopes-models that may not satisfy those who look for a new system of natural philosophy but do suffice for those who, more modestly, are satisfied by analogy, and especially mathematical analogy, that somewhat dissipates the darkness enshrouding so many phenomena of nature. (Volterra 1901, 255)

What was thus to be saved from the mechanistic view was the use of mechanical models applied to other domains by way of mathematical analogies. Consequently, and in relevance to our central theme concerning the centrality of computational templates in modeling, along the mechanical models also "a large part of the mathematical physics would still be saved from destruction" (ibid) and with it the central use of differential equations.

Summing up, Volterra's view on how biology and social sciences should be modeled on mechanics included transforming qualitative elements in quantitative measurable elements, measuring the variations, idealizing and abstracting the systems and processes under investigation—and using the mathematical tools of mechanics by way of analogies. If, for example, the phenomenon consisted of oscillations of one of the variables of the system of interest, like oscillations in the number of predators, analogies could be drawn to oscillatory systems in mechanics enabling the use of the respective mathematical tools. Moreover, according to Volterra only those components and interactions should be taken into account that were assumed to underlie the observed phenomena. This is a realistic approach in some areas of physics where comprehensive and well-confirmed background theories exist giving the resources with which to estimate the effect of distortions introduced by specific idealizations and providing guidance on how to attain particular levels of accuracy and precision. But the situation is different in such fields as biology, or social sciences, because of missing background theories and the complexity of the phenomena in question. Thus mechanical approach cannot be transferred offhand into population biology, which is shown by Volterra's subsequent attempt to apply it in modeling the predator-prey system.

The Construction of the Lotka-Volterra Model by Vito Volterra

In 1925 Volterra's son-in-law Umberto D'Ancona presented him a problem concerning the fluctuations in the number of predatory fish in the Upper Adriatic. D'Ancona had collected some statistical data on the percentages of predator fish in the total catch of fish during World War I. The data showed an unusual increase in predators during the final period of the war and immediately after, when fishing was hindered by the war. Volterra, being a theoretical physicist, embarked on finding a mechanism underlying the predator-prey dynamics that would explain the observed fluctuations in the predator population. Thus Volterra aspired to identify those component parts, component operations, and their working together, that caused the fluctuations. In this endeavor he faced two interrelated problems: First, the degree of the complexity of the system was far beyond most of the systems studied in mechanics, and second, he did not possess the appropriate mathematical tools for the description and analysis of the mechanism. The mathematical methods and techniques developed in mechanics could not be applied directly to the study of the predator-prey dynamics. Even if the variations observed in populations living in the same environment showed some well-known characteristics observed in many mechanical systems, such as oscillatory behavior, it was unclear which of the components of the system did interact and in what ways. On the one hand, the complexity of the system had to be rendered manageable enough to be modeled by mathematical tools. But on the other hand, the available mathematical tools exhibited a serious constraint on the kinds of complex structures and processes that could be studied. These two points created an essential part of the tension Volterra faced as he tried to apply the available computational templates to depict the underlying mechanism of the predator-prey dynamics.

Volterra was very much aware of this tension, as the following passage shows:

> [T]he question presents itself in a very complex way. Certainly there exist periodic circumstances relating to environment, as would be those, for example, which depend upon the changing of the seasons, which produce forced oscillations of an external character in the number of individuals of the various species.
>
> But are there others of internal character, having periods of their own which add their action to these external causes and would exist even if these were withdrawn? [...] But on the first appearance it would seem as though on account of its extreme complexity the question might not lend itself to a mathematical treatment, and that on the contrary mathematical methods, being too delicate, might emphasize some peculiarities and obscure some essentials of the question. To guard against this danger we must start from the hypotheses, even though they be rough and simple, and give some scheme for the phenomenon. (Volterra 1928, 5)

Voterra's statement: "[...] the question may not lend itself to a mathematical treatment" expresses nicely the described tension between the complexity of the phenomenon and the nature of the available mathematical methods. As a consequence, Volterra attempted to reduce the complexity of the problem by trying to set apart those components of the complex system that could be neglected and thereby rendering the problem describable by well known mathematical tools (or using Humphreys's notion, computational templates). The goal was to "isolate" those components which supposedly contributed to the variations in the number of individuals in the respective species[3]:

> Biological associations (biocoenosis) are established by many species, which live in the same environment. Ordinarily the various individuals of such an association contest for the same food, or else some species live at the expense of others on which they feed. [...] The quantitative character of this phenomenon is manifested in the variations of the number of individuals, which constitute the various species. (Volterra 1928, 4)

However, in complex environments, such as the sea, there are a lot of different factors, which have an impact on the observed variations such as changes in climate and weather, as well as seasons. This raises the question of how to identify the fundamental factors contributing to the dynamics of the predator and prey populations. What kind of assumptions had to be made and how could they be justified?

Volterra started out by distinguishing between "external" and "internal" causes. External causes are "periodic circumstances relating to the environment, as would be those, for example, which depend upon the changing of the seasons, which produce oscillations of an external character in the number of the individuals of the various species" (Volterra 1928, 5). Volterra wanted to focus on internal causes and leave aside external causes: "[...] but are there others of internal character, having periods of their own which add their action to these external causes and would exist even if these were withdrawn" (ibid.). Yet one cannot take for granted that it is possible to separate the external and internal causes, since they could be—and in fact they often are—interrelated in complex ways. Consequently, interacting species in a changing environment constituted a problem of a much higher degree of complexity than the systems studied in classical mechanics—which made it far from clear whether the modeling approach taken from mechanics could be imported to population biology.

Thus Volterra had to "start from the hypotheses" as he put it himself (see the quotation above). Since he could not isolate internal causes from the external causes due to the complexity of the interactions between the components of the system of interest, he constructed a hypothetical system with the help of some assumptions concerning them. The central assumptions made in the construction process of the model were

- The species increase or decrease in a continuous way, which makes them describable by using differential equations.

- Birth takes place continuously and is not restricted to seasons. The birthrate is proportional to the number of living individuals of the species. The same assumption is made for the death rate.
- Homogeneity of the individuals of each species, which neglects the variations of age and size.

Accordingly, Volterra concentrated exclusively on the dynamics between predators and preys leaving aside any interactions with other species or external factors. This strategy of formulating a simplified hypothetical system allowed Volterra to start out from well-known mathematical tools and computational templates and to explore their applicability, which offered at the same time some protection against the above-mentioned difficulty that the "mathematical methods, being too delicate, might emphasize some peculiarities and obscure some essentials of the question." It is important here to appreciate the difference between idealizing or abstracting away many aspects of a real system and starting out from few fundamental assumptions, which approaches Volterra himself clearly distinguished from each other.[4] Abstraction is often approached as an operation in which a theorist mentally strips away all that seems irrelevant as regards the problem at hand, in order to focus on some single property or set of properties (see Cartwright 1983, 187). Apart from the fact that it is often difficult to decompose the system of interest theoretically, let alone to decide which of the components and their interactions are irrelevant, posing the problem of abstraction in this way tends to bypass the difficulties of mathematical representation. It is as if the components and the interactions of the supposed real mechanism laid bare for the theorist to choose from, to be then described by suitable mathematical means.

In deriving the equations of the development of the two species, one of which feeds upon the other, Volterra started out from the situation in which each of the species is alone. In this situation, he assumed, the prey would grow exponentially and the predator, in turn, would decrease exponentially, because of missing food resources. Translated into the language of mathematics, the respective situations of prey and predator populations are explained by the following two differential equations, describing the change in time of the prey and predator populations.

$$dN_1/dt = \varepsilon_1 N_1, \quad dN_2/dt = -\varepsilon_2 N_2, \text{ (Equations 1a and 1b)}$$

Integration of the two differential equations leads to an exponential increase of the prey and an exponential decrease of the predator population, with N_0 referring to the numbers of individuals at $t=0$

$$N_1 = N_0 \exp(\varepsilon_1 t), \quad N_2 = N_0 \exp(-\varepsilon_2 t), \text{ (Equations 2a and 2b)}$$

Exponential growth or decrease is the most basic way of describing the development of a population in time. This first mathematical formulation of the problem does not take into account any environmental changes interacting with the population or the obvious fact that there is an upper limit in the resources provided by the environment. Rabbits do not multiply unlimitedly. Moreover, these quantitative descriptions of the predator and prey populations do not take into account any interaction between the two populations.

In the next step, Volterra introduced the interaction between prey and predator populations by introducing a coupling term. The interacting predator and prey system is described by the following set of differential equations:

$$dN_1/dt = (\varepsilon_1 - \gamma_1 N_2)N_1, \text{ (Equation 3a)}$$
$$dN_2/dt = (-\varepsilon_2 + \gamma_2 N_1)N_2, \text{ (Equation 3b)}$$

The proportionality constant γ_1 links the prey mortality to the number of prey and predators, and γ_2 links the increase in predators to the number of prey and predators. The set of differential equations is non-linear and coupled. Given that there does not exist analytical solutions to non-linear coupled differential equations, Volterra faced serious challenges in analyzing the dynamic behavior of his novel kind of model. Non-linear differential equations lie outside the classical study of oscillations, which are usually described by linear equations. Examples of such ordinary oscillations are harmonic and damped oscillators described by the following differential equations:

$$m \, d^2x/dt^2 + kx = 0 \text{ and } m \, d^2x/dt^2 + b \, dx/dt + kx = 0, \text{ (Equations 4a and 4b)}$$

An example of a harmonic oscillator is a spring, where $m \, d^2x/dt^2$ is the restoring force, which is proportional to the displacement x, and k is a constant. Damping is introduced by friction proportional to the velocity and is described by the damping term b dx/dt. The differential equations of the harmonic and damped oscillators have the advantage of being linear. This feature allows the calculation of analytical solutions. But the differential equations for harmonic and damped oscillators could not be applied to the case of interacting predator and prey systems because it does not account for the interaction between them. This interaction is described by the product N_1N_2 in the Lotka-Volterra equations, which turned a set of coupled differential equations into a set of coupled non-linear differential equations. Thus the tension between the complexity of the system under study and the available mathematical tools led to the development of a novel model, describing the dynamics of a predator-prey system by the coupled non-linear differential equations. This mathematical model of a predator-prey system became itself a computational template but interestingly only after it became mathematically tractable, a point to which we will return below.

A further critical point in the construction of the differential equations is related to the assumptions that the coefficients of increase -ε_2 and decrease ε_1 in Equations 3a and 3b) are linear in relation to N_2 and N_1. To justify this assumption, Volterra drew an analogy to mechanics by using the method of encounters according to which the number of collisions between the particles of two gases is proportional to the product of their densities. Thus Volterra assumed that the rate of predation upon the prey is proportional to the product of the numbers of the two species. The consequent mathematical analysis of the resulting equations gave Volterra some important results, including a solution to D'Ancona's problem concerning the relative abundancy of predatory fish during the war years when fishing was thwarted. Volterra summarized his results in what he called the "three fundamental laws of the fluctuations of the two species living together" (Volterra 1928, 20). The third law states that if an attempt were made to destroy the individuals of the predator and prey species uniformly, the average number of the prey would increase, and the average number of the predator would decrease. This case corresponds to heavy fishing. Next, Volterra explored small fluctuations in the two species, which corresponds to light fishing. In this case the opposite scenario would occur: The average number of the prey would decrease, and the average number of the predator to increase. This result supported the empirical findings of D'Ancona.

The Construction of the Lotka-Volterra Model by Alfred Lotka

Alfred Lotka was born in Lemberg, Austria (today Ukraine) in 1880, and he died in 1949 in the United States. In contrast to Volterra, Lotka struggled to get recognition from the scientific community for his entire life. Herbert Simon has characterized Lotka as a *"forerunner* whose imagination creates plans of exploration that he can only partly execute, but who exerts great influence on the work of his successors—posing for them the crucial questions they must answer, and disclosing more or less clearly the direction in which the answer lies" (Simon 1959, 493). It was only later in the context of general systems theory that scientists like Ludwig von Bertalanffy, Norbert Wiener, and Herbert Simon took up Lotka's work, especially his book *Elements of Physical Biology* (Lotka 1925) to which we will return below.

In 1920 Alfred Lotka published two papers. The first appeared in May in the *Proceedings of the National Academy of Sciences* entitled "Analytical \ on Certain Rhythmic Relations in Organic Systems" (Lotka 1920a), and the second paper "Undamped Oscillations Derived from the Law of Mass Actions" appeared in June in the *Journal of the American Chemical Society* (Lotka 1920b). In both articles one finds a pair of equations that have the same form as the model Volterra independently arrived at some years later. In the first paper, the equations are applied to the analysis of a biological system and in the second paper to a chemical system.[5] In the case of the biological system, Lotka began from very general considerations:

> Starting out first from a broad basis, we may consider a system in the process of evolution, such a system comprising a variety of species of

matter $S_1, S_2, \ldots, S_n$ of mass $X_1, X_2, \ldots, X_n$. The species of matter S may be defined in any suitable way. Some of them may, for example, be biological species of organisms, others may be components of the "inorganic environment". Or the species of matter S may be several components of an inorganic system in the course of chemical transformation.

We may think of the state of the system at an instant of time as being defined by statement of the values of $X_1, X_2, \ldots, X_n$; of certain parameters Q defining the character of each species (in general, variable with time); and of certain other parameters P. The parameter P will, in general, define the geometrical constraints of the system, both at the boundaries (volume, area, extension in space), and also in its interior (structure, topography, geography); they will further define such factors as temperature and climate conditions. (Lotka 1920a, 411)

On the basis of these considerations Lotka described the evolution of organic as well as inorganic systems by the following systems of differential equations:

$$dX_i/dt = F_i(X_1, X_2, \ldots, X_n; P, Q) \quad \text{(Equation 5)}$$

They describe evolution as a process of redistribution of matter among the several components X_i of the system. The function F describes the mode of physical interdependence of several species and their environment. In this general equation, the components as well as the interactions between them are not further specified. This has to be done separately for each specific system studied by using Lotka's systems approach. Defining systems in such a general way, Lotka freed his approach from any specific scientific disciplines and theories—as von Bertalanffy (1968) did later with his general systems theory. In his attempt at using methods, techniques, and concepts from statistical mechanics to describe and analyze biological systems, Lotka had realized the problems attached to the transfer of methods, techniques, and concepts from physics into biology. Frequently, the methods and techniques are not directly applicable to biology as the concepts from physics are not often apt to describe biological phenomena. Yet Lotka did not deny that the processes observed in biological systems are based on the principles of statistical mechanics and thermodynamics. He described this problem in the following passage of his book *Elements of Physical Biology* published in 1925:

> So long as we deal with volumes, pressures, temperatures, etc., our thermodynamics serve us well. But the variables in terms of which we find it convenient to define the state of biological (life bearing) systems are others than these. We may have little doubt that the principles of thermodynamics or of statistical mechanics do actually control the processes occurring in systems in the course of organic evolution. But

if we seek to make concrete applications we find that the systems under consideration are far too complicated to yield fruitfully to thermodynamic reasoning; and such phases of statistical mechanics as related to aggregation of atoms or molecules, seem no better adapted for the task. To attempt application of these methods to the prime problems of organic evolution is much like attempting to study the habits of an elephant by means of a microscope. It is not that the substance of the elephant is inherently unfitted to be viewed with a microscope; the instrument is ill adapted to the scale of the object and the investigation. (Lotka 1925, 53)

Lotka's fundamental equation (Equation 5) is an attempt to formulate a suitable, general enough instrument. The approach of Lotka is thus markedly different from that of Volterra. While Volterra is trying to specify the predator-prey dynamics in sea-fisheries by making analogies to mechanics, Lotka is making use of general templates: his analysis is not limited to biological systems but also includes "components of the inorganic environment" like chemical components. Starting out from this general outlook, Lotka deduces under assumption that P and Q are constant the equations describing a system consisting of the individuals of two hypothetical species S_1 and S_2. In this system S_2 is a herbivorous animal species that lives on the plant species S_1. The equations are of the form:

$$dS_1/dt = (A_1 - B_1 S_2)S_1, \text{ (Equation 6a)}$$
$$dS_2/dt = (-B_2 + A_2 S_1)S_2, \text{ (Equation 6b)}$$

They are of the same form as those presented by Volterra.

Having derived these equations, Lotka analyzes the stable states of the system and shows that the system under given conditions performs undamped oscillations. He concludes his paper remarking that the system of equations is identical in form with a system describing certain chemical reactions. The chemical system in question is analyzed in the paper, which was published one month later in June 1920 (Lotka 1920b). In this paper Lotka also uses the model, which he has deduced from his general considerations of how to describe the evolution of organic and inorganic systems. Consequently, he makes use of his model in the sense of a computational template: he applies the general template by adjusting it to different subject matters.

The model applying both to a biological and a chemical system can thus be regarded as a computational template. But does Lotka's set of general coupled differential equations by itself form also a computational template? Because of its generality and unspecificity, it does not lend itself directly to the modeling of specific phenomena like the Ising model or the Poisson distribution. It has to be manipulated by specifying the mechanism by identifying components and their interactions for the specific systems under study. Lotka's set of general differential equations could maybe best characterized as a very general form of a computational template linked to the general systems approach of, for example, Ludwig von Bertalanffy. Von Bertalanffy characterized systems consisting of interacting components but without any further specification of the kind of components or their interactions (von Bertalanffy 1968).[6]

MECHANISMS VS. TEMPLATES

The different approaches of Lotka and Volterra intersected at the analysis of the predator-prey interaction resulting at similar formalisms. This is in line with Paul Humphreys's claim that the process of model construction cannot be read from the formalism alone: "identical templates can be derived on the basis of completely different and sometimes incompatible assumptions" (2004, 72). The tension inherent in modeling practice between capturing the components and interactions of the supposed real causal mechanisms operative in the world and using general templates as a means in this task can be clearly seen in the opposed strategies of Lotka and Volterra. Volterra's goal was to "mathematically explain" D'Ancona's data on "temporal variations in the composition of species" (Volterra 1927b, 68). He aimed at depicting a mechanism consisting of "*the intrinsic phenomena* due to the voracity and fertility of the co-existing species" (ibid. italics of the original) and had thus clearly a certain target phenomenon in view. By contrast Lotka started his physical biology program from the general description of the interaction between species expressed as a set of simultaneous differential equations. Predator-prey dynamics, which he treated analogously to host-parasite dynamics, was just one concrete case to which his general approach could be applied. Different models of interaction could be created by choosing components and their interactions in different ways. Not all of these would necessarily correspond to real situations, but Lotka was eagerly looking for real-world situations, allowing for the application of his modeling approach. Thus, whereas Volterra approached modeling from the perspective of the causal explanation of real mechanisms, Lotka approached it from the perspective of applying a general template to specific cases.

The tension is clearly present also inside the work of both Lotka and Volterra. Not surprisingly, Volterra's endeavor was more marked by it as he had a certain target system in mind at the outset. Although he aimed at "isolating those factors one wishes to examine, assuming they act alone, and by neglecting others" (Volterra 1927b, 67), in formulating his basic model, he made intensive use of formalisms and problem solutions adapted from physics by way of analogies. And when he in his subsequent studies expanded his model to cover any number of species and to take into account different new features, he went on in this task in drawing further analogies to physics (see e.g., Kingsland 1985; Israel 1993). Consequently, in his attempt to create more *realistic* variants of the basic model, Volterra continued borrowing from mechanics some concepts and associated formalisms, which attests to the kind of tension inherent to modeling practice that we are arguing for. Volterra's excuse for making such extensive use of the principles of mechanics was that eventually his work was to be regarded as pure mathematics (see Kingsland 1985, 113). Scudo and Ziegler note that most of his original research papers "were technical and dry, especially for biologists" (1978, 58). However, being faithful to his earlier methodological pronouncements, Volterra was also interested in testing his theories on empirical data, but he typically kept the mathematical, technical treatments, and the empirical accounts separate from each other.

Lotka, in turn, perceived as problematic both the use of other disciplines such as mechanics, or statistical mechanics, as a model for biology, as well as the associated transfer of methods, techniques, and templates. Accordingly, Lotka attempted to ground his theoretical endeavor in a basic formal approach that could be used to deal with various systems, whether biological, chemical, or social. Kingsland (1985) notes his careful use of analogies and his skepticism of taking metaphorical entities for real ones, contrasting his approach with that of Volterra: "[h]e did not adopt Volterra's course. He was aware that the equations which he and Volterra had developed independently were formal statements which need not have any deeper significance" (Kingsland 1985, 125). It should be noted, however, that Lotka had also a model science in mind that guided his approach to biology. In his *Elements of Physical Biology*, he applied to biological systems the methods used in the mathematical description of the dynamics of chemical reactions. This was in line with his grand vision of evolution interpreted as mass and energy relationships cutting across species boundaries. Where chemical components exchanged matter between the components of the system, biological entities exchanged energy. Thus the tension between capturing real mechanisms and using general formal templates was present also in Lotka's work albeit in a rather programmatic level. Although he ambitiously sought for general principles guiding no less than the behavior of all organic and inorganic systems, what he eventually accomplished was a system of general templates capable of being applied to various kinds of systems thus paving way to general systems theory.

The Lotka-Volterra case also shows how, along with the formalisms, some characteristic methods, problem solutions, and concepts are transferred from one discipline to the other. It reveals an interesting linkage between Humphreys's account of computational templates and the philosophical literature on the role of analogies and metaphors in modeling (e.g., Black 1962; Hesse 1966; Bailer-Jones 2000). More often than not analogies and metaphors have been relegated to the domain of heuristics due to their supposedly vague nature. This vagueness is also affirmed by Israel (1993), who claims that the "modeling approach" employs formalisms offered by the mathematical theory on the basis of vague feelings concerning the similarities between various phenomena. We have shown that for Lotka and Volterra, this was certainly not the case. They did not take the mathematical means they used from some supposed stock of mathematical forms, instead they applied the modeling methods, as well as some techniques and concepts of some (for them) paradigmatic disciplines. All these were transferred along with the mathematical representations from one domain to another.

Consequently, we suggest that the concept of computational template gives some rigor to the discussion on modeling heuristics. The point is that the discovery and the justification of models cannot be separated clearly from each other, relegating analogies and metaphors to the side of heuristics. Analogies and metaphors can rather be treated as devices that also contribute to the justification of a model in allowing the introduction of successful computational templates and modeling methods to a new field. Through these methods and techniques, the model gets some initial built-in justification (Boumans 1999), and as a result of the subsequent construction process it gains additional credibility as the borrowed modeling methods and techniques are adapted to the new domain. This, in turn, suggests that the notion of computational template provides a convenient unit of analysis for the study of interdisciplinary exchange that is typical of modeling. As concrete "pieces of syntax," computational templates are relatively easy to follow, and in drawing together the methods, theoretical concepts, and analogies made use of in modeling, they provide a unitary perspective from which to approach the theoretical and methodological transfer between disciplines.

Moreover, our analysis of the history of Lotka-Volterra model shows that since mathematical forms such as computational templates are typically already rooted in the already established ways of modeling certain domains, their uses and interpretations in the original domain provide a clue for how to model the new domain of interest. This highlights the importance of the paradigmatic cases of how formalisms are applied at some problems. Mere templates do not carry the ways they can be used up their sleeves.

Another issue in which our account complements Humphreys's insight on computational templates concerns the question of tractability. Namely, it shows that tractability is a relative matter and in this sense historical. The mathematical model of a predator-prey system became itself a computational template, but only after it was rendered mathematically tractable. Systems exhibiting non-linear dynamics are very challenging from the mathematical point of view. Usually it is not possible to find analytical solutions for non-linear differential equations. Computer simulations are an essential tool in studying the complex dynamics represented by non-linear differential equations. The advancements of computer technologies in the 1970s sparked a new interest in the Lotka-Volterra model and its dynamic behavior. Robert May (1974) argued that especially such "simple" models as the Lotka-Volterra model offer a valuable resource for studying the basic complex behavior of non-linear systems. Turned into a computational template, the Lotka-Volterra model became a formal object of study in itself.

Although Lotka and Volterra could not employ modern computational tools, both of them introduced mathematical methods allowing for the study of the stable states of the model system. One of the big successes of the Lotka-Volterra model was that it provided tools for the study of non-linear dynamics. For instance, Lotka examined his fundamental equation (Equation 5) by showing that without knowing the precise form of the function F_i, the properties related to the stable states of the system can still be discussed. In his discussion of the stable states associated with the fundamental equation, Lotka started out by making the assumption that both the environment and the genetic constitutions are constant. By the means of a Taylor series expansion, Lotka calculated the possible stable states of the system. This mathematical procedure was taken up by many scientists dealing with non-linear differential equations, and it was especially praised by Herbert Simon (1959, 495). Consequently, we suggest, that a model *in combination with the mathematical tools enabling the study of it* can turn into a computational template: Computational templates are made productive by the methods and solutions that accompany them.

The notion of computational template solves also an interesting priority puzzle regarding the Lotka-Volterra model. Namely, Lotka claimed the priority for the model on the basis of his *Elements of Physical Biology* published in 1925, although he had presented what looked like the same model already in his 1920 articles as discussed above. Israel (1993) takes this to show that neither Volterra nor Lotka was a follower of the modeling approach but that they rather approached the use of mathematics in a classical reductionist way (see above). From the mathematical modeling conception, argues Israel, "a chemical model and a biological model described by the same equation are fundamentally the same thing, and Lotka's and Volterra's models are just two concrete examples of the same nonlinear oscillator" (1993, 500). Making a distinction between a template and a model shows that this is not the case: a chemical model and a biological model are quite different things, although the templates were the same. Lotka, the self-conscious user of templates, claimed priority not on the basis of a hypothetical possibility derived from his basic template but on what could be considered a model of a real system. Namely, in Lotka 1920a, he draws the Lotka-Volterra equations from his system of equations inspired by chemical dynamics without any discussion of *empirical* biological systems. In 1925 he lays down the equations as descriptions of a host-parasite system citing also William R. Thompson (1922) and Leland O. Howard (1897) on this topic. Moreover, later on in the same book, Lotka discusses "the interspecies equilibrium" in "aquatic life," though this time the discussion draws on other sources being unconnected to the discussion on Lotka-Volterra equations. The ensuing priority discussion shows well the differences between Lotka and Volterra regarding the tension between mechanisms and templates in modeling. Volterra acknowledges

Lotka's priority but stresses that what he had formulated were principles concerning "sea-fisheries" (Volterra 1927a) and writes to Lotka "It is the analogy between the biological case and the chemical case that has guided you" (Draft of a letter, Volterra archive, quoted in Israel 1993, 499).

THE PRODUCTIVE TENSION

We have argued that the tension between capturing the specific real mechanisms underlying the observed phenomena and using general computational templates to study them lies at the heart of contemporary modeling practice. In the case of the Lotka-Volterra model, this tension is displayed by the different modeling styles of Volterra and Lotka. Volterra sought to isolate the real mechanism, whereas Lotka aimed at developing a genuine multidisciplinary template. The tension was also felt internally in their theoretical endeavors, although it was more conspicuous in Volterra's case because of the template-orientation of Lotka's work. As we have discussed above, Lotka developed what we would call today a systems approach: mathematical techniques to model systems composed of interacting components developing in time.

The tension we have discussed is strongly present in the interdisciplinary theoretical and methodological transfer, in which computational methods and templates are applied to new domains. As such research program, making use of the tools and methods of another domain stabilizes, the tension might go unnoticed but tends still to cause conceptual problems due to the uneasy fit between the representational tools and the nature of the domain they are applied to. That this tension is highly productive can be seen from the subsequent development of the Lotka-Volterra model. For Volterra the application of the tools of mechanics to inter-species dynamics provided for a more than decade-long research program (see Volterra 1931), and since then the Lotka-Volterra model has been extended from the study of the interaction between species to the exploration of basic biological mechanisms. An example of this line of research is the investigation of the temporal organization in biological systems such as circadian rhythms, which produce oscillatory day and night rhythm in organisms (Goodwin 1963; see also Loettgers 2007). Today the Lotka-Volterra model is mostly used as a basic template for further modeling. For population biologists it serves as the "simplest imaginable model of predator-prey system" (Roughgarden 1979, 432). For scientists and mathematicians in the field of systems theory the Lotka-Volterra model has formed an ideal case for studying nonlinear dynamics.

As for the notion of computational template, the case of Lotka-Volterra sheds light on it in several ways. First of all, it shows that the distinction between a theoretical and a computational template is relative, since the tractability of a theoretical template may evolve due to new tools. Because of the non-linearity of the Lotka-Volterra equations, they are not analytically solvable, but both Lotka and Volterra used methods, which allowed them to calculate the properties of the stable states of the model system, such as the oscillatory behavior. Later on the tractability of the model was greatly enhanced due to advanced computers. Second, computational templates are important as stabilized representational tools, whose paradigmatic uses have significant impact on what can be represented and how, guiding thus theoretical development and template transfer across the disciplines. Third, we have shown how part of the tension between mechanisms and templates is due to the uneasy fit of the available representational tools to some specific subject matter, which sometimes gives a rise to a new template as in the case of the Lotka-Volterra model. This might make it seem that the tension is largely due to the limitations of cross-disciplinary templates as means of representation. But this is just one part of the story: Computational templates provide also a powerful way to generalize. Paying attention to this property of computational templates shows how the tension between specific mechanisms and general templates permeates already the aims of modeling. It is apparent, for instance, in the recent discussions in systems and synthetic biology, where identifying the mechanisms underlying specific biological functions is joined by the study of the basic design principles of the spatial and temporal organization of biological systems.

ACKNOWLEDGMENTS

We wish to thank Paul Humphreys for his support and the two anonymous referees for their constructive criticisms. We have also benefited from the numerous good comments we received from our audience in the *Models and Simulations 3* conference. This research has been supported by the Academy of Finland and the Alfred P. Sloan Foundation.

NOTES

1. A computational template that has its origin in a model of a certain system can eventually become a technique. Neural networks provide an example of subject matter independent computational templates that are used as techniques of data-analysis.
2. Michael Weisberg has recently used the Lotka-Volterra model as an example of modeling (Weisberg 2007) that for the reasons of space we cannot deal with here. Weisberg suggests that models are indirect representations in the sense of describing hypothetical entities. Although there is much to Weisberg's suggestion, it seems to us that his presentation of Volterra is somewhat too stylized in that Volterra aimed at isolating a real mechanism but had to resort to the method of hypothesis (see below).
3. On isolation as a strategy of modeling, see Knuuttila 2009.
4. This is in accord with Weisberg's (2007) claim that Volterra did not arrive at his model by abstracting away properties of real fish but rather constructed it

by stipulating certain of their properties (210). However, as we have shown, this was not what Volterra originally aimed at. Yet, in practice, he had to resort to hypotheses.

5. Lotka dealt with the rhythmic effects of chemical reactions already in his earlier writings, see e.g., Lotka 1910.

6. The complex systems studied by systems theory are sometimes called mechanisms, perhaps because they describe any group of objects interacting together to produce some result. Thus, for instance, Kuorikoski (2009) argues that there are two concepts of mechanism, the other referring to "componential causal system" and the other to "abstract form of interaction." In our sense genuinely abstract forms of interaction are better seen as templates that are used to model mechanisms and the use of them is characteristic of model-based representation. Real mechanisms can be studied also in other ways, for instance, by schematic pictures or diagrams. They are often more suitable for the tasks of decomposition and localization that are important for such disciplines as medicine and neuroscience, for example (see Bechtel and Abrahamsen 2005). As mere computational templates, the "abstract forms of interaction" comprise the object of study of formal disciplines, such as complexity theory.

REFERENCES

Bailer-Jones, Daniela M. (2000). "Scientific Models as Metaphors." In *Metaphor and Analogy in the Sciences*, ed. Fernand Hallyn, 181–198. Dordrecht: Kluwer Academic Publishers.

Black, Max (1962). *Models and Metaphors*. Ithaca: Cornell University Press.

Bechtel, William, and Robert C. Richardson (1993). *Discovering Complexity: Decomposition and Localization as Strategies in Scientific Research*. Princeton: Princeton University Press.

Bechtel, William, and Adele Abrahamsen (2005). "Explanation: A Mechanistic Alternative." Studies in History and Philosophy of the Biological and Biomedical Sciences 36: 421–441.

Boumans, Marcel (1999). "Built-In Justification." In *Models as Mediators. Perspectives on Natural and Social Science*, eds. Mary Morgan and Margaret Morrison, 66–69. Cambridge: Cambridge University Press.

Cartwright, Nancy (1983). *Nature's Capacities and Their Measurements*. Oxford: Oxford University Press.

Epstein, Joshua (1997). *Nonlinear Dynamics, Mathematical Biology, and Social Sciences*. Lecture Notes, Vol. 4, Santa Fe Institute Studies in the Sciences of Complexity, Reading, MA: Addison-Wesley.

Glennan, Stuart (2005). "Modeling Mechanisms." *Studies in the History and Philosophy of Biological and Biomedical Sciences* 36: 443–464.

Goodwin Brian C. (1963). *Temporal Organization in Cells*. New York: Springer.

Hesse, Mary (1966). *Models and Analogies in Science*. Milwaukee: University of Notre Dame Press.

Howard, Leland O. (1897). "A Study in Insect Parasitism: A Consideration of the Parasites of the White-Marked Tussock Moth with an Account of Their Habits and Interrelations, and with Descriptions of New Species." United States Department of Agriculture Technical Bulletin 5: 1–57.

Hughes, R. I. G. (1999). "The Ising Model, Computer Simulation, and Universal Physics." In *Models as Mediators. Perspectives on Natural and Social Science*, eds. Mary Morgan and Margaret Morrison, 97–146. Cambridge: Cambridge University Press.

Humphreys, Paul (2002). "Computational Models." *Philosophy of Science* 69: 1–11.
Humphreys, Paul (2004). *Extending Ourselves: Computational Science, Empiricism, and Scientific Method.* Oxford: Oxford University Press.
Israel, Giorgio (1993). "The Emergence of Biomathematics and the Case of Population Dynamics: A Revival of Mechanical Reductionism and Darwinism." *Science in Context* 6: 469–509.
Kingsland, Sharon (1985). *Modeling Nature.* Chicago and London: The University of Chicago Press.
Knuuttila, Tarja (2009). "Isolating Representations versus Credible Constructions? Economic Modeling in Theory and Practice." *Erkenntnis* 70: 59–80.
Kuorikoski, Jaakko (2009). "Two Concepts of Mechanism: Componental Causal System and Abstract Form of Interaction." *International Studies in the Philosophy of Science* 23: 143–160.
Kuhn, Thomas (1977). *The Essential Tension.* Chicago: The University of Chicago Press.
Levins, Richard (1966). "The Strategy of Model Building in Population Biology." *American Scientist* 54: 412–431.
Loettgers, Andrea (2007). "Model Organisms, Mathematical Models, and Synthetic Models in Exploring Gene Regulatory Mechanisms." *Biological Theory* 2: 134–142.
Lotka, Alfred J. (1910). "Contribution to the Theory of Periodic Reactions." *The Journal of Physical Chemistry* 14: 271–274.
Lotka, Alfred J. (1920a). "Analytical Note on Certain Rhythmic Relations in Organic Systems." *Proceedings of the National Academy of Sciences* 42: 410–415.
Lotka, Alfred J. (1920b). "Undamped Oscillations Derived from the Law of Mass Action." Journal of the *American Chemical Society* 42: 1595–1598.
Lotka, Alfred J. (1925). *Elements of Physical Biology.* Baltimore: Williams and Wilkins.
Machamer, Peter, Lindlay Darden, and Carl Craver (2000). "Thinking About Mechanisms" *Philosophy of Science* 67(1): 1–25.
May, Robert M. (1974). "Biological Populations with Nonoverlapping Generations: Stable Cycles, and Chaos." *Science* 186: 645–647.
Roughgarden, Jonathan (1979). *Theory of Population Genetics and Evolutionary Ecology: An Introduction.* New York: Macmillan Publishing Co.
Scudo, Francesco M., and James R. Ziegler (1978). *The Golden Age of Theoretical Ecology: 1923–1940.* Berlin: Springer-Verlag.
Simon, A. Herbert (1959). "Book Review: Elements of Mathematical Biology." *Econometrica* 27: 493–495.
Thompson, William R. (1922). "Théorie de l'Action des Parasites Entomophages. Les Formules Mathématiques du Parasitisme Cyclique." *Comptes Rendus de l'Acadamie des Sciences Paris* 174: 1202–1204.
Thompson, William R. (1948). "Can Economic Entomology Be an Exact Science?" *Canadian Entomologist* 80: 49.
Volterra, Vito (1901). "On the Attempts to Apply Mathematics to the Biological and Social Sciences." In *The Volterra Chronicles: Life and Times of an Extraordinary Mathematician 1860–1940*, ed. Judith R. Goodstein, 2007, 247–260. History of Mathematics, vol. 31. American Mathematical Society.
Volterra, Vito (1927a). "Letter to Nature." *Nature* 119: 12.
Volterra, Vito (1927b). "Variations and Fluctuations in the Numbers of Coexisting Animal Species." In *The Golden Age of Theoretical Ecology: 1923–1940*, eds. Francesco M. Scudo and James R. Ziegler, 65–236. Berlin: Springer-Verlag, 1978.
Volterra, Vito (1928). "Variations and Fluctuations of the Number of Individuals in Animal Species Living Together." *Journal du Conseil International pour L'exploration de la Mer* 3: 3–51.

Volterra, Vito (1931). *Lecons sur la Theorie Mathematique de la Lutte pour la Vie.* Paris: Gauthier-Villars.
von Bertalanffy, Ludwig (1968). *General System Theory: Foundations, Development, Applications.* New York: George Braziller.
Weisberg, Michael (2007). "Who Is a Modeler." *British Journal for the Philosophy of Science* 58: 207–233.
Weisberg, Michael, and Kenneth Reisman (2008). "The Robust Volterra Principle." *Philosophy of Science* 75:106–131.
Wentworth Thompson, D'Arcy (1952). *On Growth and Form.* Cambridge: Cambridge University Press.

2 Theories or Models?
The Case of Algebraic Quantum Field Theory

Tracy Lupher

INTRODUCTION

There are a plethora of philosophical issues involving the relationship between a theory and its models. What criteria distinguish a theory from its models? What is the relationship between a theory and its models? What is the role of idealization in the construction of models? When these questions are examined by philosophers in the scientific context of physics, they typically use examples drawn from Newtonian mechanics. Presumably, in classical physics, these questions receive their clearest answer. Quantum mechanics raises additional conceptual and interpretative complications. Quantum field theory (QFT), which is the successor to quantum mechanics, has its own conceptual and mathematical difficulties. Those problems might be one of the reasons that the topic of models in QFT has not received much attention in the philosophy literature (notable exceptions being Hartmann [[1998] and Krause and Bueno [2007]). Part of the problem is that there is no definitive formulation of QFT. Different formulations of QFT include Lagrangian QFT, Wightman's axiomatic formulation of QFT, effective QFT, and algebraic QFT (AQFT).

The philosophical literature on QFT has focused especially on AQFT; see Clifton and Halvorson (2001), Arageorgis et al. (2002a), Ruetsche (2003), and Kronz and Lupher (2005). The algebraic approach to QFT has two advantages for philosophers: (i) it is conceptually clear and (ii) it is mathematically rigorous. Given the conceptual clarity of AQFT, it is perhaps the best QFT framework in which to examine some of the issues surrounding models and their relationship to theories. However, no paper in the philosophical literature on models has examined the role of theories and models in AQFT. This paper fills that gap. I argue that there is a distinction between theories and models in AQFT. The theoretical content of AQFT is captured at the abstract algebraic level by imposing general constraints, while representations of the abstract algebra on a Hilbert space are models because they fill in the details about a particular system. The distinction has implications for one of the most philosophically interesting issues in AQFT, namely, unitarily inequivalent representations. Some philosophers have classified unitarily inequivalent representations as different *theories*. I will argue that unitarily inequivalent representations are examples of different *models*.

In the Theoretical Content of AQFT, I give a non-technical account of the theoretical content of AQFT. Models in AQFT explains the different types of models that can arise in AQFT. The issue of unitarily inequivalent representations will be examined in the Issue of Unitarily Inequivalent Representations. I also examine Ruetsche's (2003) "Swiss army" account of the physical content and discuss the similarities and differences between our accounts. Lastly, Implications for Theories and Models in AQFT discusses the interpretive advantages of my distinction between theories and models in AQFT and how it illuminates the relationship of a theory to its models.

THE THEORETICAL CONTENT OF AQFT

In the late 1950s, some physicists and mathematicians were dissatisfied with the conceptual and mathematical difficulties being encountered in QFT. They decided to put quantum field theory on a firm conceptual and mathematical foundation by identifying a set of fundamental physical principles for QFT, rather than focus on specific equations and their solutions. Any model of QFT should satisfy those fundamental principles. Algebraic quantum field theory is one of the outcomes of that axiomatic approach. Its emphasis on clearly formulating its basic physical principles sets it apart from perturbative QFT and the construction of phenomenological models in QFT.

The theoretical content of AQFT consists of very general constraints that require additional assumptions to apply to any specific object or system.[1] In that sense, AQFT is similar to Newtonian mechanics. Newton's theory of mechanics has three laws of motion and the law of universal gravitation, which are abstract, formal expressions. Even though force $\vec{F}$ equals mass m times acceleration $\vec{a}$ ($\vec{F} = m\vec{a}$), nothing can be inferred about specific systems without supplying additional information. No model can be deduced from the laws of motion and/or the law of universal gravitation. These laws cannot be directly applied to any specific system. To apply them to a concrete system, we need to specify a specific force function for $\vec{F}$, such as kx when the relevant force is similar to a spring force. The equations of motion can then be solved. To obtain specific values for the motion of a system, more information must be supplied about the configuration of the system such as the number of bodies, the value of the spring constant k, the mass m or masses, boundary conditions, initial conditions such as initial positions and velocities, etc. That additional information narrows down the possible states for the system and the evolution of the system through time. It also provides specific values for observables of the system such as position and linear momentum.

The theoretical content of Newtonian mechanics is incomplete—a schema of formal constraints. The methodological point is that no model of a concrete system or object can be deduced from the laws alone. A model is a specification of additional content that allows the incomplete formal schema captured in the theoretical content to be applied to a specific system or object. The theoretical content of AQFT is similar to Newtonian mechanics in that it is a formal schema that must be supplemented with additional information in order for it to be applicable to a concrete system or object. A model fills in the details. However, AQFT's incomplete schema is different from Newtonian mechanics in that equations and their solutions do not play a central role in the theoretical content of AQFT nor in the construction of models. In discussing AQFT, I will give an overview of the main mathematical structures and how they relate to each other for the purpose of clarifying the theoretical content and how models are constructed. No attempt will be made at mathematical rigor, but the mathematical details can be found in Emch (1972), Haag (1996), and Halvorson and Mueger (2007).

The observables, states, and dynamics are crucial parts of any physical theory. They form part of the theoretical content of AQFT but not all of it. The full theoretical content of AQFT is a 6-tuple: $\langle \mathcal{A}, S, A, \alpha, O, R \rangle$, where $\mathcal{A}$ are the observables, S are the states, A are the axioms imposed on $\mathcal{A}$, α describes the dynamics, O are the open bounded region(s) of Minkowski spacetime on which the observables are defined, and R is the set of relationships that the open bounded regions O of Minkowski spacetime have to each other. The connections between the elements in $\langle \mathcal{A}, S, A, \alpha, O, R \rangle$ have resulted in many deep structural relations. Sometimes, only part of the full theoretical content is necessary to prove certain theorems. The minimal theoretical content of AQFT requires an abstract algebra of observables and its set of abstract states. For example, the Weyl algebra and its states are important for the study of the canonical commutation relations, but the axioms of AQFT do not play a crucial role in the study of its algebraic structure. Each element of the 6-tuple is described in more detail below.

In AQFT, the observables are described mathematically as an abstract C*-algebra or an abstract W*-algebra.[2] An abstract algebra is essentially a list of certain relationships that hold among the observables, e.g., adding two observables yields another observable in the abstract algebra.[3] The rough reason why these algebras are called 'abstract' is that no Hilbert space is involved; the observables are not operators in a Hilbert space. The states are essentially maps from observables in the abstract algebra to the complex numbers.[4] Regardless of whether an abstract C*-algebra and its states or an abstract W*-algebra and its states are used, these mathematical structures do not apply to any concrete system. The main difference between a C*-algebra and a W*-algebra is the topology defined on them, i.e., the conditions under which sequences of observables converge. No details are filled in by choosing an abstract algebra.

28 Tracy Lupher

The axioms A impose structural relations on the abstract algebra $\mathcal{A}$ from which consequences can be derived. Though these formal structural assumptions hold in most known QFT models, the axioms do not tell us how to apply these mathematical structures to specific, concrete objects or systems. These axioms include isotony, microcausality, and covariance, though additional axioms are sometimes specified in various AQFT books and papers; see Halvorson and Mueger (2007) for more information. In AQFT, each open bounded region O of Minkowski spacetime $\mathcal{M}$ is

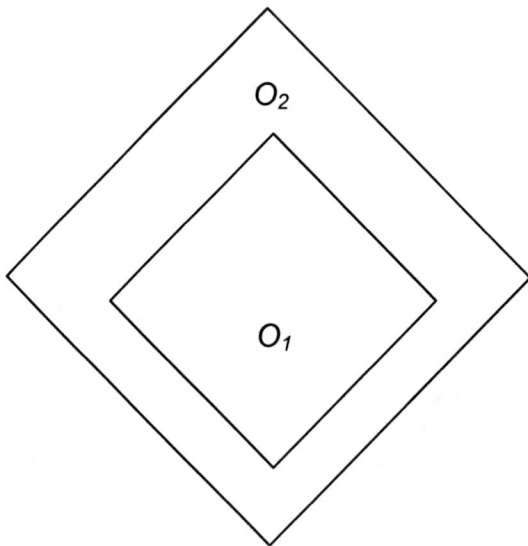

Figure 2.1 Isotony.

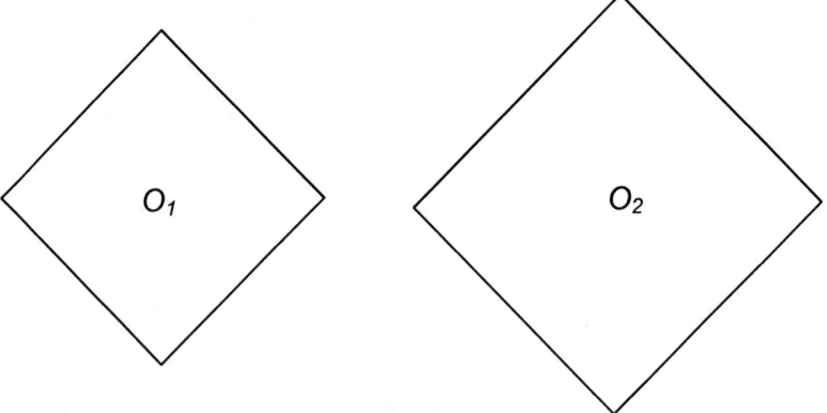

Figure 2.2 Microcausality.

associated with an algebra $\mathcal{A}(O)$ of the local observables in O.[5] The regions $O \in \mathcal{M}$ are often taken to be diamonds or double-cones, nonempty intersections of the interiors of a forward and a backward light cone. The idea behind isotony is that an observable measurable in O_1 is also measurable in a larger spacetime region O_2 containing O_1.

Isotony: If $O_1 \subset O_2$, then $\mathcal{A}(O_1) \subset \mathcal{A}(O_2)$

It does not specify what the observables in any region actually are, just that whatever observables are in O_1 will also be in the algebra of local observables in a larger region O_2. O_1 and O_2 are unspecified open bounded regions in Minkowski spacetime that have a relationship, namely, one open region is completely contained within the other open region. No actual location has been chosen.

Microcausality: If O_1 and O_2 are spacelike separated, then $[\mathcal{A}(O_1), \mathcal{A}(O_2)]=0$

Microcausality says that if two spacetime regions are spacelike separated, then any observable from the algebra associated with one region must commute with any observable from the algebra associated with the other region. This is supposed to reflect the constraints imposed by special relativity on spacetime.

Covariance: If $g \in P_+^\uparrow$ and $A \in \mathcal{A}(O)$, then $\alpha_g[\mathcal{A}(O)] = \mathcal{A}(g[O])$

Covariance says that each element g of the restricted Poincaré group $P_+^\uparrow$ (one of four disjoint classes of the Poincaré group) may be represented as an automorphism α_g of the algebra $\mathcal{A}(O)$. The automorphism transforms the algebra defined on one region, $\mathcal{A}(O)$, onto the algebra $\mathcal{A}(g[O])$ of the transformed region $g[O]$.

The dynamics α is given by a one-parameter group of automorphisms of the algebra of observables $\mathcal{A}$.[6] A quantum dynamical system can be defined for either an abstract C*-algebra or W*-algebra of observables. In the case of C*-algebras, the states are defined on the C*-algebra and the automorphisms are norm continuous in t. Such systems are called C*-dynamical systems (Sewell 2002, 27). W*-dynamical systems have a W*-algebra of observables with a set of normal states and the automorphisms are normal, i.e., weakly continuous in t. Specifying an algebra of observables, a generic set of states, and the time evolution do not provide a unique realization of a particular system. These are merely mathematical constraints.

The region(s) of spacetime O chosen is usually generic—no particular region(s) is under consideration.[7] Rather, AQFT focuses on the relations R that hold between different regions. For two regions, they could have the relation of containment (one region is completely within the other region) as in the case of isotony, disjointness (as in microcausality), partial overlap, or the sharing of a boundary. If no region or only one region is selected, then the set R is empty.

The association of open bounded regions O with local algebras $\mathcal{A}(O)$ and the axiom of isotony define a net structure of local algebras, i.e., a nesting relationship between local algebras that depends on how the various open bounded regions are contained within other larger regions. Different nets can be chosen by picking a different collection of regions. By taking the set theoretic union of all $\mathcal{A}(O)$ in $\mathcal{M}$ and closing in the norm topology, where $\mathcal{A}(O)$ is a C*-algebra defined on O, and using the isotony property, the algebra of all quasi-local observables can be defined: $\mathcal{A}_{loc} = \overline{\bigcup_{O \in} \mathcal{A}(O)}$. This inductive limit of local algebras will then cover all of Minkowski spacetime.

Additional requirements can be imposed on these theoretical structures. To quantize a system, the canonical commutation relations can be imposed. Using the Weyl form of the canonical commutation relations, an abstract C*-algebra, the Weyl algebra, can be constructed; see Clifton and Halvorson (2001) for details. Observables that satisfy this requirement do not have a unique form. In fact, there are an infinite number of different realizations of the Weyl algebra. These different realizations are unitarily inequivalent representations. There are also additional requirements that can be imposed on the abstract states, such as satisfying the Kubo-Martin-Schwinger (KMS) condition or the Hadamard condition (both of which will be discussed in the next section), and on the dynamics, such as requiring that the automorphism is unitarily implementable.

MODELS IN AQFT

A model is a partial or complete specification of the 6-tuple $\langle \mathcal{A}, S, A, \alpha, O, R \rangle$ on a Hilbert space. Which parts are specified will depend on what features are relevant for a particular physical system. Different idealizations may change how and which of the mathematical structures are implemented. In this section, I discuss how different types of models can be constructed in AQFT by specifying parts of the theoretical content. At a minimum, a model has to narrow down the possible states for a particular system. Not every abstract state in S is a possible state for a particular system. Choosing a specific abstract state for a particular system determines the other possible states for that system. Mathematically, this process involves the use of a representation of the abstract algebra of observables and a Hilbert space. The specific form that an abstract observable takes depends on the particular representation. The specific form that an abstract state takes also depends on the details of the particular representation; it may not appear at all, it may be a vector state, or it may be a density matrix.

A representation is a mapping from the abstract algebra to the set of all bounded operators acting on a Hilbert space that respects the relationships between the elements of the abstract algebra. There is a standard way of constructing a Hilbert space from an abstract algebra given by the Gelfand-Naimark-Segal (GNS) theorem.[8] Roughly, given an abstract state ω defined on the abstract algebra, a representation π_ω built on that particular abstract state can be used to build a particular Hilbert space $\mathcal{H}_{\pi_\omega}$. On $\mathcal{H}_{\pi_\omega}$, ω is a cyclic vector state $|\Omega_\omega\rangle$, which means that the other possible sates for the system can be built from $|\Omega_\omega\rangle$ by acting on $|\Omega_\omega\rangle$ with observables like $\pi_\omega(B)$, where $B \in \mathcal{A}$. The set of abstract states that can be expressed as density matrices on ω's representation is called the folium $\mathfrak{F}_\omega$ of ω. $\mathfrak{F}_\omega$ will be a proper subset of S, so not all abstract states in S will be vector states or density matrices in $\mathcal{H}_{\pi_\omega}$. This is an important feature of representations. The choice of a particular abstract state and the construction of a representation and its Hilbert space select a subset of the abstract states in S as possible states for a particular system. Abstract states do not have a specific form as a vector state or a density operator in every representation. Suppose φ is an abstract state in S, and it is a density operator in π_ω, so $\varphi \in \mathfrak{F}_\omega$. If we start with φ and use the GNS theorem, then, in its representation π_φ, φ appears as a cyclic vector state $|\Omega_\varphi\rangle$—not a density matrix. Thus, the form of an abstract state as a vector state or a density matrix will change depending on the particular representation. The abstract state may not appear as a vector state or density matrix in another representation, e.g., when π_ω and π_φ are disjoint representations of a C*-algebra in which case φ is not a possible state for π_ω.

The selection of an abstract state and the construction of its representation and associated Hilbert space coalesces the abstract algebraic structure for a particular situation. The specific form that an observable of an abstract algebra takes is not the same for every representation and its associated Hilbert space. To take an extreme example, for a nonzero element B of the abstract algebra $\mathcal{A}$, one representation π_ω might map this element to a bounded operator $\pi_\omega(B)$ on $\mathcal{H}_{\pi_\omega}$, while another representation π_φ might map the same abstract observable B to the zero operator ($\pi_\varphi(B) = 0$) on $\mathcal{H}_{\pi_\varphi}$. In that case, the expectation values, or predictions, for every state defined on $\mathcal{H}_{\pi_\varphi}$ will be equal to zero because the abstract observable was mapped to the zero operator 0, while states defined on $\mathcal{H}_{\pi_\omega}$ may have nonzero expectation values for $\pi_\omega(B)$. Different representations will map elements of the abstract algebra to different bounded operators on a Hilbert space. This can result in different physical predictions in different representations for the same observable in the abstract algebra.[9] Thus, the choice of a representation fills in the details about what states are allowed, what the form of the state is, the form of the observables in the Hilbert space, and, sometimes, the specific predictions for the states and observables in a particular representation.

Another aspect involved in the construction of models is the use of idealizations. Idealizations are supposed to simplify the model. The choice of a particular representation also introduces idealizations.

> To select a particular representation means to concentrate the attention to a particular phenomenon and to forget effects of secondary

importance... With such choice... we lose irrelevant information so that the system becomes simpler. To decide which representation to use is the same as to decide what is relevant for the problem at hand. This cannot be done *a priori* but only when the relevant patterns have been decided upon. (Primas 1983, 174–175)

[I]t seems appropriate to comment on the physical significance we ascribe to a representation of a system. Mathematically, a representation [of $\mathcal{A}$] distinguishes a certain "coherent" family of states from among [the full set of abstract states S], and at the same time, in effect, "coalesces" some of the algebraic structure... This is the effect of introducing the system into an inhibiting physical environment (compatible with it)—e.g. placing interferometers, spectrometers, polarimeters about the system, or enclosing the system within reflecting walls. (Kadison 1965, 186)

Representations "coalesce" the content of the abstract algebra for the particular abstract state we use to construct a representation and its associated Hilbert space. "Coalescing" the content means that certain conditions specific to a particular physical system have been imposed on the algebraic structure by constructing a representation of that abstract algebra.

Idealizations can require that the abstract state satisfy additional requirements, which limits the number of abstract states that can be used to construct a representation. For example, equilibrium states at a particular temperature are used in both AQFT and algebraic statistical quantum mechanics. In AQFT, they are involved in the analysis of the Unruh effect (Arageorgis et al. 2002a). Equilibrium states satisfy a mathematical relationship called the KMS condition, which is imposed on the abstract states.[10] An abstract state ω which satisfies the KMS condition and has a particular temperature T can be used to build a representation and Hilbert space. The density matrices defined on that Hilbert space all have that particular temperature. If an abstract equilibrium state φ at a different temperature T' is chosen, and its representation and Hilbert space is constructed, it has a completely different set of density matrices.[11] In the thermodynamic limit, which is taken by letting the number of particles N and the volume V become infinite while the density N/V is kept finite and nonvanishing, equilibrium states satisfy the KMS condition. They are infinite quantum systems because they have an infinite number of degrees of freedom. Though the thermodynamic limit is an idealization involving infinite volume and an infinite number of particles, it allows for different phases (e.g., a liquid or a gas) of a system to be described.

There are many other conditions that can be imposed on S, e.g., no state carrying infinite energy in an open bounded region O is allowed. For semiclassical gravity, the Hadamard condition is an important restriction on the abstract states. It allows expectation values to be assigned for the stress-energy tensor; for more information see section 4.6 of Wald (1994). These conditions are used to select a particular physical state from all of the mathematical abstract states in S. If the system under consideration does not have equilibrium states, then the KMS condition is not an appropriate restriction on the abstract states. The Hadamard condition is not necessary when there is no need to compute expectation values of the stress-energy tensor.

The topic of particles in AQFT is a very delicate and complex issue, but some of the interpretive conclusions advocated by physicists and philosophers depend on the features instantiated by representations. While a general division between bosons and fermions can be made at the algebraic level by using an abstract algebra that satisfies either the canonical commutation relations or the canonical anti-commutation relations, the form of the observables which satisfy, say, the canonical commutation relations is not unique. The form depends on the details of the representation used. Borchers (1996) defines certain representations as particle representations by imposing additional constraints such as satisfying the spectrum condition, which requires that the spectrum of the unitary representation of the translation group is contained in the forward light cone. A special case of a particle representation is a vacuum representation, which has a translation-invariant vector in the Hilbert space. One of the greatest achievements of AQFT is an account of superselection sectors. Particles with different electric charges, say an electron and a neutron, each will have a different (irreducible) representation associated with it, and those representations will be unitarily inequivalent to each other.[12] In the case of the Unruh effect, it has been argued (Clifton and Halvorson 2001) that the Minkowksi representation and the Rindler representation, which are unitarily inequivalent to each other,[13] have complementary but mutually exclusive particle content. In each of these cases, the particle content is being specified by the representation—not by the abstract algebra or its abstract states. It also suggests that the particle content changes for different unitarily inequivalent representations.

Different types of models can be constructed depending on how many parts of $\langle \mathcal{A}, S, A, \alpha, O, R \rangle$ are specified. A minimal model picks an abstract state from S and constructs a representation of $\mathcal{A}$.[14]

Minimal Model: $(\pi_\omega(\mathcal{A}), \mathcal{H}_{\pi_\omega})$

An example of a minimal model is the Fock representation that uses the Minkowski vacuum state and the abstract Weyl algebra to build a Fock space for N noninteracting boson particles; for details, see section 3 of Clifton and Halvorson (2001).

In other cases, we are interested in a model for a particular region O of spacetime, say a laboratory, and define the algebra $\mathcal{A}(O)$ on that region. A representation of $\mathcal{A}(O)$ is a regional spacetime model.

Regional Spacetime Models: $(\pi_\omega(\mathcal{A}(O)), \mathcal{H}_{\pi_\omega})$

For example, the Rindler representation is constructed on the Rindler wedge using the abstract Weyl algebra and the Rindler vacuum state.[15] Once we begin considering two or more regions of spacetime O_i, the axioms A of AQFT should hold between the regions and their geometric relationships R to each other must be specified.

Axiomatic Spacetime Models: $(\pi_\omega(\mathcal{A}(O_i)), \mathcal{H}_{\pi_\omega}, A, O_i, R(O_i))$

In some cases, the geometric relation of interest is the net of local algebras, while in other cases it might be two regions that share a boundary.

Finally, there are various dynamical models that result from specifying the automorphism α.

Dynamical Models: $(\pi_\omega(\mathcal{A}), \mathcal{H}_{\pi_\omega}, \alpha)$

When an abstract state satisfies the KMS condition, a particular (real) one-parameter automorphism α_t of the abstract algebra is chosen for inverse temperature β. If only one region of spacetime is under consideration, then there is a regional dynamic model.

Regional Dynamic Model: $(\pi_\omega(\mathcal{A}(O)), \mathcal{H}_{\pi_\omega}, \alpha)$

For example, when the Minkowski vacuum state is restricted to the right Rindler wedge, it satisfies the KMS condition, and the temperature is proportional to the acceleration; for details, see Arageorgis et al. (2002a). A complete model considers different regions of spacetime and their relationships to each other as well as providing dynamical information.

Complete Model: $(\pi_\omega(\mathcal{A}(O_i)), \mathcal{H}_{\pi_\omega}, A, \alpha, O_i, R(O_i))$

An example of a complete model would be the representation of the abstract Weyl algebra restricted to the right Rindler wedge, using the Minkowski vacuum state, and the representation of the abstract Weyl algebra restricted to the left Rindler wedge, using the Minkowski vacuum state. Both states satisfy the KMS condition and have automorphisms that describe the dynamics of possible states in each wedge.

The mathematical relationships between the theoretical part of AQFT and its models are clearly displayed by the formalism. All of these model types share a common structure: the abstract algebra. The models, which are implemented by constructing representations of that abstract algebra, preserve the mathematical relations between observables in the abstract algebra. More specifically, the representation is a *-homomorphism from the abstract algebra to the set of bounded linear operators on a Hilbert space. In the case of faithful representations (see note 9), the mapping is an isomorphism. Thus, for faithful representations, there is an isomorphism between the abstract observables, which belong to the theoretical content of AQFT, and the observables in a particular model. However, a dark cloud on the horizon is the issue of unitarily inequivalent representations, which have been taken to be physically inequivalent *theories*.

THE ISSUE OF UNITARILY INEQUIVALENT REPRESENTATIONS

The philosophical significance of unitarily inequivalent representations has been a topic of intense debate; see Ruetsche (2003) and Wallace (2006). Unitarily inequivalent representations have been involved in discussions about whether QFT can be interpreted as a theory about particles, the Unruh effect, and Haag's theorem. While unitarily inequivalent representations appear to be examples of models, they have been cited as examples of *incommsensurable physical theories* (Arageorgis et al. 2002b). The mathematical condition of unitary *equivalence*[16] has been used as a criterion of the physical equivalence of two *theories*, which implies that "unitarily inequivalent representations are physically inequivalent theories." (Arageorgis et al. 2002b, 156)

Glymour (1971) classifies two theories as equivalent only if they are intertranslatable. The philosophical criteria in the papers of Clifton and Halvorson (2001) and Arageorgis et al. (2002b) for the equivalence of two theories in AQFT have been based on Glymour's work. Clifton and Halvorson (2001, 430) interpret Glymour's notions of equivalence and intertranslatability as follows: two theories are physically equivalent *only if* there is a bijective mapping between the observables of the two theories and a bijective mapping between their states such that they make the same predictions. However, Ruetsche (2003, 1332) argues that two theories are physically equivalent *if and only if* there is a bijective mapping between the observables of the two theories and a bijective mapping between their states such that they make the same predictions. When two representations are unitarily equivalent, there is a bijective mapping between the set of observables belonging to both representations and a bijective mapping between the set of states belonging to both representations. Another consequence of unitary equivalence is that two unitarily equivalent representations are empirically equivalent, i.e., they make the same predictions. Thus, unitarily equivalent representations have been considered physically equivalent theories.[17] But why should unitarily equivalent or inequivalent representations be thought of as physically equivalent or inequivalent *theories*?

No explicit arguments are given in the papers of Clifton and Halvorson (2001) and Arageorgis et al. (2002b), but one reason that representations *could* be treated as theories is historical. The condition of unitary equivalence was used in the Stone-von Neumann theorem to classify Schrödinger's wave mechanics and Heisenberg's matrix mechanics as two mathematically and empirically equivalent versions of quantum mechanics.[18] Given that the

Stone-von Neumann theorem was used to classify wave mechanics and matrix mechanics as being equivalent physical theories, the failure of this theorem in QFT could be used to argue that there are many different quantum field *theories*. At the time of the debate between Schrödinger's wave mechanics and Heisenberg's matrix mechanics, there was no background mathematical theory against which to compare wave mechanics and matrix mechanics, but the development of von Neumann's Hilbert space formulation of quantum mechanics and the proof of the Stone-von Neumann theorem showed that they are unitarily equivalent representations of the canonical commutation relations.[19] Similarly, the development of the algebraic approach now provides a better understanding of the role of representations.

Representations and their associated Hilbert spaces are built by picking different abstract states. The selection of a particular abstract state depends on whether that state is appropriate for a particular type of system. Using a representation of a particular abstract state specifies which states are possible for the system and the form of each observable in the abstract algebra as a bounded operator in a Hilbert space. Part of the selection process can be that the abstract state satisfies some additional criteria such as the KMS condition or the Hadamard condition. The collection of abstract states in S will have equilibrium states, vacuum states, etc., but not all of these states are applicable to any situation. For example, consider two abstract states that satisfy the KMS condition but have two different temperatures. According to Arageorgis et al. (2002b), those two representations of systems in equilibrium at different temperatures would be physically inequivalent *theories*.[20] But those representations are being constructed to *model* equilibrium states at different temperatures; they are not attempts to construct physically inequivalent theories for every temperature. Treating unitarily inequivalent representations as incommensurable physical *theories* makes the use of abstract algebras obtuse. The abstract algebras would have no function if the theoretical content is completely specified by a representation. However, the proponents of AQFT constantly stress the importance of the abstract algebra for the formulation of the theory and how it unifies different unitarily inequivalent representations by exhibiting how they rely on a single abstract algebra of observables.

In a later paper, Ruetsche (2003, 1339–1342) proposes a "Swiss army approach" to specifying the physical content of a physical theory. It is based on an idea she attributes to Kadison (1965), which I discussed in Models in AQFT. Ruetsche sketches an account of the physical content of a theory in terms of possible states. The specification of physical content has two tiers. The first tier selects the broadest set of possibilities that are a priori in that no physical contingencies are taken into account. Observables are the self-adjoint elements of the abstract Weyl algebra and states are the abstract states defined on the Weyl algebra. These observables and states are supposed to be the broadest set of physical possibilities. The first tier is contained in my account of the theoretical content of AQFT by the first two elements in $\langle \mathcal{A}, S, A, \alpha, O, R \rangle$. However, I do not specify the exact abstract algebra. The abstract Weyl algebra, which is a C*-algebra, will be useful if we want to impose the canonical commutation relations and study bosons, but it is not the correct abstract algebra for fermions. Usually, AQFT just requires that the abstract algebra has observables that are bounded.

In Ruetsche's second tier ("the coalescence stage"), physical contingencies are taken into account. A contingent empirical situation has a narrower range of possible states, which is reflected in a representation because each representation has a distinguished folium of states. Each representation is like a blade in a Swiss army knife that can be used in different physical situations. Ruetsche argues that different unitarily inequivalent representations can be deployed in different physical situations, such as equilibrium states at different temperatures or in symmetry breaking, just as each blade in the Swiss army knife has a different purpose.

Ruetsche's coalescence stage is similar to my concept of a minimal model. Each blade in the Swiss army knife, which is a representation of an abstract algebra, is a minimal model, which specifies the possible states of the system. Ruetsche does not discuss the axioms of AQFT, automorphisms (dynamics), open bounded regions of Minkowski spacetime, or the relationships of those regions to each other.[21] Ruetsche discusses both tiers in terms of the possibilities of a physical theory, but her first tier has a resemblance to my account of the theoretical content of AQFT, while her second tier is similar to my notion of a minimal model. In a conversation I had with Ruetsche, she does think that representations should be thought of as models, but the focus of her papers has been on distinguishing possibilities in a physical theory as exemplified in debates about algebraic chauvinism and Hilbert space chauvinism as well as the explanatory significance of unitarily inequivalent representations.

IMPLICATIONS FOR THEORIES AND MODELS IN AQFT

The question of how to distinguish between a theory and its models in the context of AQFT can be given an unambiguous answer in part because of the mathematical and conceptual clarity afforded by working in the algebraic approach to QFT. There are other interpretive advantages to the distinction. One of the problems in the literature on models is how to decide when two models are equivalent to each other. For the models discussed in Models in AQFT, the condition of unitary equivalence provides a way to answer that question in AQFT: equivalent models are unitarily *equivalent* representations, and different models are unitarily *inequivalent* representations.[22] When two representations are unitarily equivalent, not only do they make the same predictions, they pick out the same subset of abstract states as being possible states, i.e., the folium of each representation is the same. When two representations are unitarily inequivalent, the folium of each representation will have no abstract state in common (see note 11). Unitarily inequivalent representations pick out disjoint subsets from the abstract states. Two representations that are unitarily inequivalent will also make different predictions for certain observables.[23]

The construction of representations in AQFT can rely on idealizations. For example, unitarily inequivalent representations can occur when the systems are assumed to have an infinite number of degrees of freedom, an infinite number of particles, or the thermodynamic limit is taken. These simplifying assumptions make the mathematical representation more tractable in certain cases, but more needs to be said about the nature of these assumptions and their interpretation in AQFT.[24]

The view of Arageorgis et al. (2002b) that unitarily inequivalent representations are physically inequivalent theories is conceptually incongruous when specific examples are considered, e.g., classifying two representations of systems in equilibrium at different temperatures as being physically inequivalent *theories*. Rather, these representations are the constructions of two different models of the same theory. Unitarily inequivalent representations are representations of the same abstract algebra. Viewing unitarily inequivalent representations as different models also defuses the philosophical worry that each unitarily inequivalent representation is a rival physical theory. If each unitarily inequivalent representation is a rival theory, then we might have to decide which representation is *the* theory of quantum field theory. But when unitarily inequivalent representations are viewed as different models, the choice of a representation is determined by the specific details of the system of interest. If a particular representation does not give the correct predictions for a system, we do not have to look for an alternative theoretical framework. Rather, we pick a different abstract state and see if its representation and associated Hilbert space can make better predictions.

Finally, my distinction between a theory and its models in AQFT shows in a precise way how they are related. The mathematical schema for the theoretical content of AQFT is completed through the construction of its models. A model fills in the details about a particular system and representations of an abstract algebra perform that function. A particular representation is constructed by choosing an abstract state from a set of abstract mathematical states, and that representation specifies the form of the abstract observables as bounded operators acting on a specific Hilbert space and the possible states of a system. The distinction between a theory and its models in AQFT thereby defuses the philosophical tension brought about by viewing unitarily inequivalent representations as different physical theories and illuminates how representations are used by physicists and mathematicians.

NOTES

1. Characterizing a theory as incomplete in that it does not apply to concrete situations has been articulated by Redhead (1980) and Hartmann (1998). Wightman's axiomatic formulation of QFT is what Hartmann calls a Type A Theory, which is a general background theory that requires additional assumptions in order to apply it to a concrete object or system. Neither Redhead nor Hartmann discuss AQFT.
2. Every W^*-algebra is a C^*-algebra, but every C^*-algebra is not a W^*-algebra.
3. Traditionally, the term "observable" is used to refer to any element of the abstract algebra, but this should not be taken to imply that every mathematical element of the abstract algebra is some measurable quantity. In this paper, I follow the standard practice of referring to elements of the abstract algebra as observables.
4. More precisely, a *state* ω is a positive normed linear functional. There is a difference between the states of a C^*-algebra and states of a W^*-algebra, namely the states of a W^*-algebra are normal, which means that they can be written as density matrices.
5. In some presentations of the axioms, the axioms are formulated using an abstract C^*-algebra (Haag and Kastler 1964), while at other times they are formulated in terms of concrete W^*-algebras (Buchholz 1998). Since every W^*-algebra is a C^*-algebra, it should not be surprising that the axioms can be formulated for either algebra.
6. An automorphism of an algebra is essentially an invertible linear transformation. In the case of C^*-algebras, it must be norm preserving. For W^*-algebras, it is assumed to be ultraweakly continuous.
7. O could have 2 dimensions (one space, one time), 3 dimensions (two space, one time), or 4 dimensions (three space, one time). In algebraic quantum statistical mechanics, O would be open regions of Euclidean space. The algebraic formulation of QFT has also been applied to curved spacetime. In that case, the regions are Cauchy surfaces at a particular time for globally hyperbolic spacetimes; see Wald (1994) for more details.
8. The GNS theorem can be used to build a Hilbert space for any *-algebra, so it can be used for both abstract C^*-algebras and abstract W^*-algebras.
9. The above example uses a representation that is not faithful. A faithful representation has a trivial kernel, i.e., only the zero observable in the abstract algebra is mapped to the zero operator on the Hilbert space.
10. An automorphism must also be specified. For information on the KMS condition, see Ruetsche (2003).
11. To put the point a bit more precisely, for KMS states with different nonzero finite temperatures, their respective representations are disjoint (Takesaki 1970). Disjointness implies that the intersection of the folium from each representation is empty, i.e., they have no abstract states in common.
12. Some proponents of AQFT argue that this failure of unitary equivalence makes no practical difference because the representations are in some sense physically equivalent to each other. I discuss that argument in Lupher (2008).
13. More precisely, Clifton and Halvorson prove that the Minkowkski representation and the Rindler representation are disjoint, but this implies that they are not unitarily equivalent because they are factor representations.
14. A minimal model is a representation of an abstract algebra using a particular abstract state. If the abstract algebra is a C^*-algebra, then the representation is a concrete C^*-algebra. If the abstract algebra is a W^*-algebra, then the representation is a concrete W^*-algebra, or, as it is more commonly known,

a von Neumann algebra. In the different models described in this section, the representations could be concrete C*-algebras or von Neumann algebras. Typically, models in AQFT are von Neumann algebras.
15. Around a spacetime point, there is a left and right Rindler wedge. The Weyl algebra can be defined on either wedge.
16. Let $\mathcal{A}$ be a C*-algebra and π_φ and π_ψ be two representations of $\mathcal{A}$. π_φ and π_ψ are *unitarily equivalent* to each other if there exists an isomorphism $U : \mathcal{H}_{\pi_\varphi} \to \mathcal{H}_{\pi_\psi}$ such that $\pi_\psi(A) = U \pi_\varphi(A) U^{-1}$ for all $A \in \mathcal{A}$. The relationship of unitary equivalence can also hold for representations of an abstract W*-algebra.
17. Clifton and Halvorson (2001, 431) argue for a slightly weaker condition for the physical equivalence of two representations. Two representations are physically equivalent only if they are quasi-equivalent. Unitary equivalence implies quasi-equivalence, but for some representations the distinction between unitary and quasi-equivalence collapses, i.e., in those cases quasi-equivalence implies unitary equivalence.
18. For more information on the history of the Stone-von Neumann theorem, see Summers (2001).
19. More specifically, the proof of the Stone-von Neumann theorem required that the canonical commutation relations be reformulated in the Weyl form of the canonical commutation relations. For more details, see Summers (2001).
20. More precisely, given two KMS states with different finite nonzero temperatures, their respective representations will be disjoint (Takesaki 1970). For factor representations, disjoint representations are not quasi-equivalent, which implies that they are not unitarily equivalent.
21. In a later paper, Ruetsche (forthcoming) suggests that dynamics may play a role in the selection of an abstract state. In those cases, the coalescence stage would be a dynamical model in my classification of models in Models in AQFT.
22. There is a slightly weaker condition than unitary equivalence called quasi-equivalence that might be better at ruling out artificial counterexamples for when two models are equivalent. For example, if we take the direct sum of a representation with itself $\pi_\varphi \oplus \pi_\varphi$, π_φ will not be unitarily equivalent to $\pi_\varphi \oplus \pi_\varphi$, but they are quasi-equivalent to each other.
23. The details can be found in Lupher (2008).
24. The beginning of such a discussion can be found in Ruetsche (2003, 1341–1342).

REFERENCES

Arageorgis, Aristidis, John Earman, and Laura Ruetsche (2002a). "Fulling Non-Uniqueness and the Unruh Effect: A Primer on Some Aspects of Quantum Field Theory." *Philosophy of Science* 70:164–202.

Arageorgis, Aristidis, John Earman, and Laura Ruetsche (2002b). "Weyling the Time Away: The Non-Unitary Implementability of Quantum Fied Dynamics on Curved Spacetime." *Studies in the History and Philosophy of Modern Physics* 33: 151–184.

Borchers, Hans-Jürgen (1996). *Translation Group and Particle Representations in Quantum Field Theory, Lecture Notes in Physics*. Berlin-Heidelberg: Springer.

Buchholz, Detlev (1998). "Current Trends in Axiomatic Quantum Field Theory." In *Quantum Field Theory. Proceedings of the Ringberg Workshop Held at*

Tegernsee, Germany, 21–24 June 1998 On the Occasion of Wolfhart Zimmermann's 70th Birthday, eds. Peter Breitenlohner and Dieter Maison, 43–64. Berlin-Heidelberg: Springer.

Clifton, Rob, and Hans Halvorson (2001). "Are Rindler Quanta Real? Inequivalent Particle Concepts in Quantum Field Theory." *British Journal for the Philosophy of Science* 52: 417–470.

Emch, Gérard (1972). *Algebraic Methods in Statistical Mechanics and Quantum Field Theory.* New York: John Wiley.

Glymour, Clark (1971). "Theoretical Realism and Theoretical Equivalence." In *Boston Studies in Philosophy of Science*, eds. Roger Buck and Robert Cohen, 275–288. Dordrecht: Reidel.

Haag, Rudolf (1996). *Local Quantum Physics.* 2nd ed. Berlin: Springer-Verlag.

Haag, Rudolf, and Daniel Kastler (1964). "An Algebraic Approach to Quantum Field Theory." *Journal of Mathematical Physics* 5: 848–861.

Halvorson, Hans, and Michael Mueger (2007). "Algebraic Quantum Field Theory." In *Philosophy of Physics*, eds. Jeremy Butterfield and John Earman, 731–922. Amsterdam: North-Holland.

Hartmann, Stephan (1998). "Idealizations in Quantum Field Theory." In *Idealizations in Contemporary Physics*, ed. Niall Shanks, 99–122. Amsterdam: Rodopi.

Kadison, Richard (1965). "Transformations of States in Operator Theory and Dynamics." *Topology* 3 (Suppl. 2): 177–198.

Krause, Décio, and Otávio Bueno (2007). "Scientific Theories, Models, and the Semantic Approach." *Principia* 11: 187–201.

Kronz, Frederick, and Tracy Lupher (2005). "Unitarily Inequivalent Representations in Algebraic Quantum Theory." *International Journal of Theoretical Physics* 44(8): 1239–1258.

Lupher, Tracy (2008). *The Philosophical Significance of Unitarily Inequivalent Representations in Quantum Field Theory.* Ph.D. dissertation. Austin: University of Texas.

Primas, Hans (1983). *Chemistry, Quantum Mechanics and Reductionism.* New York: Springer-Verlag.

Redhead, Michael L. G. (1980). "Models in Physics." *British Journal for the Philosophy of Science* 31(2): 145–163.

Ruetsche, Laura (2003). "A Matter of Degree: Putting Unitary Inequivalence to Work." *Philosophy of Science* 70: 1329–1342.

Ruetsche, Laura (Forthcoming). "Interpreting QM_∞." In *Handbook of Philosophy of Physics*, ed. Lawrence Sklar. Oxford University Press.

Sewell, Geoffrey (2002). *Quantum Mechanics and Its Emergent Macrophysics*: Princeton University Press.

Summers, Stephen (2001). "On the Stone-von Neumann Uniqueness Theorem and Its Ramifications." In *John von Neumann and the Foundations of Quantum Physics*, eds. Miklòs Rédei and Michael Stöltzner, 135–152. Dordrecht: Kluwer Academic Publishers.

Takesaki, Masamichi (1970). "Disjointness of the KMS-States of Different Temperatures." *Communications of Mathematical Physics* 17: 33–41.

Wald, Robert M. (1994). *Quantum Field Theory in Curved Spacetime.* Chicago: University of Chicago Press.

Wallace, David (2006). "In Defense of Naiveté: The Conceptual Status of Lagrangian Quantum Field Theory." *Synthese* 151: 33–80.

3 Modeling and Experimenting[1]

Isabelle Peschard

On the tribunal view on experimentation, experiments are the place where empirical implications or numerical simulations of theoretical models are confronted with outcomes of measurements, the data. But that supposes that it is clear what the correct theoretical implications, the correct simulation conditions, and the correct data are. By contrast and in reaction to the 'tribunal view' on experimentation, this essay will show that experimental activity, first of all, plays an essential role in a task of clarification. It will show the elaboration of an experimental system as an interactive, creative, open-ended process and highlight its constructive contribution to the processes of modeling and simulating. This constructive role of experimental activity will be explored through the analysis of a particular episode in 20th-century fluid mechanics, presented in 'Origin of the Controversy: The Discontinuity', focusing specifically on two aspects. The first, discussed from 'Conflicting Interpretations of the Discontinuity' to 'Simulation of the Wake', will be introduced with a controversy regarding the characteristics of the phenomenon under investigation ('Conflicting Interpretations of the Discontinuity'). The function of experimental activity will be to identify the conditions in which outcomes of measurements qualify as data that a putative model of the phenomenon would have to fit ('Enter the Landau Model' and 'Intrinsic Characteristics and Relevant Parameters') or a simulation would have to reproduce ('Simulation of the Wake'). The second, discussed in 'Experimental Contribution to Conceptual Understanding', is conceptual clarification and innovation. In that episode, a long controversy over the identification of some structural characteristics of the target phenomenon would finally lead to questioning implicit presuppositions about the conditions of correct measurement. The answer to these questions through experimentation would take the form of a redefinition of concepts central to the understanding of the phenomenon and the conditions of its occurrence.

WAKE, FREQUENCY AND CONTROVERSY

The Problem

Science provides us or aims to provide us with models of phenomena, for instance, a model of the dynamics of the wake that can be seen in Figure

Modeling and Experimenting 43

Figure 3.1 Wake behind a cylinder, seen from above. This figure was published in *European Journal of Mechanics B/Fluids*, 23(1): 67 by M. Provansal et al., "From the Double Vortex Street behind a Cylinder to the Wake of a Sphere." © 2003 Elsevier SAS. All rights reserved.

3.1. This figure shows the wake of a flow behind a cylinder when the velocity of the upstream flow reaches a certain critical value. The cylinder is on the left, seen from above. The flow, going from left to right, is visualized in the plane perpendicular to the axis of the cylinder, and the wake is formed by vortices emitted alternatively on each side of the cylinder and carried away with the downstream flow.

The formation of such a wake is a very common phenomenon, happening with air or liquid going over a bluff (non-streamlined) body, which can be a pole, a rock, an island. The range of application of a better theoretical understanding of wakes spreads from meteorology to the stability of bridges or platforms, from the design of cars and airplane wings to that of helicopter vanes and more generally, includes a variety of cases of development of periodic instabilities and transition towards chaotic behavior. Due to the variety of fluid structures and flow regimes that can be observed in this flow, the flow past a cylinder has served for nearly a century as a model for fundamental studies of external flows and has been characterized as "a kaleidoscope of challenging fluid phenomena" (Morkovin 1964; Karniadakis and Triantafyllou 1989).

As common as it is and as simple as it looks, this wake and the attempt to construct a theoretical model of it triggered an enormous number of studies and no less controversy. I will focus on *one*, seemingly simple, question that was the object of a debate involving experimental as well as numerical and theoretical studies in fluid mechanics in the second half of the 20th century. This question concerns the evolution of the shedding frequency of the vortices with the Reynolds number (Re), the dimensionless number representing the velocity of the flow upstream before it reaches the obstacle,[2] as Re is increased within a certain interval beyond the critical value: *is the variation of the shedding frequency with Re a continuous linear variation or is there some discontinuity?*

Again, one should not be misled by the simplicity of the question. The answer to this question is crucial not only to understanding the development of the wake but also to the general understanding of the development of fluid instabilities with all the applications one can imagine in meteorology or aeronautics. In particular, as we will see, the evolution of the frequency will be suspected of being related to the development of three-dimensional (3D) patterns of shedding, which are known to have a direct influence on the tendency for a structure to vibrate in a flow (Miller and Williamson 1994).

Modeling, Experimenting, and Relevant Parameters

Looking into the details of how this question was answered, we will see a challenge to a view that reduces the role of experimental activity in the practice of modeling to the assessment of theoretical models. It has been aptly noticed that experimental activity may also be exploratory (e.g., Steinle 2002). The following case study is an illustration of the way in which experimental activity is exploratory, but there are two temptations easily prompted by the notion of exploration that must be resisted. The first is to take exploration as implying a lack of specific direction or constraint. Exploratory experimental activity may be motivated and guided by a specific, clearly formulated question to be answered about a specific phenomenon. What makes it exploratory is that experimental activity can be seen retrospectively as a process of systematic investigation of the effects of different factors so as to identify the relations between different physical quantities that characterize the phenomenon under study. The second temptation to resist is to identify exploratory experimentation with a *type* of experimentation. Steinle explicitly avoids the first one but clearly succumbs to the second by contrasting exploratory experimentation and theory testing experimentation as two types of experimentation characterized by different types of question to be answered and different types of experimental setup used. The following case study illustrates the difficulty to sustain a contrast in type. A series of experimental procedures were conducted to clarify the origin of an unexpected effect obtained in the measurement of some physical quantities characterizing the wake. Identifying the origin of these effects was needed to decide whether the effects were really characteristic of the phenomenon, the wake, or merely artifacts. These procedures could not yet be testing a theory or a model of the wake, since it was not yet clear what would count as touchstone for a putative theoretical model.

But that this process was not about testing a theory or a model of the wake does not mean that it was not about testing at all. Steinle characterizes theory-testing experimentation by its goal "to test a clearly formulated expected effect." Such a test may well be part of a larger exploratory procedure, and an experimental investigation, which as whole can be regarded as a procedure of systematic variation needs to be composed of a series of tests of clearly formulated effects. What the experimenter sees as the goal of his/her experiment does not need to be what retrospectively, in view of what followed, will be seen as what was really at issue. In the following case study, the individuals that participate in the process had no idea they were contributing to a process of conceptual transformation of the understanding of the wake. Moreover, an exploratory process may be realized through several relevantly different experimental setups, as happens in the case that follows. Each experiment looks like one that Steinle associates with theory testing, but put together, they form a system that has the flexibility that he associates with exploratory experimentation. Rather than a difference in type between exploratory and testing, we should speak of a difference in function. The two functions are not epistemologically on a par: any empirical test necessarily has an exploratory procedure as its conditions of possibility to specify the empirical conditions in which informative, relevant measurements can be obtained.

One well-known reason why the outcomes of measurement may not be informative about the phenomenon under study is related to the proper use of the measuring instruments. The indication given by a hot-wire anemometer used to measure the velocity of a fluid in conditions where the temperature of the fluid changes will not be a relevant information. It would simply not be a measurement of velocity because the indication on the hot-wire anemometer would change with changes in temperature of the fluid independently of changes in velocity. This is why, as Margaret Morrison (2009) has recently emphasized, we need a good model of the apparatus used as measuring instrument. The model alerts the experimenter to possible sources of instrumental artifacts. But instrumental artifact is not the sort of problem that I will be concerned with. It will be admitted, in this case, that when the experimenters are reporting some measurement of frequency, the instrument functions properly and in the proper conditions of use: the instrument measures the quantity it is meant to measure, the shedding frequency. The question is whether the values of the frequency are characterizing the phenomenon or not. It is not a problem with the model of the instrument but rather with identifying what part of the experimental setup has an effect on the values of the shedding frequency and whether this effect should or should not count as an interference with the phenomenon.

To decide on whether some result is the product of an interference or characterizes the phenomenon under study is, in fact, to draw a line around the phenomenon that defines what it is and what experimental conditions need to be satisfied for the measurement of a given quantity characterizing the phenomenon.

Separating the Problem of Relevance

In the case under study, conflicting measurements will generate a controversy about whether the evolution of the shedding frequency with the upstream velocity of the wake is linear or instead displays a discontinuity. As long as this question is not answered, an assessment of a putative model of the wake, via either its analytical implications or a simulation, does not seem possible since it is not yet clear what the data are that such a model would have to account for. What is missing is the touchstone for such assessment.

What makes the case of the wake particularly interesting, however, is that it has been the target of two different types of modeling procedure. One, presented in 'Enter the Landau Model', is based on the theory of dynamical systems and builds on the periodic motion generated by the alternative emission of vortices on each side of the cylinder. It is this type of modeling that is immediately hampered by the lack of consensus on the identification of the empirical characteristics of the wake. But the wake is also a fluid mechanical phenomenon and as such falls in the domain of application of the Navier-Stokes (NS) fundamental equations. So the simulation of these equations, it was thought, should tell us what the wake is 'really' like and in particular whether or not there is a discontinuity in the evolution of the frequency with the velocity upstream. This claim could not be upheld. That is not due to a problem with the NS equations nor with the discrete model that is actually simulated, the simulation model.

The problem is with the normative (even if implicitly) assumption built in the simulation regarding the conditions in which the phenomenon normally or ideally occurs. The question is about where to draw the line, within the factors that have an effect on the phenomenon, separating those that generate artifacts or interferences and those that are characterizing the phenomenon. Morrison (2009) argues for the classification of numerical simulations as experiments on the basis of the role of models in (traditional) experimental measurements. But experimenting with a physical system is not just making measurements, it is also, and maybe primarily, identifying the influential factors and determining the conditions in which an instrumental procedure counts as a measurement of a given characteristic of the phenomenon.

Philosophical studies on experiments have mainly focused on questions of instrumental reliability (Bogen and Woodward 1988), reproducibility, or replicability (Franklin 1986; Radder 2003) and taken for granted the identification of which factors of the experimental setup 'should' and which 'should not' have an effect on the functioning of the instrument. Then the experimental problem is reduced to identifying which do have an effect and to shield those among them which should not. Epistemologically more fundamental is the problem of identifying whether a certain effect should count as an artifact or as characteristic of the phenomenon at all and on what basis.

The analysis of this function of experimental activity will call for the introduction of the concept of *relevant parameters*. I will use the term 'parameter' (rather than 'factor') to refer to a quantifiable characteristic of the experimental conditions that makes a difference in the outcomes of measurements. A relevant parameter is such that this difference is regarded as relevant to the understanding of the phenomenon, that is, as having to be

Modeling and Experimenting 47

taken into account rather than eliminated. What will count as relevant data will depend on what is recognized as relevant parameters of the system. But relevant parameters do not wear their identity on their sleeves: neither that they are parameters at all, nor that they are *relevant* parameters.

The philosophical problem at the center of this article is 'the problem of identification of the relevant parameters.' I take it to be in large part a normative problem. In practice it takes the form of a problem of interpretation of outcomes of measurement, as artifact or as characteristic of the phenomenon under study. So we will be concerned with data, data-models, and the way in which the problems of interpretation they raise are dealt with in relation to measurement. But it is not a concern with how a data-model is made out of raw data, that is, how raw data are selected, treated, analyzed, etc. Whatever one may think about what 'raw' could mean, to stay with this sort of question would still be taking for granted the identification of the experimental conditions of measurement under which the data that are representative of the phenomenon can be obtained. And it is precisely the identification of these conditions that we are here concerned with, together with the identification of the relevant parameters that it implies.

ORIGIN OF THE CONTROVERSY: THE DISCONTINUITY

Experimental Studies

The starting point of the difficulties is generally identified with the publication in 1954 of Anatol Roshko's dissertation "On the Development of Turbulent Wakes from Vortex Streets," a very detailed experimental study of the wake, still a reference in fluid mechanics, which

1) shows on the basis of measurements that for values of Re between 40 and 150, called the stable range, "regular vortex sheets are formed and no turbulent motion is developed," whereas between 150 and 300 turbulent velocity fluctuations accompany the periodic formation of vortices.
2) gives, for the stable range, an empirical formula of the linear variation of the shedding frequency with the Re.

These results had been anticipated and were shortly confirmed by other experimental studies. What makes it however, retrospectively, the starting point of the difficulties is that in 1959, a new experimental study of the wake is published that calls into question the adequacy of the distinction between only two ranges of shedding, the stable and the turbulent, and directly contradicts Roshko's results regarding the evolution of the shedding frequency with the Re.

Regarding the latter point, the author of the publication (Tritton 1959) argues for the existence of a discontinuity in the velocity-frequency curve by showing the "frequency plotted against the velocity for three different runs." He recognizes that "the discontinuity does not come in just the same position relative to the points each time," and he recognizes that "if all the runs were shown on a single Reynolds number vs. Strouhal number plot [frequency] the discontinuity would probably not be apparent" but takes that to be the reason why "it has not been noticed by other workers" (566).

Regarding the former point, the distinction between two ranges of shedding, one stable and the other turbulent, Tritton argues, on the basis of visualizations of the wake that had not been made by Roshko, that the dynamics of the wake is not what Roshko's identification of the main ranges of shedding suggested. Beyond the discontinuity, that is, for values of Re greater than the one for which the discontinuity occurs, the shedding of the vortices *along* the cylinder is not simultaneous, the invariance by translation is broken: the imaginary lines joining side-by-side vortices along the cylinder, are not parallel to the axis of the cylinder; they are oblique. And that cast serious doubt on Roshko's previous results.

The Conflict

Roshko's simple distinction between two ranges of shedding suggested that the dynamics of the wake in the stable range would be two-dimensional (2D), contained in the plane perpendicular to the cylinder. "In the stable range," he wrote, "the vortex street has a periodic spanwise structure" (1954, 1), suggesting that the vortices emitted on the same side of the cylinder are emitted simultaneously and that the line of vortices, if observed, would be parallel to the axis of the cylinder. That the successive lines of vortices are or are not parallel to the axis of the cylinder translates in terms of the dimension of the dynamics of the wake: parallel lines of vortices corresponds to a 2D dynamics of the wake, whereas non-parallel lines of vortices testifies to the existence of a dynamics in the direction of the cylinder, which would make the total dynamics of the wake 3D. But 3D effects on the dynamics were thought to be associated with the development of turbulence, which according to Roshko took place beyond the stable range.

This conflict between Roshko's and Tritton's experimental results will start a controversy of 30 years about "whether the discontinuity [and the oblique shedding] is an intrinsic, fluid-mechanic phenomenon, irrespective of the experimental set-up" (Williamson 1989, 580). And it is the way in which this controversy developed and was resolved that I am about to analyze. I just quoted one of the participants in the debate. More than just a participant, he will offer an unexpected solution to resolve the controversy that will transform the understanding of the problem.

Modeling and Experimenting 49

CONFLICTING INTERPRETATIONS OF THE DISCONTINUITY

Numerous publications followed those by Tritton and Roshko arguing for one side or the other. Rather than giving an exhaustive account of the literature, I will focus on studies that explored original explanations of the observed discontinuity, presenting evidence for these explanations, and often motivating, in their turn, new sets of experimentation and arguments. What is philosophically remarkable here is that the panel of explanations and supportive evidence, that were advanced, pretty well exhausted the different forms of justification of the reliability or significance of experimental observations, like reproducibility, replicability, controlled and quantifiable modification of the observed effects, without however being able to put an end to the polemic.

The reason, as I will show, is that some basic assumptions regarding the conditions of measurement were implicitly made. Calling these assumptions into doubt will later on lead to the refinement and redefinition of concepts central to the understanding of the phenomenon and the conditions of its occurrence.

Claim: The discontinuity is an artifact

In 1969, Tritton's claims for a non-linear evolution of the shedding frequency were objected to on the basis of new experimental results, which supported an explanation of the discontinuity and the 3D effects in terms of effects of non-uniformities. The main sources of non-uniformities were identified as being in the upstream flow and irregularities in the diameter of the cylinders. In effect, some experiments conducted by Gaster (1969) showed that increasing artificially the non-uniformities in the flow, or irregularities in the cylinder diameter, had the effect of increasing the discontinuities and, in fact, generated additional discontinuities.

What made Gaster confident that non-uniformities were responsible for the discontinuity is the observation that reducing the span of the cylinder, and reducing thereby the total amount of irregularities along the cylinder, had the effect of suppressing the 3D dynamics along the cylinder, allowing a parallel shedding, with lines of vortices parallel to the axis of the cylinder. These results strongly suggested that the discontinuities and the 3D effects presented by Tritton as features of the wake should rather be regarded as an artifact of the experimental setup.

Tritton had supported his argument that the discontinuity could not be an artifact by the reproducibility of the discontinuity with different setups. But if the factor responsible for the discontinuity is so basic that it is present in all the setups used by Tritton, then this reproducibility is exactly what one would expect. And the irregularity of the diameter of the cylinder is such a factor. So by showing the influence on the measurements and observations of such a basic feature of the experimental setup,

Gaster could counter Tritton's argument from reproducibility. Or so it could seem.

In fact, Gaster's results related to the effect of the length of the cylinder would prove significant, but for reasons quite different from what he thought. It is true that he showed the fragility of the argument from reproducibility. But the causal argument for the role of irregularities, based on the observed consequences of a variation in length on the discontinuity and 3D dynamics will prove to be flawed as well. On the one hand, what happens when the span is reduced, no more discontinuity and 2D dynamics, will be explained independently of the decrease in the amount of irregularities; and on the other hand, the increase of the span, followed with additional discontinuities and 3D dynamics, will be shown not to be necessary for the existence of a discontinuity and 3D dynamics. But I am running ahead.

Claim: The Discontinuity is not an Artifact

In 1971, new studies were made by Tritton where again he argued in favor of the 'reality' of the discontinuity. He then proposed to understand the discontinuity as a transition between two distinct instabilities requiring two different explanations of the development of the wake but without offering more to support their existence than the need to account for the discontinuity and without saying more about these explanations than what could be observed in the experiments.

This did not prevent new experimental arguments for the opposite interpretation of the discontinuity, this time as resulting from micro-vibrations of the cylinder itself as the velocity of the flow increases (van Atta and Gharib 1987). Tritton's explanation probably seems too ad hoc to be right. But, on the other hand, that new discontinuities arise with an increase of non-uniformities does not show that without non-uniformities there would be no discontinuity. The observations made when the length of the cylinder is reduced seemed to partly fill the gap: not only are there more discontinuities when there are more non-uniformities, but there are less discontinuities when there are less non-uniformities. But even then, as I mentioned, Gaster was too hasty in identifying reduction of the length with reduction of the non-uniformities due to irregularities of the diameter, and another interpretation will be given of the relation between length and mode of the shedding vortices.

And finally, importantly, no attempt to attribute the origin of the discontinuity to some aspects of the experimental setup yielded a convincing account of the specific value or range of values of Re at which the discontinuity appears. That was one of Tritton's strongest points for requiring an explanation in terms of two instabilities involving a point of transition.

In this section, I have presented several experiments that were made to clarify the effects of different characteristics of the experimental setup on the phenomenon under study, the evolution of the frequency of the wake with the Re of the flow. No experimental results were yet compelling enough to close the controversy about the form of this evolution. But several important points can already be made at this point.

First, each of these experiments was specifically directed at testing one particular factor and specifically designed for this task. But each of them is also, retrospectively, part of a process which, as a whole, constitutes a systematic investigation of the effect of different parameters akin to what Steinle (2002) identifies as characteristic of exploratory experimentation. So testing and exploring should not be seen as two types of experimentation but as a distinction in function. And what the function of an experiment turns out to be depends, in part, on what happens after it.

Second, there was apparently no theoretical modeling of the phenomenon involved in this experimental activity. Still, as will become clear in the next section, we are already engaged with an activity of modeling. The investigation of the effects of different factors is embedded in a dispute about whether the discontinuity observed in the evolution of the frequency was or was not merely an artifact: what is at issue is what a model would have to account for, which form of evolution of the frequency, and what it would have to take into account, which experimental factors, in order to qualify as a model of the phenomenon.

ENTER THE LANDAU MODEL

Model Implications vs. Experimental Measurements

It is not until 1984 that a model of the wake was proposed to account for its temporal dynamics, i.e., the temporal evolution of the amplitude of the vortices and of the frequency at which they are emitted (Mathis et al. 1984). The model in question is obtained by applying to the wake situation the general model proposed in 1944 by Landau to describe the development of a periodic instability—which he viewed as the first step towards turbulence. It was then introduced as 'the Landau model of the wake':

$$dU/dt = (\sigma_r + i\sigma_i)U - (l_r + il_i)|U|^2 U \text{ (Equation 1)}$$

where U is the complex amplitude of the wake, and $(\sigma_r + i\sigma_i)$ and $(l_r + il_i)$ are, respectively, the linear and non-linear coefficients.

From this model can be derived the evolution of the real component of the amplitude of the instability with the control parameter, here the Re, and its critical value Re_c

$$U^2_{max} = \sigma_r/l_r (Re - Re_c) \text{ (Equation 2)}$$

The measurements of the amplitude that were made showed that, in this respect, the model works beautifully –even better than expected. So for the evolution of the amplitude, at least one and the same model can account for the development of the instability on the whole range of the Re. This result contradicts Tritton's claim that two different instabilities are at play on two ranges of Re. From the same model can also be derived the evolution of the frequency with the Re, and it shows a linear variation, with no discontinuity.

When measurements are made on the frequency, however, once more, what they show is the existence of a discontinuity. And additional measurements made along the cylinder indicate the existence of a 3D dynamics, an oblique shedding. The results are quite clear. What is not clear is what to do with them.

Interpreting the Result of the Comparison

Does the discrepancy between the model's prediction of the evolution of the frequency and the outcomes of measurement show or even indicate that the Landau model is not an adequate model for the wake? It will depend on whether that discontinuity has to be accounted for by a model of the wake. If the discontinuity is an artifact, the model of the wake not only does not have to account for it but should not account for it. On the other hand, if it is an intrinsic feature of the wake, a model that does not account for it cannot, in that context, count as a model of the wake.

Ronald Giere writes that "If what is going on in the real world is similar in structure to the model of the world then the data and the prediction from the model should agree." (Giere 2005, 30) But what are the data in question? The problem is not one of data analysis or of construction of what is usually referred to, after Suppes (1962), as 'models of the data.' Even if some procedures of analysis are assumed to be in place, and a data model is produced, like the data-model of the evolution of the shedding frequency with Re, we are still left with the open question of whether this data-model is one that the model of the wake *should* agree with.

INTRINSIC CHARACTERISTICS AND RELEVANT PARAMETERS

The predictions of a theoretical model of the wake will have to agree with the data only if these data are informative about the features of the wake that the theoretical model aims to provide information about, instead of being non-informative or maybe even misleading,[3] as an artifact is. Just as the theoretical aim is not just any theoretical model but a theoretical model *of* the dynamics of the wake, so it is not just any data-model but a data-model *of* the dynamics of the wake that experimental measurements aim at.

As quoted above, Williamson described the controversy as a search to determine *'whether the discontinuity is an intrinsic, fluid-mechanic phenomenon, irrespective of the experimental set-up.'* The idea of being irrespective of the experimental setup seems to offer an empirical criterion to distinguish genuine data, ones that are informative about the target phenomenon, from non-informative data, including artifacts. If the discontinuity is intrinsic, it should not depend on the experimental setup, so if it is shown to depend on the experimental setup, then it is not intrinsic. This is obviously what motivated the experimental studies of the effect of an increase of the non-uniformities in the flow or in the diameter, as well as of the effect of making a cylinder vibrate. In each case, the idea is to show that the discontinuity is generated by some specific features of the experimental setup and consequently is not a feature of the wake itself.

It is not sufficient, however, to show that the discontinuity is only the effect of non-uniformities or vibrations. For that, it should be shown that without non-uniformities or vibrations, there is no discontinuity. That is, indeed, as we will see, the challenge that some numerical studies were going to try to address.

But the notion 'irrespective of the experimental set-up' raises a fundamental question that will prove crucial to the interpretation of the simulations that were carried out. "Irrespective of the experimental set-up" cannot be taken to mean "completely independent" of the characteristics of the experimental setup. After all, the shedding itself does depend on the value of the velocity of the upstream flow, and other characteristics of the experimental setup, like the diameter of the cylinder or the viscosity of the fluid, do influence the dynamics of the wake. The influence of these measurable characteristics will be part of our understanding of the wake; they will be taken into account through the Re, a control parameter of the system. The effect of a change in value of Re counts as an intrinsic feature of the wake and has to be accounted for by a model of the wake. The Re is therefore a *relevant parameter* of the system.

The experimental investigation of the wake takes place against a theoretical background where lines have already been drawn between what is to count as relevant parameter and what as a source of mere interference and artifact. All the factors whose effect was carefully investigated in experiments presented in Conflicting Interpretations of the Discontinuity, like non-uniformities or vibrations of the cylinder, were regarded as possible sources of artifact. If they turned out to have an effect, it would not have to be accounted for by a model of the wake, and they themselves would not have to be taken into account by the model. Whether they had an effect was what had to be established, not their status as non-relevant parameters.

The next section will bring to light the crucial, even though implicit, normative function of the contour of these lines in relation with simulation. And the last section will show that this contour is not written in stone. Their alteration, even if it is empirically justified, needs to be seen, first of all, as a normative transformation of the investigation, with a new conception of what is relevant and what the phenomenon under study is.

SIMULATION OF THE WAKE

To demonstrate that there are more discontinuities when the non-uniformities and vibrations are increased is not enough, as we saw, to show that the discontinuity is merely an effect of non-uniformities or vibrations of the cylinder and therefore is an artifact. What should be shown is that when there is *no* non-uniformity or *no* vibration, there is no discontinuity. But this is not an easy thing to show. A flowing fluid as well as the diameter of a cylinder keep a certain level of non-uniformities however carefully they are prepared. This is where one would like to make a thought experiment starting with 'imagine a flow with no non-uniformities. . . . ' Fortunately, the situation of the wake, by the end of the 1980s, lent itself to the modern alternative: the numerical simulation, which one should think will be less subject to ambiguous interpretation. But that is not so. The ambiguity stems from how the line is drawn that separates the characteristics of the experimental set up qualifying as relevant parameters from the others.

What the Simulation Shows

A simulation of the equations of NS, fundamental equations in fluid mechanics, was performed to find out how the flow behind a cylinder develops when there are no non-uniformities of any sort and no vibration. The results of the simulation were presented as pointing to a definite answer to the question of the nature of the discontinuity (Karniadakis and Triantafyllou 1989). And the answer was that the evolution of the frequency with Re is linear, with no discontinuity.

These results certainly show that the occurrence of the discontinuity in the experiments results from the influence of some factors that are not taken into account as parameters of the system in NS. And the parameters of NS are those the effect of which is constitutive of fluid-mechanical phenomena. So, if one trusts the method used for the simulation (a spectral-element method, used successfully in previous studies) and does not envisage calling into question the validity of the fundamental equations, it is most likely that the effect of these factors constitutes an artifact and should be, consequently, neutralized.[4]

Interpreting the Results of the Simulation

This conclusion about the discontinuity only holds, however, under certain assumptions. Imagine that the simulation is not an imitation of the phenomenon we are interested in; then the results are not relevant to the understanding of this phenomenon. This is where the problem of identification of the relevant parameters sneaks in. When one speaks of simulating the fundamental equations, this is not exactly right in at least two main respects. First of all, the simulation requires the construction of a system of discrete equations and a method of discretization for time and space to obtain the discrete model that is actually run by the computer, the simulation model (Lenhard 2007). Lenhard has shown how the construction of the simulation model may become a modeling process 'in its own right,' when the agreement with the observations of the phenomenon that the simulation intends to imitate is the norm that guides and regulates the construction. The main normative requirement being the successful imitation, the simulation model may be as far from the theoretical model as what is called a phenomenological model may be.

But there is something else that may be overlooked when one speaks of simulating the fundamental equations. It is something that is independent of the way in which the simulation model is obtained. What will turn out to be critical here is that the fundamental equations are abstract (Cartwright 1999; Giere 2010) and going from the fundamental equations to the simulation of a target phenomenon must involve some specifications that determine what particular situation is the target of the simulation. At closer inspection then, what raises doubt as to the significance of the result of the simulation is the geometry of the simulated situation: it is a 2D geometry representing a plane containing a cross section of the cylinder.

Remember that in this case, there is no observation that can play the role of guide or regulator. The simulation is meant to tell what the observation should be, what the phenomenon is really like, whether the discontinuity is part of it or not. But how could this simulation of the development of a flow in a 2D plane tell what it is like when a flow goes around a cylinder? In the latter case, the flow develops in a space that contains not only the plane perpendicular to the axis of the cylinder but also the plane that contains the axis of the cylinder.

There is an answer to this question: suppose that, with respect to the phenomenon under study, the frequency of shedding of the vortices forming the wake, all points on the cylinder are equivalent to one another and that the same thing happens everywhere. Then, no need to simulate the wake in each point of the cylinder; any cross section will suffice. What the 2D simulation shows then is how the wake develops, according to NS equations, in conditions where all the points on the cylinder are interchangeable. But why should we think that all the points are interchangeable? The presence of the ends obviously creates an asymmetry contradicting the assumptions of the simulation!

To this question too there is an answer: suppose that a cylinder that is long enough can be regarded as infinite and that an infinite cylinder can be regarded as cylinder that has no end. If there is no end, then we are in the situation where all points are interchangeable. All that is needed to satisfy this assumption of an infinite cylinder is that, for a long enough cylinder, what happens in the middle part of the cylinder be independent from what happens at or near the ends. And it can then be admitted that the 2D simulation will, at least, show what should happen in a 'long enough' cylinder, far enough from the ends.

Taking the simulation as relevant is, consequently, taking the ends of the cylinder as not being relevant to the understanding of the fluid mechanical features of the wake, amplitude, or frequency. The ends of the cylinder are treated in the same way as non-uniformities of the flow or vibrations of the cylinder: if they have an effect on the outcomes of measurement this effect will be classified as artifact and should be neutralized. *The ends are taken not to be a relevant parameter of the system.*

EXPERIMENTAL CONTRIBUTION TO CONCEPTUAL UNDERSTANDING

The applicability of the results of the simulation rests on the assumption that the effects that the ends of a cylinder of finite length may have on the dynamics of the wake are not intrinsic characteristics of this dynamics. This assumption will be, temporarily, supported by some measurements that have shown that for a cylinder that is long enough, the frequency of shedding in the middle of the cylinder is different from that found near the ends of the cylinder. But that should not mislead us into thinking that the assumption was an empirical assumption.

This assumption is clearly normative in that the normal conditions of development of the wake, where it has its 'pure' form, are taken to be those where the cylinder is sufficiently like one with no end. Conditions under which the ends would have an effect on the measurements would not count as right conditions of measurement.

In addition, the distinction between cylinder with ends and cylinder-like one with no end is taken to hinge on the length of the cylinder. The way to neutralize the effect of the ends, it is assumed, is to have a sufficient length.

These two assumptions will be called into question in a thoroughgoing experimental study of the evolution of the shedding frequency of the wake (Williamson 1989). This study was to represent a turning point on the issue of the discontinuity and the development of 3D effects.

Checking the Influence of the Non-uniformities

To begin with, a series of measurements was carried out to test the attribution of the discontinuity and 3D effects to the existence of non-uniformities or vibrations. On the basis, on the one hand, of measurements comparing the spectra of the wake and of the cylinder vibrations and on the other hand, of measurements made, first, with rates of turbulence and flow uniformity lower than those usually obtained, and then with 'the cylinder towed through the quiescent fluid' (instead of being fixed in a moving flow), both attributions were rejected.

Checking the Influence of the Ends

The next and main part of the investigation focuses on the study of the influence of the ends of the cylinder.

Measurements of the shedding frequency with a probe moving along the span of the cylinder showed the existence of different regions characterized by different shedding frequencies. In particular, a region of lower frequency was found near the ends.

More precisely, for a cylinder with an aspect ratio (the ratio L/D of the length to the diameter) beyond a specific value, the frequency near the ends is different from the frequency in the central region. "This suggests," Williamson writes, "that the vortex shedding in the central regions of the span is unaffected by the *direct* influence from the end conditions" (590, italics added). Note however, it will be of crucial importance, that Williamson only recognizes the absence of a *direct* influence.

For smaller values of the aspect ratio, however, the frequency of shedding is found everywhere the same as that found near the ends, as the size of these regions is now sufficient to cover the whole span. Remember that Gaster interpreted the reduction of the span of the cylinder in terms of reduction of the non-uniformities of the diameter and took that to be the explanation for the shedding being parallel in this condition. Williamson will propose the existence of a single frequency over the whole span as alternative explanation. But before getting there, another set of measurements and observations is still needed, leading to a new conceptual understanding of the notion of ends, independent of the notion of length, and to the inclusion of the ends among the relevant parameters of the system.

Redefining the Notion of Ends

Why did Williamson underline the absence only of a *direct* influence of the ends on the wake in the central region? In the case in which there is a difference in frequency between the ends and the central part of the cylinder, visualizations of the temporal development of the wake along the cylinder were made. They showed that, initially, the lines of vortices traveling downstream are parallel to the cylinder, and that progressively the parallel pattern is transformed into a stable oblique pattern which propagates from the ends of the cylinder towards the central region. These observations suggested that the ends did have some influence on the way in which the wake develops in the central region. But that does not mean that this influence, indirect in that it is not on the value of the frequency itself, *should* be part of our understanding of the wake. The decisive measurements and observations in favor of a new conception of the ends, their role, and consequently the wake itself, are still to come.

Early in the experimentation, it was noticed that the repeatability of the measurements was greatly improved by the use of endplates, little disks fixed at the ends of the cylinder perpendicular to the axis of the cylinder, to shield the span of the cylinder from the boundary layer along the test-section walls. So far, all the observations and measurements had been made with endplates *perpendicular* to the axis of the cylinder, in the same way as in previous experimental studies. Some further measurements, however, were made for different values of the angle between the axis of the cylinder and the plates. And it was then realized that, for a certain angle, the shedding becomes parallel, that is, the dynamics becomes 2D, and the discontinuity disappears even though *the length did not change*.

After further measurements, the explanation that was eventually given for the effect of change of angle was the following. Changing the angle of the plates has the effect of changing the *pressure* conditions responsible for the existence of a region of lower frequency towards the ends. When there is such a region of lower frequency, a phase difference propagates from the ends towards the central region and this propagation creates the pattern of oblique shedding. For a certain interval of angles of the endplates, when the pressure and the vortex frequency match those values over the rest of the span, there is no region of lower frequency, no propagation of phase difference, and the shedding is parallel. And the discontinuity only appears in the oblique mode of shedding and is found to correspond to the transition of one oblique pattern to another with a slightly different geometry.

New Relevant Parameter and Reconception of the Wake

Williamson takes his results to "show that [the oblique and parallel patterns] are both intrinsic and are simply solutions to different problems, because the boundary conditions are different" (579). In effect, the two forms of shedding simply correspond to different values of the angle between the endplates and the axis of the cylinder. If no special status is bestowed on certain values of this angle by contrast with others, there is no reason to take only one of the shedding patterns as being normal or intrinsic. In this new perspective, the parallel and the oblique pattern are not two distinct phenomena with only one being the normal form of the wake, but two possible configurations of the flow corresponding to different values of a parameter of the experimental system, two possible solutions for the same system in different conditions.

But this new way of seeing implies that the two assumptions, on which the relevance of the simulation depended, must be rejected. First, a new parameter should be added to the set of relevant parameters of the system; and this parameter characterizes the end conditions of the cylinder. That means that the effect that the end conditions have on the development of the wake is now part of the structural characteristics of the wake, rather than being a source of interference or artifact. Second, this parameter is independent of the length of the cylinder. The concept of end needs to be redefined in terms of pressure difference and value of the angle of the end plates that determined the value of this pressure difference.

By integrating this parameter among the set of relevant parameters, the gain is one of conceptual unification: what were seen as two distinct phenomena have been unified under the same description. Steinle (2002) is right to associate exploratory experimentation with the introduction of new concepts that allow the formulation of empirical general regularities. But what generalities are the relevant ones is not an empirical claim.

To integrate the ends among the relevant factors, through the definition of a new relevant parameter and not to bestow a special status on a particular range of values of the angle are normative transformations of the investigation. There is a price to pay. On this understanding of the wake, the Landau model, only able, as we saw, to predict a linear variation of the frequency with the Re, and consequently to account for a parallel pattern, does not account for the phenomenon. A different model will be needed able to account both for parallel and oblique patterns (Albarède and Monkewitz 1991).

CONCLUSION

The elaboration of an experimental system, I said to begin, is an interactive, creative, open-ended process and contributes constructively to the processes of modeling and simulating.

The constructive contribution is mediated by the identification of the relevant parameters. The relevant parameters are characteristics of the experimental system such that their variation not only has an effect on the phenomenon, but this effect is also constitutive of the phenomenon, instead of being an artifact. The identification of the relevant parameters is required for determining the conditions in which measurements provide the empirical touchstone of a putative model of the phenomenon. Before that, a putative model is non-testable. The specification of the relevant parameters involves a systematic empirical investigation of the effects of different factors, but the line that is drawn, between which effects are relevant and which are not, is normative. The effects of the relevant parameters are those a model of the phenomenon should account for.

When the route to the origin of certain effects is scattered with ambiguities and controversies, the simulation of fundamental equations may seem to offer a sure way to determine whether a certain effect is constitutive of the phenomenon. But, as we saw, only if some normative assumptions as to what factors need to be taken into account are built in the simulation. By establishing the relevance of additional factors, experimental activity may contribute to the reformulation of what factors need to be represented in the simulation.

We had an interactive process in that both the predictions of a still non-testable model, and the results of a prejudiced simulation contributed to shaping the experimental search for the relevant parameters. The new relevant parameter that was introduced in the conception of the phenomenon amounted to a conceptual innovation. It is why I speak of a creative process.

And it is open-ended. As a new model was formulated in response to the reconception of the phenomenon, already, the exactitude of some of Williamson's measurements was called into question on the basis of an analysis of the solutions of this model. New measurements were to follow, and new simulations, and a modified version of the model, and so it goes on.

NOTES

1. The author wishes to acknowledge support for this research by National Science Foundation (NSF) grant SES-1026183, and to thank two anonymous referees for profitable comments, as well as Paul Teller, Joseph Rouse, Ronald Giere, and the participants in colloquia where preliminary versions were presented at University of California (UC) Santa-Cruz, UC Davis, UC Berkeley, and at 'Models and Simulations 3' in Charlottesville, VA. for helpful discussions and correspondence.
2. The definition of the Reynolds number is Re= Ud/v where U is the velocity upstream of the flow, d is the diameter of the cylinder, and v the viscosity of the fluid. For a given fluid and cylinder, increase in Re expresses increase in the velocity upstream.
3. Thanks to Paul Teller (private correspondence) for pointing that out.
4. What counts as artifact is relative to the definition of the target of the experimental investigation. If vibrations of the cylinder were shown to have an effect on the outcome of measurement of the frequency, it would be regarded as an artifact if the target of the measurement is the 'fluid mechanics phenomenon' of development of the wake. But the effect of the vibrations of the cylinder on the development of the wake is also a common subject of study in which the phenomenon is that of aeroelastic coupling between the vortex street and the cylinder vibrations, and the effect of the vibrations is what is measured.

REFERENCES

Albarède, Pierre, and Peter Monkewitz (1991). "A Model for the Formation of Oblique Shedding and 'Chevron' Patterns in Cylinder Wakes." *Physics of Fluids A* 4(4): 744–756.

Bogen, James, and James Woodward (1988). "Saving the Phenomena." *Philosophical Review* 97: 303–352.

Cartwright, Nancy (1999). *The Dappled World: A Study of the Boundaries of Science*. Cambridge: Cambridge University Press.

Franklin, Allan (1986) *The Neglect of Experiment*. Cambridge: Cambridge University Press.

Gaster, Michael (1969). "Vortex Shedding from Slender Cones at Low Reynolds Numbers." *Journal of Fluid Mechanics* 38: 365.

Gaster, Michael (1971). "Vortex Shedding from Slender Cones at Low Reynolds Numbers." *Journal of Fluid Mechanics* 46: 749.

Giere, Ronald (2005). *Understanding Scientific Reasoning*. New York: Holt, Rinehart, and Winston.

Giere, Ronald (2010). "An Agent-Based Conception of Models and Scientific Representation." *Synthese* 172: 269–281.
Karniadakis, George E., and George S. Triantafyllou (1989). "Frequency Selection and Asymptotic States in Laminar Wakes." *Journal of Fluid Mechanics* 199: 441.
Landau, Lev (1944). "On the Problem of Turbulence." *C. R. Acad. Sci. U.S.S.R.* 44: 311–314.
Lenhard, Johannes (2007). "Computer Simulation: The Cooperation between Experimenting and Modeling." *Philosophy of Science* 74(2): 176–194.
Mathis, Christian, Michel Provansal, and Louis Boyer (1984). "The Benard-Von Karman Instability: An Experimental Study near the Threshold." *Journal de Physique Lettres.* 45(10): 483–491.
Miller, G. D., and Charles H. K. Williamson (1994). "Control of Three-Dimensional Phase Dynamics in a Cylinder wake." *Experiments in Fluids* 18: 26–35.
Morrison, Margaret (2009). "Models, Measurements and Computer Simulation: The Changing Face of Experimentation." *Philosophical Studies* 143: 33–57.
Morkovin, M. (1964). "Flow around Circular Cylinder –A Kaleidoscope of Challenging Fluid Phenomena." *American Society of Chemical Engineers (ASME) Symposium on Fully Separated Flows* 102–118.
Provansal, Michel, Lionel Schouveiler, and Thomas Leweke (2003). "From the Double Vortex Street behind a Cylinder to the Wake of a Sphere." *European Journal of Mechanics B/Fluids* 23(1): 65–80.
Radder, Hans (2003). *The Philosophy of Scientific Experimentation.* Pittsburgh: University of Pittsburgh Press.
Roshko, Anatol (1954). "On the Development of Turbulent Wakes from Vortex Streets." Report No 1191. National Advisory Committee on Aeronautics (NACA).
Steinle, Friedrich (2002). "Experiments in History and Philosophy of Science." *Perspectives on Science* 10(4): 408–432.
Suppes, Patrick. (1962). "Models of Data." In *Logic, Methodology and Philosophy of Science: Proceedings of the 1960 International Congress*, eds. Ernest Nagel, Patrick Suppes, and Alfred Tarski, 252–261. Stanford: Stanford University Press.
Tritton, David J. (1959). "Experiments on the Flow Past a Circular Cylinder at Low Reynolds Numbers." *Journal of Fluid Mechanics* 6: 547–567.
Tritton, David J. (1971) "A Note on Vortex Streets behind Circular Cylinders at Low Reynolds Numbers." *Journal of Fluid Mechanics* 45: 203–208.
Van Atta, Charles W., and Morteza Gharib (1987). "Ordered and Chaotic Vortex Streets behind Circular Cylinders at Low Reynolds Numbers." *Journal of Fluid Mechanics* 174: 113–133.
Williamson, Charles H.K. (1989). "Oblique and Parallel Modes of Vortex Shedding in the Wake of a Circular Cylinder at Low Reynolds Number." *Journal of Fluid Mechanics* 206: 579–627.

4 Model Dynamics
Epistemological Perspectives on Science and Its Educational Practices[1]

Michael Stöltzner

The aim of this paper is to foster a constructive interaction between philosophers of science and educational scientists in regard to the study of models. Not that such interactions would be entirely absent. To the contrary, in the works of several educational scientists, one finds quite a few references to the recent philosophical debates about models.[2] Sometimes these citations are emblematic, but in other cases educational scientists have undertaken detailed empirical studies into questions that are close to the philosopher's heart, among them Kuhnian gestalt switches in the history of models, the role of models in Lakatosian research programs, or a model's representative features. It is thus unfortunate that there is less transfer of ideas in the other direction, if only to argue that educational scientists, in the theoretical analysis of their empirical findings, might profit from a somewhat broader perspective that includes recent debates in philosophy of science. What is more, educational scientists' findings may shift philosophers' focus from issues of representation to an epistemic dynamics of models.

Since the purpose of my paper is to motivate further interactions by specifying issues of common interest, rather than to lay out a theory embracing cognitive and historical model dynamics, I will largely ignore the question to what extent the findings of educational scientists that I am going to discuss, have been influenced by their diverging conceptions of developmental psychology and cognitive science. As Ronald Giere's (1988) *Explaining Science: A Cognitive Approach* has undoubtedly shown, philosophy of science profits from a solid understanding of psychological theory. But only some investigations undertaken by educational scientists provide a theoretical underpinning to their investigations on models; others mainly report empirical findings and discuss their implementation in instructional practice. Moreover, both in educational science and philosophy of science, we are today witnessing a certain plurality of approaches; developmental psychology and cognitive science have come under strong influence from biology and physiology. Finally, the philosophical debates about models have recently moved slightly away from issues of cognitive representation, without educational scientists denying the relevance of these philosophical debates for their own work.

A DAPPLED CONCEPT IN BOTH FIELDS

There is at present no agreement among philosophers of science about the concept of model, let alone a universally accepted definition. Not unlike the classic encyclopedia entries by Ludwig Boltzmann (1902) and Mary Hesse (1967), today a comprehensive analysis would diagnose different uses of 'model' throughout the single sciences and the various philosophical approaches. Yet within the last decade or so models have moved center-stage largely by abandoning the previously dominating syntactic perspective—which understood models in the sense of formal logic as intended (or unintended) interpretations of an axiomatized theory—and by endowing models with a certain autonomy with respect to scientific theories.

The semantic view even turned the tables by understanding models as non-linguistic entities that "can be characterized in many different ways, using many different languages. This makes it possible to identify a theory not with any particular linguistic formulation, but with a *set of models* that would be picked out by all the different linguistic formulations" (Giere 1988, 48). While some proponents of this view have advocated a kind of isomorphism between a model and the real world, Giere holds that "models need only be similar to particular real-world systems in specified respects and to limited degrees of accuracy" (1999, 92–93). Accordingly, "the primary representational relationship is not the truth of a statement relative to the facts, but the similarity of a prototype to putative instances" (ibid., 123). Not only theoretical statements and equations are able to define such a model but also diagrams and visual renderings. A visual model, e.g., of continental drift, then "serves as an organizing template for whatever other potentially relevant information the agent may possess" (ibid., 132).

In their influential collection *Models as Mediators. Perspectives on Natural and Social Science,* Mary Morgan and Margaret Morrison have criticized the semantic view for dealing mostly, if not exclusively, with "'models of theory' or 'theoretical models'" (1999, 7), as in Giere's (1988) example of a mathematical pendulum, where the framework of Newtonian mechanics allows for a certain type of correction terms that yield a real pendulum by successive de-idealization of the model, e.g., by introducing friction or the mass of the bob. Against this conception, Morgan and Morrison understand models as mediators between theory and empirical data and endow them with autonomous representative features. This "autonomy is the result of two components (1) the fact that models *function* in a way that is partially independent of theory and (2) in many cases they are *constructed* with a minimal reliance on high level theory" (Morrison 1999, 43). Sometimes, they even function as measuring devices for specific parameters of the problem at hand that are typically not given by the framework theory.

Morrison's (1999) primary example for a model, whose corrections are not prescribed by a theoretical framework, is the boundary layer model of a flow in a pipe.[3] Ludwig Prandtl's water tunnel was both an autonomous model and a real experimental device that united the ideal flow in the center of the pipe and the two-dimensional friction-dominated boundary layer. It is true; both regimes can be mathematically expressed as appropriate simplifications of the Navier-Stokes equations, simplifications that allow scientists to tract the problem at all. The model's representative features, however, transcend such mathematical unifications.[4] But, representation comes at a price because it restricts a model to a certain class of phenomena. The boundary layer model is unable to explain turbulent flow around obstacles or bows in the pipe.

Similar restrictions to a specific domain of explananda exist for the classical models of the atomic nuclei (cf. Morgan and Morrison 1999, 23–24). While the liquid drop model is able to explain most instances of nuclear fission—in analogy to the splitting of a droplet of liquid—by a phenomenological formula for the energy, the shell model adapts the quantum mechanics of the atomic hull into a model of the nucleus. It can explain—at least at the phenomenological level—why nuclei consisting of certain numbers of neutrons and protons that correspond to closed shells, are more stable than others, such that there could be islands of stability for super-heavy nuclei. Even if one departs from the theory of quantum chromodynamics, one can build autonomous phenomenological models of the nucleus emphasizing distinct theoretical properties (cf. Hartmann 1999).

Arguably, the 'Models as Mediators' approach is not irreconcilable with Giere's conception of models. Revisiting the pendulum, Giere, for one, holds that "to some extent the models themselves provide guidelines for the relevant similarity judgments" (1999, 123). Moreover, his adoption of the investigations of Rosch, Mervis, Gray, Johnson, and Boyes-Braem. (1976) on natural categories for establishing a two-dimensional map of the families of models in classical mechanics is precisely intended to abandon the syntax-oriented statement view of scientific theory then prevailing among philosophers of science. Thus to my mind, the differences between Giere's semantic approach and 'Models as Mediators' rather lie in the fact that Giere, first, emphasizes the prototypical aspects of models instead of their autonomy and, second, that he introduces a gradation that globally separates the vertical levels on his model map. Thus he emphasizes that, according to Rosch, "basic-level terms are more quickly learned that either superordinate or subordinate terms" (Giere 1999, 104). Similarily, George Lakoff's (1987) work on radial structures shows that "the graded structure results from overlaps and other connections among the various models, with the focal model being at the center of a radiating structure. It is the focal model that provides whatever unity the whole family of models may exhibit" (Giere 1999, 106).

To the hierarchy of categories thus established, Giere applies an empirical study of Chi, Feltovich, and Glaser (1981) according to which "novices classify problems on the basis of surface features while experts classify on the basis of underlying physical principles" (Giere 1999, 113). Thus, "part of becoming an expert in a science like mechanics is learning to categorize systems at levels of abstraction *above* the Roschian basic level" (ibid., 113–114). Later investigations of Chi, Hutchinson, and Robin (1989), however, have led to qualifications of the distinction between experts and novices. Using the example of children's knowledge about dinosaurs, they found that even four-year-old children, if they have a sufficient knowledge base, can become experts in a given domain and organize their knowledge in a more integrated and coherent way than novices, while they fail to apply this knowledge to analogous domains about which they have no information. This and subsequent studies led developmental psychologists to believe that the development of knowledge and the distinction between experts and novices is significantly domain specific. To the extent that Giere insists on the global nature of categorical distinctions, rather than a local and domain-specific mode of gradation, the dinosaur study and its rich offspring favor the context-relative autonomy of models characteristic of the 'Models as Mediator' approach.

Not surprisingly, the issue of representation has stood in the center of cognitive scientists' treatment of models. The historical development since the 1970s was characterized by a similar shift away from linguistic entities and formal logic to entities representing in their own right: the mental models advocated most influentially by Philip N. Johnson-Laird (1983).[5] In Stella Vosniadou's concise formulation: "Mental models are dynamic and generative representations which can be manipulated mentally to provide causal explanations of physical phenomena and make predictions about the state of affairs in the physical world" (1994, 48). Despite the significant literature on the topic and its fruitfulness for empirical studies, "the theory was born—and remains—highly controversial, and has attracted admirers and critics in roughly equal numbers" (Evans 1996, 321).[6] In the consideration of mental models, one needs to distinguish "four different things: the *target system*, the *conceptual model* of that target system, the user's *mental model* of the target system, and the *scientist's conceptualization* of that mental model" (Norman 1983, 7), which is in effect a model of a model. Moreover, a mental model arises, and is constantly modified, though the interaction with the target system—which might be a physical model.

Although I believe that mental models bear promising perspectives for the philosophical model debates, I will not pursue this line further here. Instead, the aim of the present paper is to establish a direct dialogue between educational scientists—whether they argue on the basis of a specific psychological theory or not—and philosophers of science. The reason is not so much that the mental model approach has been complemented, or supplemented, by a new paradigm, the investigation of neuronal structures, but, more importantly, that the present paper intends to promote model *dynamics* rather than provide another detailed investigation of model *representation*. The former does not necessarily presuppose a solution of the latter. For scientists and philosophers since Boltzmann's (1902) encyclopedia entry have generally learned to live well with a certain multifacetedness in the meaning of 'model,' embracing internal (mental) as well as external (physical) aspects.

Finally, philosophically minded educational scientists who call for a robust terminology might nevertheless prefer a working functional classification instead of one that is based on deep, yet not fully settled theoretical disputes. Rosaria Justi and John Gilbert, for instance, take a model "to be a representation of an idea, object, event, process, or system" (2000, 994), and distinguish (not directly accessible) mental models, expressed models (in the public domain), scientific models (tested by the community of scientists), consensus models (at the forefront of research), historical models (superseded by the course of science), and curricular models (a simplified version of any consensus or historical model included into an educational curriculum). As convincing as this might seem as a operational classification, the distinction between scientific models and consensus models, for instance, is problematic in a philosophical perspective because poorly tested models will hardly be consensual within the scientific community, and the distinction between the forefront of research and the established knowledge may be drawn either sociologically or by rational reconstruction, yielding not necessarily consistent results.[7]

Themes of Contact—A Sketch

In a survey paper that concludes a special issue of the *International Journal of Science Education*—a main source for the present paper—John Clement has identified a series of

> unanswered questions at the cognitive level that are centrally important and that could make a significant contribution to theories of instruction if we can make progress on them. Among these are: What is the role of mental models in science learning? What is the nature of these models as knowledge structures? What learning processes are involved in constructing them? What teaching strategies can promote these learning processes? (2000, 1041)

The ideal aim of science education is to arrive at a "deeply understood model that the student can use to reason with and make inferences from" (ibid., 1044f.). It is less clear, however, what kind of mental models in the students and in their instructors are most likely to fulfill that aim, and how they are related to physical or virtual classroom models.

Learning scientific facts typically is not a one-shot interaction between teacher and student but extends over a longer period from elementary school to university. Science educators accordingly investigate the process from a preconceived model before instruction—be it the primary model of an infant, the model used at a lower grade, or some socially well-entrenched lay conception—to a target model that is either set out by the instructors or

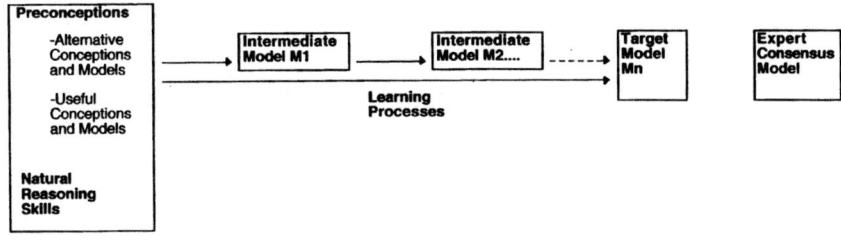

Figure 4.1 Scheme of a learning process from Clement (2000, 1042).

corresponds to the model presently accepted by scientists. Instruction typically proceeds stepwise over the course of an instructional unit or a whole learning career, such that scientific facts and theories—especially those that are complex and far removed from our lifeworld—will be learned through a series of intermediate models. Clement (2000) has provided the above scheme for these steps (see Figure 4.1).

As simple as this scheme might appear at first glance, it contains already some of the basic problems to be addressed in the present paper. (i) What is the status of the intermediate models? (ii) Are they simply pieces of, or dispensable steps towards, the target model, or do they contain necessarily contain ontological commitments? (iii) Are parts of them conserved in the target model or will they be fully replaced, the motivation for teaching them at all stemming from the fact that they make students understand the process of model building?

As Clement emphasizes,

> developing such [mental] models through instruction can turn out to be surprisingly difficult, for several reasons. Hidden explanatory models cannot be observed directly. Students may be accustomed to learning at a more superficial level. New qualitative models may conflict with pre-existing intuitive models, requiring a conceptual change or reorganization to take place. (2000, 1042)

Apart from the problem to access the mental models—rather than students' strategies to prevail in an exam situation—there are different views about the target model. "This may not be as sophisticated as the *expert consensus model* currently accepted by scientists. In addition to (or at some age levels, instead of) logical relationships in formal treatments of the topic, an educator's view of the target model may need to reflect qualitative, simplified, analogue, or tacit knowledge that is often not recognized by experts" (ibid., 1042). After all, the vast majority of students will never become socially recognized experts but rather educated laypersons.

Clement emphasizes the parallels between a "portrayal of learning as involving recurring model construction cycles and "cognitive studies of scientists" (ibid., 1049)—a heading under which he subsumes historical and epistemological investigations.

> Much of the recent progress in history and philosophy of science can be seen as a struggle to move away from a simplistic view of how theories are formed in science as being either pure induction upward from observations or pure deduction downward from axioms followed by testing. Instead, we see movement toward a view that involves both top-down, bottom-up, and horizontal (e.g., analogical) processing in a cycle of conjecture and modification. (ibid., 1050)

Citing earlier investigations from this field, among them Nersessian (1984), Giere (1988), and Darden (1991), Clement argues

> that there is now considerable empirical grounding for progressive model construction via revision in science. This is not to deny Kuhn's claim that major revolutionary paradigm shifts and the ensuing fractures in scientific theory are sometimes required; but even there the development of the new view is found to require many cycles of criticism and revision rather than being a single insight. (2000, 1050)

Clement, accordingly, locates the interaction between educational scientists and philosophers of science in the domain of model and theory dynamics rather than model statics and representation.

It is the aim of the present paper to take up this line of thought and elaborate on five issues that have emerged from philosophically informed studies of educational scientists.

1. *Discontinuous versus stepwise revision.* Since Kuhn's *The Structure of Scientific Revolutions* (1962), and even before, it has become a commonplace that conceptual change proceeds steadily or discontinuously. Although the philosophical discussion about paradigms has mainly focused on scientific theories and their relationship with empirical data, the dichotomy 'continuity versus fracture' emerges for models as well, particularly if they are endowed with explanatory features and are not just considered as didactical instruments. Science educators accordingly have investigated why certain intermediate models can be modified by instruction less easily than others, why certain paths of intermediate models work out in practice while others do not, and how we can understand that some modifications go smoothly while others require a significant gestalt switch.

2. *Are intermediate models mere instruments or necessary steps?* The role and status of intermediary models poses ontological and epistemological questions above and beyond those in the center of the present model debates. Are they formed merely in an ad hoc fashion and

do they vary across the individual students, in their different attempts to get hand at a newly introduced target model that is difficult to understand? Or are there necessary intermediate steps leading from the initial model to a target model that expresses new and difficult facts? If so, do such intermediate steps correspond to models actually advocated in the history of science? If not, are they merely instrumentalist learning tools without ontological commitments on the part of the learner?

3. *Experimenting with models.* Imre Lakatos's (1976) conception of mathematical proofs as thought experiments has been highly influential on classroom didactics (cf. Hanna 1997 and Hanna and Jahnke 1999), and the famous polyhedron problem around which *Proofs and Refutations* was built, can well be understood as a model.[8] This suggests inquiring whether students can experiment with models in more than a playful fashion, and whether such experimentation comes close to Morgan and Morrison's (1999) understanding of scientific models as measuring instruments.

4. *Hierarchy versus autonomy, or the semantic view versus 'Models as Mediators.'* If, in the sense of 2., there is some degree of internal necessity in the sequence of intermediate models, one can investigate whether such a sequence has affinities to Giere's (1999) hierarchy of models extending from the concrete to the abstract or whether these intermediate models enjoy an explanatory and representative autonomy in the sense of Morgan and Morrison (1999). Do these different views on the relationship between model and theory relate to the difference between target model and expert consensus model outlined by Clement? For as mentioned above, expert consensus models are more likely to be incorporated into a mathematized scientific theory aimed at generality, while the target of instruction might well consist in a comprehensible simplified model that suffices for all practical purposes.

5. *Bottom-up empirical approaches.* In the age of experimental philosophy, empirical investigations into an epistemological concept seem no longer outlandish. In this perspective it is promising to continue a line of research launched by the late Daniela Bailer-Jones (2002), by combining more refined methods of social research and philosophical classifications. Science educators in turn should be interested in such analyses since they have shown that instruction is more effective if educators are aware of the epistemology of scientific models.

In the subsequent sections, I will not treat these issues one by one but steer a middle course between keeping them neatly apart and preserving the integrity of the discussions among educational scientists, even if they involve multiple issues from the above list.

ON THE SIGNIFICANCE OF EPISTEMOLOGICAL REFLECTION—SOME EMPIRICAL EVIDENCE

Let me begin with the fifth issue and compare the empirical investigations, mainly on the basis of interviews and questionnaires, of the philosopher Bailer-Jones and two educational scientists about the use of models in scientific practice and in the classroom. They reveal a great variety of views and starkly different usages of a genuinely philosophical terminology, both among scientists and science teachers, that bring mainly the fourth issue to the fore: hierarchy versus autonomy.

Even though Bailer-Jones restricted her expert interviews about "Scientists' Thoughts on Scientific Models" to a highly homogeneous sample involving nine scientists from the United Kingdom, who were working mainly in physics at the Open University, she obtained a pretty variegated set of responses in which terminological and thematic issues were intermingled. Thus she could draw only general conclusions.

> (1) Models are perceived as central to doing science. . . . (2) Just as in philosophy, there is no uniform, and certainly no sharp, distinction between theory and model that is shared by scientific practitioners. . . . (3) Definitions of what a scientific model is are diverse. It therefore transpires that any philosophical definition of what a model is would have to be broad in order to cover at least some of the main uses of 'model' by scientists. (2002, 297f.)

While the second point reaffirms a core feature of the present debates about models, the third recaps the challenges any descriptively oriented encyclopedia entry on 'model' would face.

From a philosophical perspective, Bailer-Jones diagnosed three dichotomies: The first, capturing the essence by simplification and idealization versus accuracy, is exemplified by the difference between computational models that are accurate once appropriate data correlations are specified and physical models that reveal (at least in part) the laws and causal processes involved.[9] "Another dichotomy is that models are expected to be subject to and live up to empirical tests, yet, they are not necessarily expected to be correct or true" (ibid., 298). On the basis of the first two dichotomies, there emerges a third of being about reality versus only capturing the essence. Dealing with this dichotomy, Bailer-Jones expresses skepticism about the concept of representation. "Representation does not meet the criteria of empirical adequacy or scientific realism, although there is an aspiration to both" (ibid., 299). This dilemma, to my mind, was a substantial motivation for the Morgan-Morrison program to emphasize that models both involve an autonomous representation—yielding a certain dose of realism in a sufficiently large domain between theory and empirical data—and act as measurement devices—providing thus a standard of accuracy. But it is not only the bottom-up studies of philosophers that show that the fourth theme, the distinction between semantic approaches and 'Models as Mediators,' arises within daily practice.

The educational scientists Harrison and Treagust (2000) have put forward a pretty complex and fine-grained typology of school science models that is based on the observation of science teachers in class and interviews conducted with them. They suggest that there is an overall scheme according to which students begin with concrete scale and iconic models before advancing to more abstract or mathematical models and that the respective levels can be empirically distinguished—as has already been advocated by Grosslight et al. (1991). Such a leveling seems to support Giere's hierarchy of models, but Harrison and Treagust's series of models is not targeted to any theory representing the envelope of all these models. Instead, they point out "that no single model can ever adequately model a science concept; therefore, students should be encouraged to use multiple explanatory models wherever possible. This is best done by teachers modeling multiple modeling in their lessons. In its simplest form, this requires teachers to avoid early closure in discussions by asking the students for 'another model please.' It also asks teachers to socially negotiate model meanings with their students and to regularly remind students that all models break down somewhere and that no model is 'right'" (2000, 1023).

This suggestion for the classroom combines three distinct aspects. First, many models have indeed a limited domain of applicability because they emphasize a certain feature, for instance of the atomic nucleus, and are explanatory for certain observable effects (See Section 'Themes of Contact—a Sketch'). Second, "model-based thinking is a sophisticated process that should be an explicit part of learning in science [. . . , and] teachers should be sensitive to the familiarity of, and similarities and differences between the models they use to explain science phenomena" (ibid., 1023). Only then can they properly guide students in generating their own model that will substantially differ from the expert model. Thus in the end, a teaching that increases student ownership, rather than trying to water down an expert model, requires two levels of epistemological reflection: in the teacher and in the student thus encouraged by the teacher. No doubt, an important aspect of instructional practice is to convey the ability to build and critically apply scientific models. But, third, the combination of pluralism and ownership bears the danger to lose sight of the representative features of models and advocate an approach that is either radically constructivist or undervalues the ties with the history of science and the validity claims inherent in scientific knowledge. At least at first glance, it seems to me that the slogan "child as a researcher" runs such risks by combing children's autonomy with a perspective that is strongly informed by philosophical hermeneutic (cf. Kellet 2005; Kinash and Hoffmann 2008). The autonomy of scientific models does not imply the nomological autonomy of the modelers; both experimenters and modelers navigate within the ocean of natural laws—even though these laws may be unknown or predictively worthless.

There exists even direct empirical evidence for the importance of a solid epistemological background for model instruction. Once again the fourth theme comes to the fore. Two Dutch studies (van Driel and Verloop 1999; Justi and van Driel 2005) have found that instructing science teachers about the epistemology of models—the authors here explicitly mention Giere (1999), Morgan and Morrison (1999), and Nersessian (1999)—improves classroom performance. "The most significant changes were those related to how: (i) they introduced and/or used the idea of models as partial representations, (ii) they introduced and conducted modeling activities, (iii) they used teaching models, and (iv) they started to pay attention to students' ideas" (Justi and van Driel 2005, 564). The context of this and earlier studies was a new Dutch high school curriculum that introduced public understanding of science as a separate subject, designed "to make students aware of the ways in which scientific knowledge is produced and developed" (van Driel and Verloop 1999, 1144).

On the basis of two types of questionnaires and their statistical analysis, Jan van Driel and Nico Verloop diagnosed "a large degree of variety of the teacher's knowledge of models and modeling in science" (1999, 1146) that is not so different from Bailer-Jones's aforementioned findings.

(1) The teachers held different beliefs with respect to the *representational modes* of scientific models. When asked to respond to seven specific examples, one teacher classified all these, including a picture of a house and a toy car, as a model, referring to each example' s potential to represent specific aspects of reality. On the other hand, other teachers rejected almost all the examples, including a molecule of water. In the view of these teachers, explanatory potential appeared to be an important criterion for an example to qualify as a model.

(2) The teachers emphasized different *functions* of models. Specifically, the explanatory function and the descriptive function of models were stressed. However, some important functions (e.g. using models to make predictions) were rarely mentioned.

(3) The teachers mentioned different *characteristics* of models. All ... mentioned the relation between model and target.... [Some additionally said that a model is a research tool, that it bears analogies to the target, that it is simple, and that it is developed through an iterative process.] The latter group was expected to have rejected most of the examples of models presented earlier, arguing that these examples would not comply with all the characteristics. No relation could be found, however, between the number of characteristics the teachers mentioned, and the number of examples (see (1)) they classified as models. . . .

(4) [As regards epistemology, m]ost of the teachers displayed a constructivist orientation, indicating, for instance, that different models can co-exist for the same target, dependent on the researchers' interest or theoretical point of view. A minority of the teachers, however, reasoned in terms of logical positivism. These teachers stated that a model should always be as close to reality as possible and that a model may become 'outdated' when new data are obtained. However, it was not possible to identify relations between the teachers' epistemological orientations and the other themes discussed above. (ibid., 1146f.)

Item (1), the representative modes, and (2), the different functions, on the above list correspond to the items (3), the diversity of scientists' understanding of models, and (2), the model-theory distinction, from Bailer-Jones's list respectively. Item (3) above, teachers' diverging views about the characteristics or the essential features of models, to my mind and owing to the difference between the instructional and the scientific context, does not match Bailer-Jones's second dichotomy, the question as to how realistic a model is. Item (4), the general epistemological outlook, erroneously identifies logical positivism with some brand of convergent realism and thus fails to appraise that all logical positivists adopted a conventionalist position according to which there could well be competing theories because theory is always underdetermined by the data. Subscribing to the underdetermination thesis, however, does not presuppose constructivism. This muddled philosophical terminology may be responsible for the surprising lack of correlations between (4) and (1)–(3); this is unfortunate also because the tension between a constructivist approach and one that stresses the rationality of science, either on a realist or an empiricist basis, sits at the bottom of many discussions about model dynamics. Pursuing this matter further brings us to the oft-discussed relationship between rational reconstruction and historicity of science that is of vital importance for the status of intermediate models, especially in cases in which these are more than mere teaching instruments. Accordingly, the next section mainly deals with issue 2.

TARGET OR PATH? HISTORICAL PRESENTATION OR RATIONAL RECONSTRUCTION?

Using the example of models, Rosaria Justi and John Gilbert investigate how "the systematic inclusion of the history and philosophy of science (HPS) in science education" can contribute to "making the latter more closely related to the process and conduct of science *per se*" (2000, 993). Historical models, on their account, are situated in the core of a Lakatosian scientific research program. Our conception of the atom, they argue, proceeded in six distinctively separated research programs: from the Ancient Greek model, Dalton's (chemical) model, Thomson's 'embedded mass' (or plum pudding) model, Rutherford's 'nuclear' model, Bohr's 'orbit' model, and the quantum mechanics model. At each stage, they identify the respective hard core and: "(1) The deficiencies in the explanatory capability of a given model; (2) The features of that given model that were modified and incorporated into the new model; (3) The way by which the new model overcame the explanatory deficiencies of its antecedents; (4) The subsequently emergent explanatory deficiencies of the new model" (ibid., 995).

Against this backdrop, Justi and Gilbert argue that the school curricula, both in Brazil and the United Kingdom, neither clearly refer to, nor make appropriate use of, historical models. Moreover, there is a lack of distinction between 'model' and 'theory,' "the theoretical background on which each model is based is generally not clearly discussed," and "[s]ome textbooks assume that the introduction of a different positive heuristic signals the introduction of a new model, rather than being either an improvement to a given model or a step towards the refutation of that model" (all ibid., 1001). Unfortunately to their mind, most textbooks analyzed by them presented hybrid models. "A hybrid model . . . is not a curricular model, that is a simplified version of one distinct historical model [included into an educational curriculum]. It is instead something in which the characteristics of several distinct historical models have been merged" (ibid., 1002). As a consequence of teaching hybrid models,

> students think that: there is an absolute and unchanging conception of the atom; the production of experimental data, and not the hypotheses on which they are based or the interpretations derived from such data, are the most important part of a scientist's work, and; if a given model is changed by another scientist, the scientist who had proposed that model had made experimental mistakes. (ibid., 1005)

These students obviously fail to understand the dynamics of models and theories and because of their inadequate philosophy of scientific progress erroneously charge great scientists of mistake—a well-known leitmotif of Lakatos's (1978) methodology of scientific research programs (MSRP).

Accordingly, Justi and Gilbert call "for the elimination of hybrid models" (2000, 1007). But, to my mind, this needs qualification. First, as Clement rightly observes,

> if a hybrid intermediate model appears to be more readily understood than a historical model—should it be disallowed? And more generally, if a successful learning pathway through certain intermediate models is different than the historical pathway, but that pathway appears to be more natural and expedient for learning, which should be used? If the first pathway is not misrepresented as history, can it still convey the spirit of model improvement in science? . . . It may be a difficult trade off decision where it arises. (2000, 1050f.)

Clement's valid criticism can be complemented by the epistemological observation that historians and philosophers of science occasionally embark on counterfactual histories, not only to accentuate the underdetermination thesis but to better understand the course of events and its rationale. A model thus construed would accordingly be classified as a hybrid model, even though it occurs within a historical-critical analysis.[10]

Second, Clement rightly points out that students' idea of a 1:1 correspondence between model and reality might have more practical roots than a false epistemology.

> The idea that future models will supersede the current expert consensus model is not intuitive [for students]. They may think of learning as memorization of the expert model rather than model revision. And they may be impatient with the idea of learning one or more intermediate models as stepping-stones to a more sophisticated model. They might say: 'Why not save time and just memorize the final target model?' (Clement 2000, 1047f.)

As the previous section has demonstrated, instructing students and teachers about model building and model dynamics is both rewarding and difficult.

There are also three epistemological caveats. First, quite a few historians of science consider Lakatos's rational reconstruction as precisely the way how not to understand a model as a historical entity, but to disassociate it from how it actually emerged, how it historically gained acceptance, and why it was factually abandoned in the end.[11] This difference leads back to Clement's criticism because rational reconstruction typically buttresses a certain epistemological stance in the same vein as a series of intermediate hybrid models is constructed under the auspices of a certain psychology of learning. It is true that Justi and Gilbert might retort with the suggestion to include the troubling interaction between history and philosophy of science into the curriculum as well.

Second, Lakatos's own approach differs significantly from Justi and Gilbert's insofar as he did not subsume models under the negative heuristics safeguarding the hard core against untimely refutation but placed them into the protective belt that "saves the scientist from becoming confused by the ocean of anomalies. The positive heuristic sets out a programme which lists a chain of ever more complicated *models* simulating reality. . . . [The scientist thus] ignores the *actual* counterexamples, the available '*data*'" (1978, 50). And Lakatos defines: "A '*model*' is a set of initial conditions (possibly together with some of the observational theories) which one knows is *bound* to be replaced during the further development of the programme" (ibid., 51). This implies that one factually experiments with models, not only in thought—as mentioned above when I introduced issue (3.) with reference to Lakatos's philosophy of mathematics– but also as an attempt to discover new empirical confirmations for the research programme under the guidance of the positive heuristic. Models, in the perspective of rational reconstruction, come in directed series. Within Newton's research program, for instance, Lakatos identifies Newton's initial model of a fixed sun and one revolving planet that, due to its inconsistency with the force law, was replaced by both bodies revolving around their common center of mass, and finally led into perturbation theory. Within the early history of quantum physics, to give a second example, Bohr's 1913 model of the atom represented only as a first step that was superseded by the model of ionized helium and then the introduction of a reduced mass (cf. ibid., 61–63). Lakatosian models, it becomes clear, are subservient to the hard core and the theoretical assumptions contained in it. The examples clearly show that Lakatonsian models correspond rather to a theoretical model in the style of the mathematical pendulum and downgrade precisely the representative features of models and their relative autonomy from theory stressed by Morgan and Morrison (1999).

Third, as is well known since the work of Laudan (1977), MSRP awards only predictive success and not theoretical progress. Thus models can only count as progressive if, when functioning as a measuring device, they are able to predict new data, and not if they provide a unifying representation of a previously known domain of facts. For the aims of science education, this is not particularly helpful, especially because this discipline, in contrast to psychology and cognitive science, has a normative side: empirically justified recommendations for the classroom. It thus appears that Lakatos's MSRP fares worse in the classroom and among science educators than has his *Proofs and Refutations*.[12]

ABSTRACTION, SEMANTICS, REPRESENTATION

While MSRP neither describes nor rewards any tendency towards abstraction, two other empirical studies by science educators address precisely that point when studying the dynamics of intermediate models. Apart from being thus a contribution to problem (2.), they are both relevant for the semantic approach (4.). Additionally, I wonder whether the first of them does not reveal a discontinuity in the learning of models in the sense of (1.).

Janice D. Gobert (2000) had 47 fifth graders read a text describing plate tectonics and subsequently asked them to draw diagrams of the spatial, causal, and dynamic processes inside the earth that make continents drift. In doing so, she treated spatial, causal, and dynamic information as corresponding to an increased sophistication in forming conceptual models and to a progress from left to right in Figure 4.1 (p. 67). The study identified two types of models of the inside of the earth:

> spatially incorrect models, . . . in which the spatial arrangement of the layers is not correct; and spatially correct models, . . . in which the

> inside of the earth is depicted in concentric circles. . . . Four types of models of the causal and dynamic processes involved in volcanic eruption were identified, ranging from local models including heat-related mechanisms only, local models including movement-related mechanisms only, mixed models including some heat- and movement-related mechanisms, and integrated models (the most sophisticated type of model observed at this age level). (2000, 960)

There were also students who formed models that became a barrier to a deeper understanding. One failed to conceive the inside layers of the earth as concentric circles. Until this spatial arrangement is remedied, "she will not be able to use her model of the earth to engage in model-building and inferencing about the causal and dynamic processes involved in plate tectonics" (ibid., 947). On the other hand, correct spatial models became a useful tool for progressive model building through a series of constant modifications and improvements. Gobert's investigation both supports a conception like Giere's and the 'Models as Mediator' account because in the sequence of model building, the models become more abstract, and they are themselves used as devices to construct further models. That an inappropriate model turns into an impediment speaks for it containing representative features.

Another study by Jennifer L. Synder (2000) directly addresses Giere's (1999) above-discussed claim that models are dwelling on certain levels of a Roschian categorical hierarchy and that novices and experts categorize problems at different levels. Snyder's focus was to investigate the transition between these levels and the relationship between models and theory. She asked a group of nine experts (university professors), nine intermediates (graduate students), and nine novices (who had only one introductory course in the field) to classify a set of problems of classical mechanics according to model-based attributes, among them coefficient of friction and inclined plane, versus theory-based attributes, among them Newton's laws and the equations of motion. Comparing the proportions of theory- and model-based categories, she found that novices predominantly created hierarchies of model-based categories, while both intermediates and experts created theory-based categories at the highest, most abstract level but combined model- and theory-based categories at middle and lower levels. "It was expected that intermediate subjects' categorizations would be a hybrid of expert and novice subjects' categorizations. Instead, intermediates categorized in generally the same way as experts did in terms of using theories to categorize the problems more than did the novices" (Snyder 2000, 988).

Snyder takes the absence of a region of transition as a significant problem for Giere's approach and concludes that her "results support a representation of physics knowledge in which a hierarchy of models is organized by theory-based categories" (ibid., 979). If one retorts that this was to be expected in a domain as theory-laden as classical mechanics, one implicitly assents to Morgan and Morrison's above-quoted criticism that the semantic view is limited to theoretical models. A stronger limitation to Snyder's criticism of Giere is, I think, that one may wonder whether the separation into model-based and theory-based categories is really so rigid as she claims. Why, for instance, does velocity range under model and force under theory (cf. 2000, 983, Table 1)? To my mind, they both stand on a par if one does not only look at the Newtonian axioms but at the fact that the treatment of any mechanical problem requires the specification of the forces acting and the initial positions and velocities of the mass points. And there are mathematical reformulations of Newtonian mechanics in which the velocities disappear and such in which forces do not exist.

DISCONTINUOUS VERSUS STEPWISE REVISION: THEORIES OF CONCEPTUAL CHANGE

There exist a number of studies providing empirical evidence that some concepts or models are learned smoothly and in small increments while others, among them atomic models, require substantial changes in the learner's knowledge structure, such that they lead to a larger number of misconceptions and occasionally meet stubborn resistance on the part of the students. The two studies discussed in this final section provide diverging psychological explanations for this phenomenon and involve quite specific philosophical presuppositions that concern the ontological status of intermediate models (issue 2.) and the relationship between conceptual fractures and scientific revolutions (issue 1.).

Chi, Slotta, and de Leeuw have elaborated a theory of conceptual change that is based on a rather rigid ontological classification of concepts into categories and subcategories (See Figure 4.2). For their analysis of the rich empirical material, they assume "that the meaning of a concept is determined by the category to which the concept is assigned. Conceptual change occurs when the category, to which the concept is assigned, changes" (1994, 27f.). Changes between ontological subcategories on the same tree, e.g., Natural Kinds and Artifacts, often can be accomplished by invoking "higher order ontological attributes that span them both, such as 'is brown' and 'can be contained' for objects such as a dog and a knife.... [However,] ontological attributes of trees are mutually exclusive. There is no higher order ontological attributes that can span entities on two different trees" (ibid. 30). Thus, conceptual change between concepts belonging to different trees is most difficult to achieve.

Students' naïve categorizations of some basic physical concepts, such as electric current or force, are often situated on the matter tree, while in actual fact they represent constraint-based interactions, i.e., processes determined by a known or knowable set of constraints.

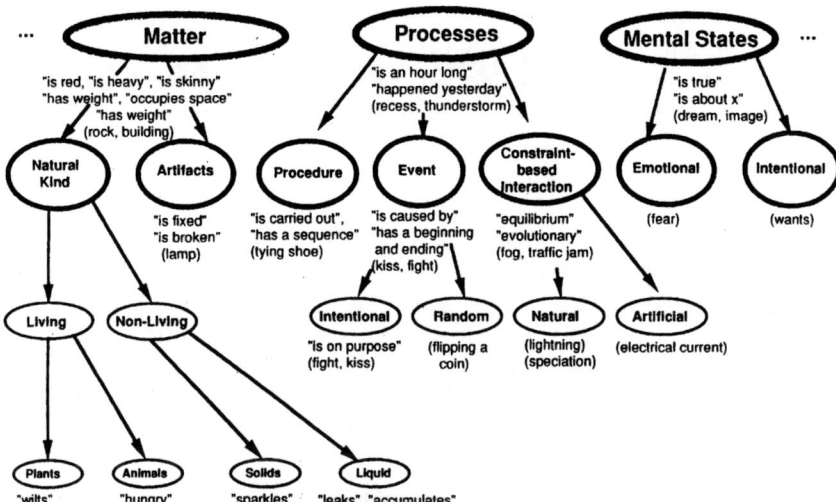

Figure 4.2 An epistemological supposition of the nature of our conceptions about the entities in the world. The three primary categories of MATTER, PROCESSES, and MENTAL STATES are ontologically distinct. Courtesy of Elsevier Science and Technology Journals from *Learning and Instruction*, Chi, Slotta, and de Leeuw, 29.

Current exists only when electrically charged particles are moving, usually because of an electric field. A field fills all space, but an electrical current exists only when a charged particle is introduced into the field. Hence, an electrical current is neither MATTER nor properties of MATTER, but a PROCESS that is fundamentally constraint-based and has no causal agents. (ibid., 31)

Confusion may arise, because Constraint-Based Interactions involve components of other ontological categories, especially MATTER. Returning to the example of electrical current, the MATTER components include moving particles, wires, batteries, and so on. But the involvement of these components does not imply that electrical current belongs in the same category with them. (ibid., 32)

On the basis of this conception, Chi et al., posit an incompatibility hypothesis: "When a student's initial representation of the concept is incompatible with the concept's veridical ontological status, then learning the concept requires *conceptual change,* meaning that the concept's categorical membership has to be re-assigned *across trees*" (ibid., 34).

In such cases, students' misconceptions show characteristic traits. The naïve conceptions are robust, consistent (the same student displays it in different contexts over time), persistent across different ages and schooling levels, homogeneous across different students, and recapitulated across historical periods "(i.e., the medieval scientists and contemporary naive students tend to hold the same misconceptions)" (ibid., 35). If on the other hand, students remain within one tree, misconceptions should be randomly distributed. As a counterexample, Chi et al., were "tracking learning of a 102-sentence passage about the human circulatory system by eighth grade students" (ibid., 36). Since,

> [i]n junior high school textbooks, [in contrast to university textbooks] discussion of the human circulatory system consists almost entirely of conceptions which belong to the MATTER category, because the topics covered pertain strictly to physical connections between the blood vessels and the heart, lungs and other organs, as well as the direction and paths of blood flows, (ibid., 33)

none of the above regularities was statistically significant. Most interesting in the present context is that their misconceptions, in contrast to the examples from physics, did not significantly repeat the conceptions of medieval scientists. "Only 30% of the false beliefs mentioned more than once were at all related to historical misconceptions" (ibid., 38).

Although Chi et al., do not explicitly talk about models, it is clear that the ontological characterizations of electric current or the circulatory system are exemplified in terms of mental models, say of electric fluids or a circuit of blood vessels. On their account, models rigidly belong to an ontological category, and although complex models may involve other categories, the primary category determines whether they are easily abandoned or resilient to change. In virtue of the above characteristics, Chi's conception parallels the history of science and learning, and thus distinguishes normal development within a tree and revolutionary chance across trees. However, this parallel comes at a high price because from the perspective of a philosopher of science, Chi's ontology appears pretty Aristotelian and too static for a historical account of model dynamics. Moreover, if university textbooks, for instance, present the circulatory system as a process, should eight graders not somehow have heard of that, feeling no surprise with their matter-based high-school textbook? The categories thus seem to require too hermetic a separation of social contexts.

Stella Vosniadou (1994) accordingly criticizes that some of Chi et al.'s categories "seem rather arbitrary" (ibid., 65) and that her theoretical approach is syntactic instead of semantic. While I agree with the first criticism, it seems to me, that Chi et al.'s (1994) concepts are endowed with too rigid a semantics through the categories they fall into. It is true, such semantics is unable to incorporate the representative features of mental models that are the center of Vosniadou's investigation and that, to my mind, provide a more adequate picture of model dynamics that allows one to distinguish two tiers of learning, continuous and discontinuous, without previously making strong ontological commitments. What remains to be achieved is to relate the mental models she builds upon to the philosophical model debates.

> It is assumed that conceptual change proceeds through the gradual modification of one's mental models of the physical world, achieved either through *enrichment* or through *revision*. Enrichment involves the addition of information to existing conceptual structures. Revision may involve changes in individual beliefs or presuppositions or changes in the relational structure of a theory. Revision may happen at the level of the *specific theory* or at the level of the *framework theory*. Revision at the level of the framework theory is considered to be the most difficult type of conceptual change and the one most likely to cause misconceptions [because it requires to abandon the most fundamental presuppositions]. Misconceptions are viewed as students' attempts to interpret scientific information within an existing framework theory that contains information contradictory to the scientific view. (Vosniadou 1994, 46)

A key point in Vosniadou's conception is that, already in infants, there exists a

> *framework theory of naive physics* which is not available to conscious awareness and hypothesis testing. This framework theory of physics constrains the process of acquiring knowledge about the physical world in ways analogous to those that research programs and paradigms have been thought to constrain the development of scientific theories (Kuhn, 1977; Lakatos, 1970). (Vosniadou, 1994, 47)

Since Vosniadou is after distinguishing two types of change, those involving the framework (revision) and those involving specific theories only (enrichment), the reference to Kuhn is the more basic one. Lakatos's MSRP, however, comes into play in the description of the intermediate models which represent strategies to integrate new features that are inconsistent with the previous model.

From a psychological perspective, the main place where new information is incorporated, requiring revisions of different degree, are mental models. "It is assumed that most mental models are created on the spot to deal with the demands of specific problem-solving situations. Nevertheless, it is possible that some mental models, or parts of them, which have proven useful in the past, are stored as separate structures and retrieved from long-term memory when needed" (ibid., 48). Since mental models are representations, if internal ones, they can be reconstructed from empirical inquiry as depicting a certain set of beliefs the student verbally expressed. In a previous

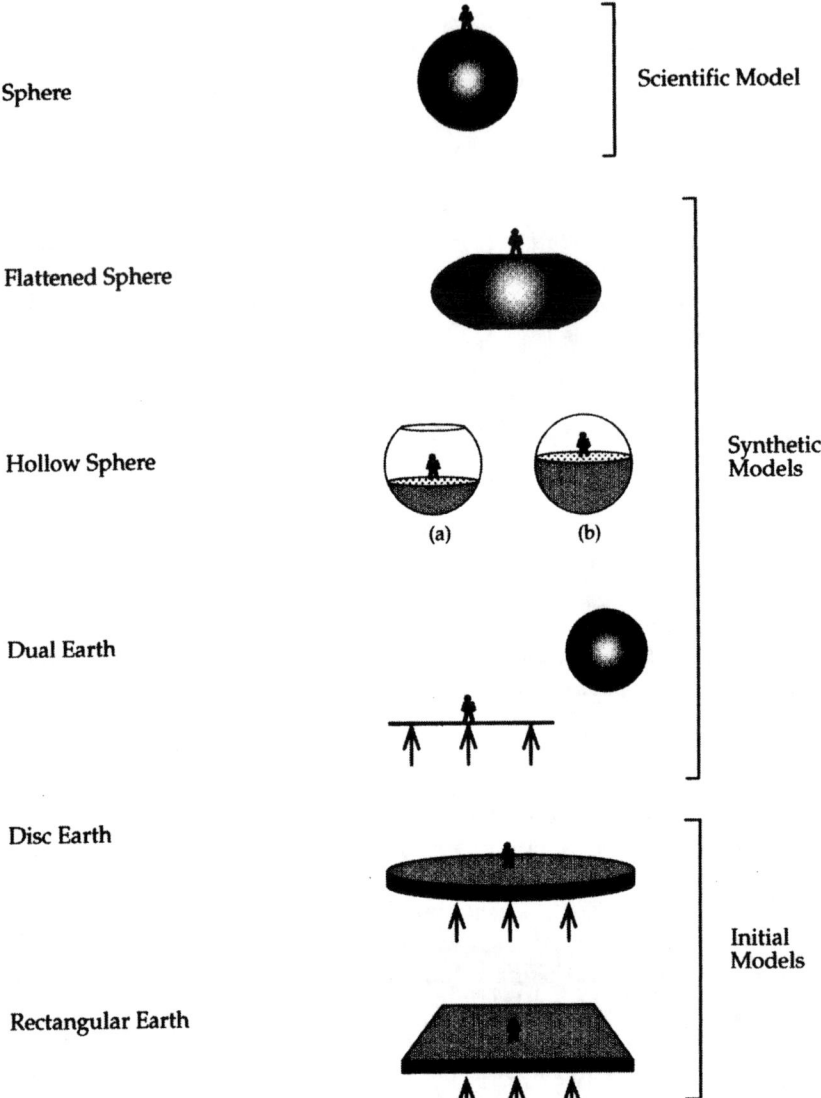

Figure 4.3 A series of mental models of the Earth. Courtesy of Elsevier Science and Technology Journals from *Learning and Instruction*, Vosniadou, 53.

study, Vosniadou and collaborators had asked school children from grade one to five about their concept of the earth. Testing for the internal consistency of their responses, i.e., that they not just memorized knowledge parts, the investigators found that the transition from the naïve initial model to the scientific mode proceeded through a variety of synthetic intermediate models that revealed various misconceptions (Figure 4.3) because students simply added or integrated a novel feature that was in actual fact inconsistent with the previous framework.

Children "seem to start by categorizing the Earth as a physical object—rather than as an astronomical object—and apply to it all the presuppositions that apply to physical objects in general" (ibid., 54). Dual earth is a way to solve the problem by simple accretion, while the hollow sphere and the flattened sphere correspond to only partial changes in the underlying beliefs and presuppositions. As expected, younger children are more likely to form synthetic models, and at fifth grade all children understand Earth as an astronomical body. Students' explanations of the day/night cycle revealed a similar sequence of mental models that were correlated with the more basic assumptions about the Earth.

In both cases it was found that the synthetic models, on the one hand, were formed in an ad hoc fashion and in themselves did not correspond to deeply held beliefs, but that, on the other hand, no direct transition from the initial to the scientific model was observed. Accordingly, concepts and models act as constraints and some are prerequisites to others. This indicates the representative features of synthetic models, even though students do not actually experiment with them. Synthetic models are not only a feature of instructing young children, but they also occur in the history of science, if new features, e.g., the quantization of radiation or Bohrian orbits, are introduced into a previous model, e.g., of electrodynamic processes, even though one knows that they contradict the latter.

In discussing the possible implication of her findings for instruction, Vosniadou recommends, among a general epistemological awareness, to "[t]ake students' mental models seriously and create environments that allow students to express their representations of situations, to manipulate them, to test them, and to have the experience of revising them successfully" (ibid., 67f.)—in short, to experiment with them. This shows that despite the problem of epistemic access to our mental models, they have both representative and explanatory features that are constrained, but not deductively implied, by the framework theory. Since the misconceptions are instantiated by inconsistent synthetic models, Vosniadou's model dynamics cannot be thought of as a hierarchy in the sense of the semantic view because theories thus conceived as an envelope of models, it seems to me, still have to be consistent. Rather can Vosniadou's conception of model be understood as a psychological correlate to the 'Models as Mediators' conception that emphasizes their (context-dependent) representative features.

OUTLOOK

To conclude, a lot has still to be fleshed out on the way towards a philosophical model dynamics. If one could establish that students can actually experiment with models, intermediate models might obtain a certain degree of autonomy as required by the Morgan-Morrison program. Otherwise they would rather be helpless attempts of the learner to come to terms with the learning situation. Another point concerns the difference in education levels. To what extent is an investigation of university students and professors like Snyder's—which is, to my mind, reconcilable with Giere's outlook– at all comparable to Vosniadou's studies of elementary school children that supports the 'Models as Mediator' approach? Moreover, as Chi, Hutchinson, and Robin (1989) above-mentioned dinosaur study shows, there will perhaps not be a monolithic answer. Philosophical model dynamics, to be sure, cannot be restricted to adult experts. Another promising approach might be to start empirical investigations into the teaching of some of those models that have stood in the center of the philosophical debates, among them the shell model and the liquid drop model of the atomic nucleus. Pursuing such questions would foster the interaction between philosophers of science and educational scientists in the study of model dynamics.

NOTES

1. This paper has emerged from a joint seminar with Cornelia Gräsel (University of Wuppertal), to whom I am indebted for many suggestions and critical remarks. I also want to thank the editors and the anonymous referees for their thoughtful comments and suggestions.
2. A case in point is a 2000 special issue of the *International Journal of Science Education*, in which some of the papers discussed here have been published.
3. Although most scientific authors, it is true, prefer the term boundary layer *theory*, I consider Morrison's classifying Prandtl's approach as *model* to be essentially correct. The sometimes fluid terminology is not only a consequence of scientists' rather diverging distinctions between theory and model—see the study of Bailer-Jones (2002) discussed in Section 'On the Significance of Epistemological Reflection—Some Empirical Evidence'—but it is also a product of history and highly domain-specific. Here is a converse example: the standard *model* of elementary particle physics is as much specific a theory as one can get in this field, at least as long as quantum field *theory* is able to provide only a conceptual framework.
4. Still, there are contexts where this mathematical unification should be taken into account, see (Stöltzner 2004).
5. Giere (1988, 6) mentions mental models alongside schemata, cognitive maps, and frames. For a history of mental models, see (Johnson-Laird 2004).
6. Evans's paper is an afterword of a collection of essays assessing the various aspects of the debates about mental models.
7. As we shall see in Section 'Target or Path? Historical Presentation or Rational Reconstruction?', Lakatos's methodology of scientific research programs is indeed the authors' philosophical background. But even within this framework, the problem of distinction remains.
8. Cf. Lakatos (1976, 8), where the term 'model' however does not appear. In the methodology of scientific research programs, Lakatos's use of 'model' does not involve representative features (see in Section 'Target or Path? Historical Presentation or Rational Reconstruction?').
9. Note that this dichotomy is not limited to computational models. For, Hughes (1999) and Batterman (2002) have shown that mathematical models

in statistical mechanics can also capture the essence, even though they are known to be both numerically inaccurate and physically wrong.
10. Ernst Mach (1989) was one of the first to emphasize the value of alternative historical pathways for both the history and philosophy of science.
11. Cf. Kuhn (1971) and his constant efforts to properly distinguish his roles as a philosopher and a historian of science (1977, 3–20).
12. See the above quoted (Hanna 1997) and (Hanna and Jahnke 1999).

REFERENCES

Bailer-Jones, Daniela M. (2002). "Scientists' Thoughts on Scientific Models." *Perspectives on Science* 10: 275–301.
Batterman, Robert W. (2002). "Asymptotics and the Role of Minimal Models." *The British Journal for the Philosophy of Science* 53: 21–38.
Boltzmann, Ludwig (1902). "Model." In *Encyclopedia Britannica*, 10th ed. (Reprinted in *Theoretical Physics and Philosophical Problems. Selected Writings*, 1974, 213–220, Dordrecht: Reidel).
Chi, Michelene T. H., Paul J. Feltovich, and Robert Glaser (1981). "Categorization and Representation of Physics Problems by Experts and Novices." *Cognitive Science* 5: 121–152.
Chi, Michelene T. H., Jean Hutchinson, and Anne F. Robin (1989). "How Inferences about Novel Domain-Related Concepts Can Be Constrained by Structured Knowledge." *Merrill-Palmer Quarterly* 35: 27–62.
Chi, Michelene T. H, James D. Slotta, and Nicholas de Leeuw (1994). "From Things to Processes: A Theory of Conceptual Change for Learning Science Concepts." *Learning and Instruction* 4: 27–43.
Clement, John (2000). "Model Based Learning as a Key Research Area for Science Education." *International Journal of Science Education* 22: 1041–1053.
Darden, Lindley (1991). *Theory Change in Science: Strategies from Mendelian Genetics*. New York: Oxford University Press.
Evans, Jonathon S. B. T. (1996). "Afterword. The Model Theory of Reasoning: Current Standing and Future Prospects." In *Mental Models in Cognitive Science. Essays in Honor of Phil Johnson-Laird*, eds. Jane Oakhill and Alan Garnham, 319–327. Hove, East Sussex: Psychology Press.
Giere, Ronald N. (1988). *Explaining Science: A Cognitive Approach*. Chicago: University of Chicago Press.
Giere, Ronald N. (1999). *Science without Laws*. Chicago: University of Chicago Press.
Gobert, Janice D. (2000). "A Typology of Causal Models for Plate Tectonics: Inferential power and Barriers to Understanding." *International Journal of Science Education* 22: 937–977.
Grosslight, Lorraine, Christopher Unger, Eileen Jay, and Carol Smith (1991). "Understanding Models and Their Use in Science: Conceptions of Middle and High School Students and Experts." *Journal of Research in Science Teaching* 28: 799–822.
Hanna, Gila (1997). "The Ongoing Value of Proof." *Journal für Mathematik-Didaktik* 18: 171–185.
Hanna, Gila, and Hans N. Jahnke (1999). "Using Arguments from Physics to Promote Understanding of Mathematical Proofs. In *Proceedings of the 23rd Conference of the International Group for the Psychology of Mathematics Education, July 25–30, 1999*, vol. 3, ed. Orit Zaslavsky, 73–80. Haifa, Israel.
Harrison, Allan G., and David F. Treagust (2000). "A Typology of School Science Models." *International Journal of Science Education* 22: 1011–1026.

Hartmann, Stephan (1999). "Models and Stories in Hadron Physics." In *Models as Mediators. Perspectives on Natural and Social Science*, eds. Mary Morgan and Margaret Morrison, 326–346. Cambridge: Cambridge University Press.

Hesse, Mary (1967). "Models and Analogy in Science." In *The Encyclopedia of Philosophy*, Vol. 5, ed. Paul Edwards, 354–359. New York: Macmillan.

Hughes, R. I. G. (1999). "The Ising Model, Computer Simulation, and Universal Physics." In *Models as Mediators. Perspectives on Natural and Social Science*, eds. Mary Morgan and Margaret Morrison, 97–145. Cambridge: Cambridge University Press.

Johnson-Laird, Philip N. (1983). *Mental Models. Towards as Cognitive Science of Language, Inference, and Consciousness*. Cambridge: Cambridge University Press.

Johnson-Laird, Philip N. (2004). "The History of Mental Models." In *Psychology of Reasoning: Theoretical and Historical Perspectives*, eds. Ken Manktelow and Man Cheung Chung, 179–212. New York: Plenum.

Justi, Rosária, and John Gilbert (2000). "History and Philosophy of Science through Models: Some Challenges in the Case of 'the Atom.'" *International Journal of Science Education* 22: 993–1009.

Justi, Rosária, and Jan van Driel (2005). "The development of science teachers' knowledge on models and modelling: promoting, characterizing, and understanding the process." *International Journal of Science Education* 27: 549–573.

Kellett, Mary (2005). "Children as Active Researchers: An New Research Paradigm for the 21st Century." National Centre for Research Methods (*NCRM*) *Methods Review Papers* CRM/003. Economic and Social Research Council (ESRC) NCRM.

Kinash, Shelley, and Madison Hoffman (2008). "Child as Researcher: Within and Beyond the Classroom." *Australian Journal of Teacher Education* 33: 76–93.

Kuhn, Thomas S. (1962). *The Structure of Scientific Revolutions*. Chicago: University of Chicago Press.

Kuhn, Thomas S. (1971). "Notes on Lakatos." In *Philosophy of Science Association (PSA) 1970 (Boston Studies in the Philosophy of Science*, Vol. 8), eds. Roger C. Buck and Robert S. Cohen, 137–146. Dordrecht: Reidel.

Kuhn, T. S. (1977). *The Essential Tension*. Chicago: University of Chicago Press.

Lakatos, Imre (1976). *Proofs and Refutations*. Cambridge: Cambridge University Press. (Originally in *British Journal for the Philosophy of Science* 1963).

Lakatos, Imre (1978). "Falsification and the Methodology of Scientific Research Programmes." In *Philosophical Papers*, Vol. 1, eds. John Worrall and Gregory Currie. 8–101. Cambridge: Cambridge University Press, (Originally In *Criticism and the* Growth *of Knowledge*, eds. Imre Lakatos and Alan Musgrave, 91–195. Cambridge: Cambridge University Press, 1970).

Lakoff, George (1987). *Women, Fire, and Dangerous Things: What Categories Reveal about the Mind*. Chicago: Chicago University Press.

Laudan, Larry (1977). *Progress and Its Problems. Towards a Theory of Scientific Growth*. Berkeley: University of California Press.

Mach, Ernst (1989). *The Science of Mechanics. Account of Its Development*. La Salle, IL: Open Court. (First published in 1883).

Morgan, Mary S., and Margaret Morrison eds. (1999). *Models as Mediators. Perspectives on Natural and Social Science*. Cambridge: Cambridge University Press.

Morrison, Margaret (1999). "Models as Autonomous Agents." In *Models as Mediators. Perspectives on Natural and Social Science*, eds. Mary Morgan and Margaret Morrison, 38–65. Cambridge: Cambridge University Press.

Nersessian, Nancy J. (1984). *Faraday to Einstein: Constructing Meaning in Scientific Explanation*. Dordrecht: Martinus Nijhoff.

Nersessian, Nancy J. (1999). "Model-Based Reasoning in Conceptual Change." In *Model-Based Reasoning in Scientific Discovery*, eds. Lorenzo Magnani, Nancy J. Nersessian, and Paul Thagard, 5–22, New York: Plenum.

Norman, Donald A. (1983). "Some Observations on Mental Models." In *Mental Models*, eds. Dedre Gentner and Albert L. Stevens, 7–14. Hillsdale, NJ: Lawrence Erlbaum.

Rosch, Eleanor, Carolyn B. Mervis, Wayne D. Gray, David M. Johnson, and Penny Boyes-Braem. (1996). "Basic Objects in Natural Categories." *Cognitive Psychology* 8: 382–439.

Stöltzner, Michael (2004). "Application Dominance and the Model Web of Plasma Physics." In *Selected Papers Contributed to the Sections of GAP.5, Fifth International Congress of the Society for Analytical Philosophy*, eds. Roland Bluhm and Christian Nimtz, 128–139. September 22–26, 2003, Bielefeld, Germany. (CD-ROM) Paderborn: Mentis.

Synder, Jennifer L. (2000). "An Investigation of the Knowledge Structure of Experts, Intermediates and Novices in Physics." *International Journal of Science Education* 22: 979–992.

Van Driel, Jan H., and Nico Verloop (1999). "Teachers' Knowledge of Models and Modelling in Science." *International Journal of Science Education* 21: 1141–1153.

Vosniadu, Stella (1994). "Capturing and Modeling the Process of Conceptual Change." *Learning and Instruction* 4: 45–69.

Part II
Simulations

5 Weak Emergence and Computer Simulation

Mark A. Bedau

INTRODUCTION

There is a revival today of the belief that some metaphysically innocent form of emergence plays an important constructive role in the empirical sciences that explain complex natural systems, although much controversy remains (Bedau and Humphreys 2008). This paper contributes to this revival by explaining how emergence is connected with computer simulations and with complex systems in nature.[1]

Recently, emergence has been viewed with disdain, for a variety of reasons. For example, it was thought to be an admission of scientific ignorance (Hempel and Oppenheim 1965) or to be disproved by contemporary empirical science, such as quantum mechanics (McLaughlin 1992), or to be excluded from explanations by more fundamental micro-explanations (Kim 1999). In a series of papers, I have argued that a certain form of emergence deflects all those traditional worries (Bedau 1997, 2003, 2008, 2010). This form of emergence (which I call *weak emergence*) is closely associated with computer simulations and scientific explanations of complex systems. This paper focuses on two important connections: (i) that certain simulations produce artificial examples of weak emergence and (ii) that the same kinds of simulations play a crucial role in helping us understand natural examples of weak emergence. My explanation of these two connections will be illustrated with simulations of a very simple abstract model of an evolving population: Norman Packard's Bugs.[2]

Emergence involves the relations between wholes and their parts. It is colloquially described as the whole being more than the sum of its parts, but it is not clear exactly what this means. Examples sometimes better convey the idea of apparent emergent phenomena. Although this paper concentrates on an example involving an artificial life computer model, there are plenty of examples of apparent emergence in the real, physical world. One very simple and familiar example is a traffic jam. The "whole" in this case is the jam, and its "parts" are the individual cars with their individual drivers.[3] The relationship between the whole and its parts has a distinctive quality. At any given time, the jam is nothing more than the aggregation of cars that comprises it.[4] The composition of that aggregation changes over time, as individual cars enter and exit the jam. Many cars are part of a jam over its lifetime. So, the jam's continued existence and overall behavior is in some sense autonomous from the individual cars that are part of the jam at any given time.

Another very simple example of an alleged emergent phenomenon is the spontaneous formation and subsequent growth and division of tiny structures, such as vesicles. Vesicles are microscopic hollow spherical bodies, formed from a thin membrane, with a watery fluid inside and out. The vesicle membrane forms spontaneously when certain special molecules, called amphiphiles, are mixed in water. Amphiphiles are polar molecules; they have a hydrophilic end that tends to move toward water, and a hydrophobic end that tends to move away from water. So, when amphiphiles are mixed in an aqueous solution, they spontaneously clump together in certain shapes so that the hydrophilic heads are in contact with water, and the hydrophobic tails are not. One such aggregation is a bilayer membrane. These membranes can be millions of amphiphiles wide, but they are only two amphiphiles thick. The hydrophilic ends of the amphiphiles are all facing outside the membrane, toward the water, and the hydrophobic ends are all facing inside the membrane, between the two layers. In this example, the vesicle is a "whole," and the amphiphilic molecules are its "parts."

Before explaining weak emergence and illustrating it in Packard's Bugs, I should explain how I propose to evaluate conceptions of emergence. Many different conceptions of emergence exist in the literature today. I hold that the general conditions for being a conception of emergence are captured by the twin hallmarks of *dependence* and *autonomy*: A system's emergent macro-states in some way *depend on* the system's micro-states, and those same macro-states in some way are *autonomous from* the system's micro-states. These two hallmarks apply to every conception of emergence; every conception involves wholes that both depend on their parts and are autonomous from them.

The twin hallmarks concern how wholes are related to parts. For terminological consistency, I will refer to "macro" states (properties, events, phenomena, and the like) when speaking about wholes, and I will refer to "micro" states (properties, events, phenomena, and the like) when speaking about their parts. Note that my terms "macro" and "micro" are relative terms that refer to wholes and their parts, no matter what their physical size. A "micro" state might apply to a macroscopic object the size of an automobile (in the traffic jam example). A "macro" state might concern a microscopic object (such as a vesicle) that is too small to be seen with the most powerful light microscopes.

Elsewhere I show how these hallmarks fit three different broad kinds of conceptions of emergence, which I term *nominal*, *strong*, and *weak* emergence (Bedau 2003), though here my main concern is weak emergence. These two hallmarks can be taken in many different ways, so there are a corresponding number of different conceptions of emergence. In general, the hallmarks define the whole diverse field of all the different conceptions of emergence. Not all of those conceptions of emergence are viable; some are but others are not, and this changes over time. By *viable* I mean a conception of emergence that is a leading candidate, given everything we know, for capturing a kind of emergence that actually exists and is important. Importance obviously varies from context to context; here, what is especially important is whether a conception of emergence contributes to our scientific understanding of the natural world. Whether a given conception is viable at a given time depends on what evidence is possessed at that time. It is possible in principle for many different, independent conceptions of emergence to be viable at the same time, each for their own reasons.

I judge a conception of emergence viable only if it passes the following three tests:

Is the conception clear and explicit, especially with regard to the interpretations of dependence and autonomy?
Is the conception coherent and consistent with everything else we know philosophically and scientifically?
Is the conception useful in empirical science?

I think that any conception of emergence that passes all three tests deserves our consideration and respect. This paper will conclude that weak emergence passes all three tests.[5]

This methodology for evaluating conceptions of emergence is pluralistic. Different conceptions of emergence are not necessarily competing for our affirmation; many could all be viable at the same time. For example, I think that traffic jams and vesicles exemplify both nominal and weak emergence. (I will say more about these examples below.) The upshot of this pluralism is that each conception of emergence must be evaluated independently and on its own merits. It is not competing against other conceptions for the top spot on the podium. Every viable conception would share space on the podium with the other leading candidates for actual and important kinds of emergence.

WEAK EMERGENCE AS EXPLANATORY INCOMPRESSIBILITY[6]

Weak emergence (as I understand it) can be explicated in various ways.[7] My explanation here uses the concept of *explanatory incompressibility*. Previously I have defined weak emergence using the concept of underivability except by simulation (Bedau 1997, 2003). These two definitions are essentially equivalent (Bedau 2008), and I support them both. Here I express the argument in terms of explanatory incompressibility in order to highlight that weak emergence applies not just to computer simulations. It is true that weak emergence has special connections with computer simulations, as I will explain, but it also applies equally well to many natural systems, especially those that motivated the original discussions of emergence by the British emergentists (e.g., Broad 1925).

Emergence always involves a certain kind of relationship between global or macro-phenomena and local or micro-phenomena, and emergent macro-phenomena somehow both depend on, and are autonomous from, micro-phenomena. The characteristic feature of weak emergence, in general, is that the macro is dependent on the micro because it is ontologically and causally reducible to the micro *in principle*, but the reductive micro-explanation is autonomous from the macro-explanation because it is *especially complex*. Different kinds of explanatory complexity create different versions of weak emergence.[8] My definition of weak emergence is this, specifically:

> If P is a macro-property of some system S, then P is *weakly emergent* if and only if P is explainable from all of S's prior micro-facts but only in an incompressible way.[9]

This defines weakly emergent macro-properties by the distinctive incompressible way in which we explain how they are generated from underlying micro-states. It is easy to generalize definitions for weakly emergent macro-states, macro-events, macro-phenomena, macro-states of affairs, and macro-facts. It is also easy to generalize definitions for weakly emergent patterns in macro-properties, states, events, phenomena, states of affairs, facts, and the like. To finish explaining my definition of weak emergence, I need to say more about generative explanations and the important subset of them that are incompressible.

By a generative explanation from underlying micro-states I mean an explanation that exactly and correctly explains how macro-events unfold over time, i.e., how they are generated dynamically. The generative explanation assumes complete information about both the micro-causal dynamics that drive the system and the system's earlier micro-states and boundary conditions. The explanation works simply by tracing through the temporal details in the complex web of micro-level causal interactions that ultimately generate the macro-events. This kind of explanation is appropriate for any system with (global) macro-features that depend on (local) micro-features in certain complex ways. In particular, it is appropriate if we can describe the system's macro-features *at a given time* by appropriately conjoining, aggregating, or summing the (local) micro-features that exist at the same time. This is a synchronic generative reduction of macro to micro. Now, by starting with a completely specified initial condition, and by tracing forward in time through the network of local micro-level causal interactions, the system's macro-features (which are an aggregation of its micro-features at a given time) can be explained from immediately preceding aggregations of micro-features. Explaining the generation of a system's macro-behavior by aggregating and iterating the earlier local micro-interactions over time provides a dynamic generative reduction of the macro to the micro; I shall describe these explanations as *crawling the micro-causal web* because the explanations proceed by aggregating micro-causal influences and iterating them over time.

Next, in order to understand what incompressibility is, let's turn to compressibility. An explanation of a macro-property is *compressible* if and only if it can be carried arbitrarily far into the future with some fixed and finite amount of explanatory effort. The amount of explanatory effort needed is capped; explaining behavior farther into the future takes no more explanatory effort than explaining earlier behavior. A compressible explanation can achieve these explanatory economies because it somehow avoids the incrementally growing cost of crawling the micro-causal web.[10]

Now, an *incompressible* explanation is simply one that is not compressible. That is, an explanation of some macro-property of a given system is incompressible just in case there is no shortcut generative explanation of that macro-property that is true, complete, accurate, and avoids crawling the micro-causal web.[11] Incompressible generative explanations have no alternative but to crawl the web. They cannot be replaced without explanatory loss by fundamentally shorter explanations that ignore the web. So, incompressible explanations have a characteristic temporal signature: Explaining behavior farther in the future requires proportionally more explanatory effort (Crutchfield et al. 1986). This kind of explanatory incompressibility is the key feature of my paradigm case of weak emergence. How much additional explanatory effort is required for calculating the system's future states depends on a constant of proportionality. Perhaps these constants of proportionality would provide a way to compare and rank different cases of weak emergence.[12] But any constant of proportionality would still signal explanatory incompressibility in principle.

One might ask exactly how much micro-causal complexity is necessary and sufficient for explanatory incompressibility in principle. This question would make sense if some bright line separated weakly emergent properties from merely resultant properties, but the truth is more complicated. Emergence comes in degrees. Some kinds of emergence require explanations that are so complex that they are more than incompressible, and others require explanations that are less than compressible but still complex. Assad and Packard (1992, 232) describe one scale for degrees of emergence, ranging from behavior that is "immediately deducible upon inspection of the specification or rules generating it" to behavior that is "deducible in hindsight from the specification after observing the behavior," and continuing to behavior that is "deducible in theory, but its elucidation is prohibitively difficult," and finally reaching behavior that is "impossible to deduce from the specification." Explanatory incompressibility can be arrayed into similar stages: The macro-states might be explainable immediately upon inspection, or in hindsight only, or in principle but not in practice, or they might be inexplicable in principle. So, since weak emergence depends on explanatory incompressibility, weak emergence also comes in different strengths or degrees (more on this below).[13] In this paper I will not propose a precise way to measure degrees of weak emergence, but it surely depends on the amount of effort required for something's generative explanation, i.e., the amount of effort that must be expended in crawling the micro-causal web in order to generate or explain some macro-state.

Many philosophers and scientists assume that emergence and reduction are incompatible; it is worth correcting this mistake.[14] It is certainly true that strong emergence is incompatible with some kinds of reduction (e.g., Kim 1999), but weak emergence is *consistent* with other kinds of reduction. Each concrete physical embodiment of weak emergence is ontologically nothing more than some kind of aggregation of smaller embodied objects. For example, the ontological substance of a traffic jam is nothing more than an aggregation of cars on a road, and the ontological substance of a vesicle is nothing more than a certain kind of closed bilayer aggregation of amphiphilic molecules in water. Furthermore, the causes and effects of each concrete instance of any kind of weakly emergent macro-phenomenon are reducible to the iteration of the aggregation of the causes and effects operating at the micro-level, at least in principle. So, each example of macro-level weak emergence is ontologically and causally reducible to micro-level phenomena. However, in practice, the only way we know how to discern the consequences (generic macro-behavior) of such systems is by studying a generative model that crawls the micro-causal web on a computer; as far as we know, a certain kind of micro-reductive computer simulation is necessary to explain the generic macro-behavior of these systems.

Thus, weak emergence depends on the distinction between explanations or reductions that hold only *in principle,* and those that also hold *in practice.* A reductive generative explanation of macro from micro might exist, in principle, but be unhelpful for explaining weakly emergent phenomena, in practice, for a variety of reasons. One is that some relevant micro-level details required for the explanation might be unknown and inaccessible. Furthermore, even if all those details were known, the explanation might still be too complex and tedious for anyone to work through without the aid of a computer simulation. Nevertheless, given enough time and patience, anyone could work through all those micro-level explanatory details, as least in principle.[15] And working through those details is exactly what a computer simulation does. So, weakly emergent phenomena always have complete and accurate explanations solely from micro-phenomena, at least in principle. These explanations rely on complete prior micro-level information, and they necessarily proceed by crawling the micro-causal web. It is easy to see why it is typically impossible for anyone to grasp or understand how the emergent phenomena unfold from the micro, in practice, without resorting to computer simulations. This leads to the special connection between weak emergence and computer simulations.

The distinction between explanation and reduction *in principle* and *in practice* helps explain how weak emergence fits the two hallmarks of emergence mentioned earlier: the dependence of the macro on the micro and the autonomy of the macro from the micro. In cases of weak emergence, the macro depends on the micro because, in principle, each instance of the macro ontologically and causally is nothing more than the aggregation of micro-causal elements. For example, the ontological and causal state of a cellular automaton macro structure is nothing more than the aggregation of the ontological and causal states of its micro constituents. At the same time, weak emergence exhibits a kind of macro autonomy because of the incompressibility of the micro-causal generative explanation of the macro structure. Because the explanation is incompressible, it is useless in practice (except in so far as it serves as the basis for a good simulation of the system). Thus, my conception of weak emergence passes the first of my three tests for emergence, for it is clear how autonomy and dependence are to be interpreted.

The subtle way in which weak emergence balances reduction in principle and in practice can be summarized with the awkward but apt notion of *in principle irreducibility in practice*. Although weakly emergent phenomena have true, complete, and exact micro-level generative explanations, at least in principle, incompressibility makes the explanations of little use, in practice. In practice, we have no alternative but to simulate the system's micro-level behavior, if we want to observe what macro-behavior will emerge. This is a practical limitation, a limitation on irreducibility in practice. Furthermore, this limitation holds *in principle* for any naturalistic epistemic agent that is trying to explain the behavior of complex systems. We can put these points together by saying that weakly emergent phenomena are in principle irreducible in practice.

Another point to appreciate about weak emergence is that it is ubiquitous. People sometimes expect emergence to be a rare and precious thing, but this is certainly false for weak emergence. Most instances of weak emergence have no special philosophical or scientific significance. The macro-behavior of many kinds of complex natural systems is generated and can be explained only by crawling the micro-causal web. All those natural systems are, then, full of emergent macro-properties. But many of those macro properties are contingent and reflect no general, robust macro-pattern. Nevertheless, they still are in principle irreducible in practice, so they are weakly emergent. These highly contingent weakly emergent phenomena typically lack any individual philosophical or scientific significance.

Weakly emergent phenomena become more interesting when they are *robust*, that is, when they involve causally salient law-like patterns in weakly emergent macro-properties (Bedau 2003). Many features of emergent patterns are insensitive to the details of the local micro-interactions that produce the patterns, so the emergent patterns have multiple realizations.

When multiply realizable emergent patterns are generic or robust, they can have significant explanatory force. The Edge of Disorder law observed in Packard's Bugs model (see below) is an example of a robust emergent pattern with significant explanatory force.

Since weak emergence is found throughout nature, and so many natural complex systems exhibit weak emergence, emergent properties are sometimes viewed as a hallmark of complex systems (e.g., Bar-Yam 1997; Morowitz 2002; Sawyer 2005; Mitchell 2009). We often have more or less complete knowledge of the micro-level rules that generate a natural system's behavior. Given those rules and given a system's initial and boundary conditions, in principle we often can work out exactly what would happen in the system at any scale arbitrarily far into the future. But in practice we need to use computer simulations to figure out almost anything about what will happen in the system. That is why computer simulations are such crucial tools in current scientific attempts to understand robust emergent patterns in complex systems. I say more about this connection between weak emergence and computer simulations below.

The two examples with which this paper started are great examples of weak emergence in the real world. The spontaneous chemical self-assembly, growth and division of vesicles in an aqueous solution containing amphiphiles is a well-known process. We can understand the micro-scale chemical and physical forces at work in a soup of amphiphiles (Deamer 2008; Stano et al. 2008), and we can simulate the cycle of vesicle growth and division on computers today (Jiang et al. 2008; Mouritsen and Jakobsen 2008). But the only way we know to explain the details of vesicle growth and division is by means of these computer simulations. So, everything we know today about vesicle growth and division is consistent with the hypothesis that it has no compressible explanation. The second example of weak emergence in the real world is traffic jams (Nagel and Rasmussen 1996; Sugiyama et al. 2008). One striking and robust pattern about traffic flow is that jams will spontaneously form if the traffic density crosses a certain critical value. Another robust pattern is that traffic jams move slowly backwards against the direction of the flow of traffic. This happens because cars enter the jam at one end and leave at the other, so the location of the jam moves slowly backwards. These macro-patterns involving traffic jams are easy to explain by means of simple micro-causal generating processes. But, as far as we can tell, those generating processes are so complex and incompressible that we can explain the details of traffic jams only by using micro-causal computer simulations.

We can now see how my position on emergence contrasts with the view of Hempel and Oppenheim (1965). For Hempel and Oppenheim, there is no objective emergence in the world, and any appeal to emergence merely signals our ignorance about some true underlying, non-emergent explanation. For me, on the other hand, many natural systems really do have objectively incompressible micro-causal explanations, but we might be mistaken about exactly which systems those are. Even if we have no mathematical proof that there is no compressible explanation of how micro-processes generate certain macro-patterns, we can still (provisionally and defeasibly) identify which systems have only incompressible explanations.

Some people might think that apparent micro-causal complexity is never good evidence for incompressibility, on the grounds that the complexity might be *merely* apparent.[16] Now, it is true that beliefs about weak emergence (complexity, incompressibility) can be empirically well grounded and yet still be false. But this does not entail that we never have good evidence for weak emergence. We often have good empirical evidence that a specific kind of macro-pattern (traffic jams, vesicles) is weakly emergent, because often the best scientific explanations of the macro-pattern rely on apparently incompressible computer simulations. Note that this evidence is not mathematically irrefutable; it is empirical. In general our empirical evidence that a macro-pattern is weakly emergent will always be consistent with the hypothesis that the pattern is actually *not* weakly emergent. In general, we have no mathematical proof that the macro-pattern has no compressible explanation. But that does not mean that we cannot have defeasible evidence that the macro-pattern has no compressible explanation. It does not mean that there is no empirical evidence for or against the weak emergence of traffic jams or vesicles.

My argument for weak emergence in vesicles and traffic jams can be adapted to make similar arguments involving many other complex systems (e.g., Bar-Yam 1997; Morowitz 2002; Sawyer 2005; Mitchell 2009). In what follows, I describe one further example in greater detail. This example is an artificial life computer simulation, so it is a paradigm case of a system that is artificial rather than natural. Studying this artificial life simulation enables us to describe and explain some interesting weakly emergent macro-patterns observed in the simulation. This simulation example illustrates the connections between weak emergence and computer simulations.

SIMULATING PACKARD'S BUGS REVEALS THE EDGE OF DISORDER LAW[17]

Norman Packard's Bugs model (1989) illustrates how artificial life simulations can be used to explain certain fundamental emergent macro laws about evolving systems. In this model evolving populations display various macro-level patterns on an evolutionary time scale. Adaptive innovations that arise through genetic changes tend, *ceteris paribus*, to persist and spread through the population and increase the population's adaptive fit with its environment. Of course, these sorts of patterns are usually not precise and exceptionless universal generalizations; instead, they are rough rules of thumb that hold only for the most part. Context-dependent micro-level fluctuations buffet the system's evolutionary dynamics and sometimes these fluctuations affect macro-scales; nevertheless, overall generic macro-level patterns can still emerge statistically from the micro-level contingencies of natural selection. Even though they are imprecise and have exceptions, these patterns can nevertheless be real and important. The Edge of Disorder law, described below, is a candidate for that status.

The Bugs model is a classic artificial life computer system with interesting evolutionary dynamics. The model consists of an evolving population of sensory-motor agents and demonstrates how various macro-level evolutionary patterns emerge implicitly from explicit micro-level generative rules. As the population of agents evolves, the environment it encounters changes, so the agents' sensory-motor functionality can continue to adapt and evolve in an open-ended fashion (Bedau and Packard 1992). Packard's Bugs exhibits a very simple form of open-ended adaptive evolution.

The model consists of a finite two-dimensional world with a fixed resource distribution and a population of agents. An agent's survival and reproduction depend on its ability to find enough resources, and this ability in turn depends on its sensory-motor genes, which encode the way in which the agent's perception of its contingent local environment affects its behavior in that environment. The genes can mutate when an agent reproduces. Thus, on an evolutionary time scale, the process of natural selection implicitly adapts the population's sensory-motor strategies to the environment. Furthermore, the agents' actions change the environment because agents consume resources and compete with each other for space. This entails that the mixture of sensory-motor strategies in the population at a given moment is a significant component of the environment that affects the subsequent evolution of those strategies. Thus, the fitness function in Packard's model—what it takes to survive and reproduce—is constantly buffeted by the contingencies of natural selection and unpredictable changes.

All macro-level evolutionary dynamics produced by this model ultimately result from explicit micro-level mechanisms acting on external conditions. The model starts with a population of agents with randomly chosen sensory-motor strategies, and the subsequent model dynamics explicitly controls only local micro-level states. Resources are locally replenished, genes randomly mutate during reproduction, and an agent's genetically encoded sensory-motor strategy determines its local behavior, which determines its internal resource level, which, in turn, determines whether it survives and reproduces. Each agent is autonomous in the sense that its behavior is determined solely by the environmentally sensitive dictates of its own sensory-motor strategy. On an evolutionary time scale these strategies are continually refashioned by the historical contingencies of natural selection. The model generates macro-level evolutionary dynamics only as the indirect product of an unpredictably shifting agglomeration of directly controlled micro-level events (individual actions, births, deaths, mutations). The model has no provisions for explicit control of

Weak Emergence and Computer Simulation 101

macro-level dynamics. Moreover, macro-level evolutionary dynamics are weakly emergent in the sense that, although constituted and generated solely by the micro-level dynamic, they can be derived only through simulations that crawl the micro-causal web.

It should be noted that Packard's model is not intended as a realistic simulation of some actual biological population. Rather, it is an "idea" model, aiming to capture the key abstract principles at work in evolving systems generally. Packard's model is in effect a thought experiment—but an *emergent* thought experiment (Bedau 1998). As with the armchair thought experiments familiar to philosophers, Packard's model attempts to answer "What if X?" questions; but emergent thought experiments are distinctive in that what they reveal can be discerned only by simulation. This kind of thought experiment can still help us to understand complex systems, both those found in nature and those that humans have constructed. This is one of the connections between weak emergence and computer simulations, discussed further below.

I will illustrate the emergent supple dynamics in Packard's Bugs by describing some simulations in this model concerning the evolution of evolvability. The ability to successfully adapt depends on the availability of viable evolutionary alternatives. Enough alternatives can make evolution easy; too many or too few can make evolution difficult or even impossible. In short, the system needs a capacity for evolutionary "novelty." At the same time, the population's sensory-motor strategies can adapt to a given environment only if strategies that prove beneficial can persist in the gene pool; in short, the system needs a capacity for evolutionary "memory." The need to balance evolutionary novelty and memory is a well-known fact about evolving systems, and in the field of evolutionary computation, this is known as the tradeoff between evolutionary "exploration" and "exploitation" (Holland 1975).

The simplest mechanism that simultaneously affects both evolutionary memory and novelty is the mutation rate. The lower the mutation rate, the greater the number of genetic strategies remembered from parents. The higher the mutation rate, the greater the number of novel genetic strategies introduced with children. Successful adaptability requires that these competing demands for memory and novelty be suitably balanced. Too much mutation (not enough memory) will continually flood the population with new random strategies; too little mutation (not enough novelty) will tend to freeze the population at arbitrary strategies.

Successful evolutionary adaptation requires a mutation rate suitably balanced between these extremes. Furthermore, this balance point might shift as the context of evolution changes. An evolutionary process that could continually support evolving life should have the capacity to adapt automatically to this shifting balance. So, in the context of Packard's model, it is natural to ask whether the mutation rate that governs first-order evolution could adapt appropriately by means of a second-order process of evolution. If the mutation rate could adapt in this way, then the Bugs model would yield a simple form of the evolution of evolvability and, thus, might illuminate one of evolving life's fundamental prerequisites.

If you vary the mutation rate in the Bugs model, you see two robust effects (Bedau and Packard 2003). The first observation is that the mutation rate governs a transition between genetically ordered and genetically disordered phases of the gene pool. When the mutation rate is too far below the phase transition, the whole gene pool is frozen at a small number of given strategies; when the mutation rate is too far above the phase transition, the gene pool is a continually changing plethora of randomly related strategies. At a certain intermediate mutation rate, which can be experimentally measured, the gene pool flips from being highly ordered to highly disordered. The second observation is that the process of evolution produces the most fit population (measured by efficiency of extracting available energetic resources) when the mutation rate is a little below this transition—or, as I will say, "at the edge of genetic disorder." These two observations show that in the Bugs model, the capacity to evolve (or evolvability) is maximized when the gene pool is at the edge of disorder.

Earlier we discussed how the capacity to evolve depends on properly balancing the demands for memory and novelty. Putting this together with our observation that the Bugs evolve best at the edge of genetic disorder suggests the following hypothesis about the Bugs:

> **Balance hypothesis:** Evolutionary memory and novelty *ceteris paribus* are balanced at the edge of genetic disorder.

This hypothesis claims that a system's ability to evolve is maximized at the edge of genetic disorder. This hypothesis is not exact and exceptionless; instead, it describes where memory and novelty are balanced if everything else is equal, and this holds only typically but not universally. So far, the only evidence presented here in favor of the Balance hypothesis results from simulations of the Bugs model. But the hypothesis is interesting because it is quite general and could be applied to many other evolving systems. If the hypothesis holds for the Bugs, then it might turn out to be a fundamental property of a broad class of evolutionary systems that includes much more than the Bugs.

To test the Balance hypothesis in the Bugs model, the model was modified in such a way that each agent has an additional gene encoding its personal mutation rate (Bedau and Packard 2003). In this case, two kinds of mutation play a role when an agent reproduces: The child inherits its parent's sensory-motor genes, which mutate at a rate controlled by the parent's personal (genetically encoded) mutation rate; and the child inherits its parent's mutation rate gene, which mutates at a rate controlled by a population-wide meta-mutation rate. Thus, first-order (sensory-motor) and second-order (mutation rate) evolution happen simultaneously. So, if the Balance hypothesis is right and mutation rates at the critical transition optimize evolvability because they balance memory and novelty, then we would expect second-order evolution to drive mutation rates into the critical transition. It turns out that this is exactly what happens; below I term this effect the Edge of Disorder law.

An examination of hundreds of simulations confirms the predictions of the Balance hypothesis: Second-order evolution tends to drive mutation rates to the edge of disorder, increasing population fitness in the process. If natural selection cannot affect mutation rates, the rates wander aimlessly due to random genetic drift. But the mutation dynamics are quite different when governed by natural selection. Even if the population is initialized with quite high mutation rates, as the population becomes more fit the mutation rates in the population drop back into the ordered phase but near the edge of disorder.

If the balance hypothesis is the correct explanation of this second-order evolution of mutation rates into the critical transition, then we should be able to change the mean mutation rate by dramatically changing where memory and novelty are balanced. In fact, the mutation rate can be observed to rise and fall along with the demands for evolutionary novelty. For example, when we randomize the values of all the sensory-motor genes in the entire population so that every agent immediately forgets all the genetically stored information accumulated in a lineage over its evolutionary history, the population must restart evolutionary learning from scratch. There is no immediate need for memory because the gene pool contains no information of proven value; instead, the need for novelty dominates. Under these conditions, we observe a typical sequence of events: (i) the population immediately becomes much less fit; (ii) the mean mutation rate dramatically rises; (iii) after the mean mutation rate peaks and the population's fitness dramatically improves, the mutation rates fall back toward their previous equilibrium levels.

The same sequence happens when the Bugs face environmental catastrophes (Buchanan et al. 2004). We created hundreds of catastrophic changes to the global resource environment in the Bugs model, and the mutation rate evolved up and down in such a way that evolutionary novelty and memory were fluidly balanced. (See Figure 5.1).

We are unsure whether this balance was "optimal," but we are quite confident that mutation rates evolved up when evolutionary novelty became more valuable, and they evolved down when evolutionary memory became more valuable.[18] Environmental catastrophes typically triggered the same characteristic (i)-(iii) cycle of changing mutation rates; mutation rates rise when evolutionary novelty is important right after a catastrophe, and they fall again after the population adapts to the new environment and evolutionary memory becomes important again. Furthermore, the changing mutation rates follow a characteristic trajectory, with a distinct trajectory for distinct kinds of environments. In other words, the temporal trajectory of the mutation rates is a characteristic signature of a given kind of environment. In general, different kinds of environments could elicit an open-ended variety of different typical mutation rate trajectories. These trajectories illustrate some of the open-endedness in the ability to adapt mutation rates in the Bugs model.

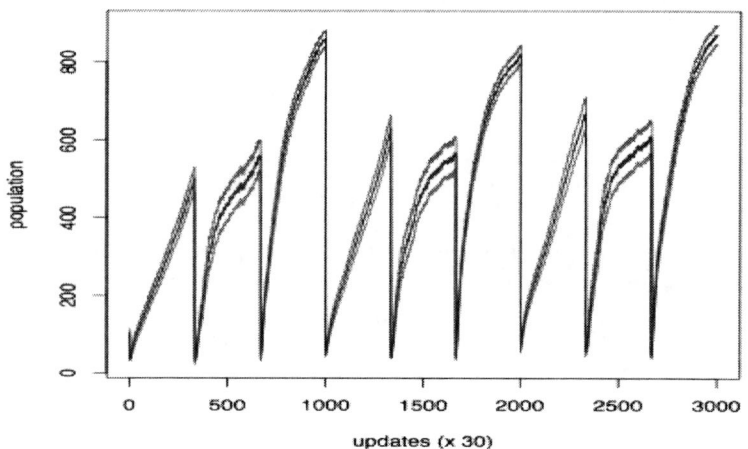

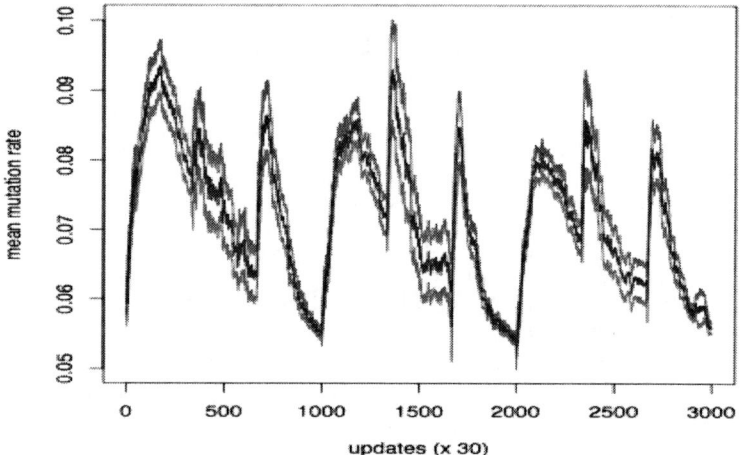

Figure 5.1 Population dynamics and mutation rate. Top: dynamics of the population level (dark grey), plus and minus the standard error (light grey), averaged across 80 runs of Packard's Bugs model. Bottom: dynamics of the mean mutation rate in the population (dark grey, plus and minus the standard error in light grey), averaged across the same 80 runs. The effects of eight catastrophic changes in the environment are clearly visible. In response to environmental catastrophes, the population level and mutation rate exhibit a characteristic robust weakly emergent macro-level pattern: Immediately after an environmental catastrophe, the population level plummets and the average mutation rate skyrockets. Next, as the population adapts to its new environment, the population steadily rises and the average mutation rate drops back towards its equilibrium level (adapted with modification from Buchanan et al. 2004).

These results show that the mutation rate distribution shifts up and down as the Balance hypothesis would predict. A change in the context for evolution can increase the need for rapid exploration of a wide variety of sensory-motor strategies and thus dramatically shift the balance toward the need for novelty. Then, subsequent sensory-motor evolution can reshape the context for evolution in such a way that the balance shifts back toward the need for memory.

This all provides evidence for the following empirical generalization about second-order evolution in Packard's Bugs:

> **Edge of Disorder law:** Mutation rates evolve *ceteris paribus* to the edge of disorder.

According to the Balance hypothesis, the Edge of Disorder law is simply a manifestation of the deeper regularity that, *ceteris paribus*, mutation rates evolve in such a way that evolutionary memory and novelty are optimally balanced. This balance point is typically at the edge of genetic disorder, but an indefinite variety of environmental contingencies can shift the point of balance. In other words, the Edge of Disorder law has exceptions that prove the rule.

As with all *ceteris paribus* laws, the Edge of Disorder law is not precise and exceptionless. Furthermore, when *ceteris* is not *paribus*, it is generally unpredictable how contextual contingencies will relocate the balance between evolutionary memory and novelty. No evident fixed mechanism could be guaranteed to find the appropriate mutation rate. Yet it remains a lawful regularity that mutation rates evolve to the point of balance, *ceteris paribus*. This is because the process of evolution is continually changing the system's micro-level mechanisms. The micro-level mechanisms in this case consist of the agents with their genetically encoded sensory-motor strategies and their genetically encoded mutation rates. As the environmental context of this mechanism changes, the mechanism itself is subject to continual alteration by first- and second-order evolution. There is no algorithm for determining what adaptation will prevail in which context or which alteration in the mutation rate genes will emerge, but trial and error and natural selection can be counted on—*ceteris paribus*—to continually create appropriate mutation rates as novel situations unfold. This open-ended supple flexibility in the dynamics of the evolution of evolvability is the deep reason why the statistical regularity about the balance of memory and novelty resists any precise and exceptionless formulation.

CONNECTIONS BETWEEN COMPUTER SIMULATIONS AND WEAK EMERGENCE[19]

Weak emergence has an especially close connection with computer simulations and computational systems. Some of the best examples of weak emergence come from computer simulations, and my previous definitions of weak emergence rely centrally on the notion of "underivability except by simulation" (Bedau 1997, 2003). But the link between simulations and weak emergence can be misunderstood.

My first point concerns not simulations but complex systems found in the real world. The fact is that a great many complex natural systems have incompressible micro-explanations and so exhibit weak emergence. A trivial consequence is that weak emergence does *not* apply merely or primarily to computer simulations or other kinds of computational systems. Once again, consider the spontaneous formation of traffic jams when the traffic density exceeds a critical value and the slow backwards movement of jams. Scientists explain these macro-behaviors by iterating and aggregating all the simple local interactions among individual vehicles (Nagel and Rasmussen 1996; Sugiyama et al. 2008), but they have no idea how to give a shortcut explanation of this behavior from complete information about micro-states and boundary conditions. The same holds for the spontaneous self-assembly, growth, and division of vesicles; scientists can explain these macro-behaviors with incompressible simulations, but they know of no shortcut substitute (Deamer 2008; Jiang et al. 2008; Mouritsen and Jakobsen 2008; Stano et al. 2008). In fact, explanatory incompressibility typically characterizes a vast number of global properties of complex systems in molecular and cellular biology, including regulatory gene networks, metabolic networks, and the process by which proteins fold into three-dimensional (3D) structures. The same can be said for many psychological and social systems. So, our best current explanations of complex systems imply that weak emergence is rampant throughout nature. It is not limited only or even mainly to computational systems and computer simulations created and studied by humans.

This brings me to my second point: Certain computer simulations and computational systems produce some of the most striking examples of weakly emergent phenomena. Computer simulations of complex, non-linear, dynamical, hierarchical systems in nature comprise one class of computational embodiment of weak emergence. Another class consists of complex computational systems that are not simulations of something else but are studied in their own right, such as Conway's Game of Life (Bedau 1997). Packard's Bugs falls into this category, too; it is not designed as a simulation of something else but as a computational embodiment of a certain kind of evolving system. That kind of evolving system is captured in the details of the micro-level rules that drive the model. The Game of Life and the Bugs both have a massively parallel architecture governed by non-linear local interactions. Wolfram has used cellular automata and the like to produce some of the most vivid and striking images of weakly emergent patterns in space and time (Wolfram 2002).

It is worth emphasizing that these examples are all artificial systems that exist only because of conscious human intention and design. These examples of weak emergence are real, but they exist only because someone created the software needed to run the cellular automata or the artificial life system and extensively studied it through various simulation experiments. In this respect, they differ dramatically from examples of weak emergence found in the non-digital world (traffic jams, vesicles, etc.), which exist whether or not anyone intended or designed them.

Note that the macro-level patterns and regularities created in a bottom-up simulation are not *mere* simulations of emergent phenomena. Rather, they are computational *embodiments* of actual emergent phenomena. That is, there are computation states that have incompressible explanations, because they are generated by a complex micro-causal process. In other words, those computational states meet the definition of weak emergence. When the simulation is running, those computational states really exist, and they really involve weak emergence. For example, a simulated traffic jam or a simulated vesicle corresponds to certain software data structures being in certain kinds of software states. The information in the data structures can be "visualized" in various ways, such as scatter plots of traffic density and velocity, or as 3D computer-generated movies of simulated vesicle growth and division. Those visualizations were produced by crawling the micro-causal webs involved, and there is no compressible shortcut process for producing them. So, those scatter plots and computer-generated movies have weakly emergent macro properties.

This brings me to my third point: Computer simulations can provide important evidence for and explanations of weak emergence in the real world.[20] Those sciences that study the behavior of complex systems are undergoing a methodological revolution (e.g., Wolfram 2002). They are increasingly relying on massive computer simulations because they typically have no other way to explain various complex behaviors. Computer simulations fill the study of complex natural systems today. Often they provide our only solid and useful evidence about how real complex systems will behave, and what global patterns will emerge through the course of the myriad micro-interactions that drive them. That is why it is possible to define weak emergence as that which is "underivable except by simulation" (Bedau 1997, 2003).

Of course, it is possible to simulate a system that has a compressible explanation, so the mere fact that traffic jam formation and movement, and vesicle growth and development, can be studied by simulation does not prove that those processes are weakly emergent. The signature of weak emergence is not that something can be simulated but that it can be explained *only* by an incompressible micro-causal process. We do have some evidence that there is no compressible explanation for traffic jams and vesicles, because the best scientific experts today have found no such compressible explanations.

Like any other empirical evidence, the evidence from computer simulations for weak emergence is defeasible. Empirical evidence can be misleading, and we can be mistaken about what natural systems are weakly emergent. Thus, we might form mistaken beliefs about which systems are weakly emergent. My claims that traffic jams and vesicle reproduction are weakly emergent are defeasible; I have no mathematical proof that jams and vesicles are weakly emergent. It is possible that I am mistaken. But there is no special reason to think that I am mistaken about them, and there is special reason to think that I am correct. For our best current scientific accounts of traffic jams and vesicle growth and reproduction rely on computer simulations in the way that is characteristic of systems generated in an incompressible way.

If we later discover a compressible explanation for some complex behavior that we thought was weakly emergent, this shows merely that we were wrong about an example of weak emergence. It does not show that weak emergence is "merely epistemological" any more than any other defeasible, empirical scientific claim is "merely epistemological." The possibility that we are wrong about which systems exhibit real weak emergence is to be expected, since the evidence for emergence is typically the empirical result of many computer simulations.

The indirect epistemological role of computer simulations in explaining weak emergence might be what fuels the common belief that weak emergence is merely epistemological emergence (Silberstein and McGeeve 1999), but that belief is a mistake (Bedau 2008). The weak emergence exhibited in jamming traffic and dividing vesicles is not merely epistemological. Traffic jams and vesicles are not just in the mind; they are a real part of the causal fabric of everyday life. And it is because of their objective, intrinsic microcausal complexity that they have incompressible explanations.

Packard's Bugs model nicely illustrates how micro-causal complexity can generate striking macro-patterns. First, the model produces many examples of macro-properties that are weakly emergent. Almost any precise and interesting macro-property that you might define over the population in the model is impossible to predict because it is generated in such a complex manner. The Edge of Disorder law (which states that mutation rates tend to evolve toward the edge of genetic disorder) is an especially interesting kind of case to consider. This is a macro-pattern describing how the Bugs model typically behaves when the mutation rate can evolve. This pattern exists not in a single state or a single run of the Bugs model but rather in the space of all possible runs of the model under different mutation rates. The pattern concerns the dynamics of mutation rates, specifically, the changes in the mutation rate distributions in the gene pool. These mutation rate distribution dynamics are generated by aggregating and iterating the fundamental micro-level processes, i.e., they percolate up out of the generative process that crawls step by step through the micro-causal web.

Striking and robust patterns in mutation rate distributions can be discovered by examining patterns in how the mutation rate distribution changes.

One striking pattern is the Edge of Disorder law, but other patterns can also be detected. For example, mutation rates evolve up with increasing need for evolutionary novelty and down with increasing need for evolutionary memory. As far as anyone knows, there is no way to accurately predict or explain these patterns without running the simulations, even given complete information about the model's micro-dynamics and its initial and boundary conditions. These explanations are incompressible, so the patterns are weakly emergent.

Furthermore, at least in the ideal case, study of the Edge of Disorder law in the Bugs model can teach us something fundamental about certain evolving populations found in nature. At a minimum, it shows how a simple evolving system with evolving mutation rates can have an open-ended ability to evolve its own evolvability. The Bugs model is a constructive demonstration of a simple example of the open-ended evolution of evolvability. Furthermore, the behavior of the Bugs lends support to the Balance hypothesis, that the ability to evolve is typically maximized at the edge of genetic disorder. Thus, the model could teach us something quite fundamental about the ability to evolve in general, by providing evidence for sweeping conjectures like the Balance hypothesis.

This illustrates my third point above: Studying computer simulations is often the key to *understanding* certain kinds of weakly emergent behavior, such as the evolution of mutation rates. Natural evolving systems must strike some more or less appropriate balance between evolutionary memory and novelty, if life is to evolve and flourish. Not much is known about how life strikes this shifting balance. Evolving mutation rates in the Bugs model can help illuminate how natural systems achieve a similar balance.

CONCLUSIONS

Contemporary biology rejects vitalism and instead holds that the macro-phenomena in biology happen because of how things are at the micro-level. Ernst Mayr, the archenemy of reductionism in biology, accepted at least that much micro-causal reductionism when he said: "Every biologist is fully aware of the fact that molecular biology has demonstrated decisively that all processes in living organisms can be explained in terms of physics and chemistry" (1982, cited in Weinberg 1987, 352). The huge advances in molecular biology in the past 60 years are great successes that support this kind of micro-causal reductionism. But it is important to recognize that those micro-causal reductions are typically context-sensitive and involve synergistic interactions among the micro-causal constituents. This makes the micro-causal webs very complex, and that makes their explanations incompressible. So weak emergence abounds in biology. Weak emergence is the macro-level mark of incompressible complexity in the micro-causal web. When the micro-causal web is sufficiently complex, generative explanations of the resulting macro-properties are inevitably incompressible. These incompressible explanations are the root of weak emergence.

I began this paper by listing three tests for viable conceptions of emergence: (i) clear and explicit interpretation of dependence and autonomy, (ii) coherence and consistency with everything else we know, and (iii) utility in empirical science. We noted above that weak emergence passes the first test. We now can see that weak emergence passes the second test. The bottom-up micro-causal generative processes at work in traffic jams, vesicle self-assembly, and mutation rate dynamics in Packard's Bugs are certainly coherent and consistent, and they illustrate the general type of generative processes that are behind many evolving systems. A great many other examples of weak emergence are driven by similarly coherent and consistent bottom-up generative processes that are complex enough to have incompressible micro-causal explanations. This is the case for both traffic jam formation and movement, and vesicle growth and division. The micro-causal processes that generate and explain traffic jams and vesicles are incompressible, but they still provide coherent explanations that are consistent with everything else we know.

Packard's Bugs illustrates a number of important properties of weak emergence. For example, in contrast with many forms of emergence discussed in contemporary philosophy of mind, weak emergence is dynamic rather than static, because it concerns how macro-properties are generated over time, rather than how macro-states are realized at a given instant. More central here, the weak emergence in the Bugs is consistent with any reasonable form of naturalism. Rather than being mysterious or a sign of our ignorance, weak emergence is the natural and expected outgrowth of a specific, context-sensitive micro-causal process that generates complex macro-behaviors. These micro-causal generative processes are precisely specified in the computer code that implements the model.

The third test concerns the scientific usefulness of a conception of emergence. I take my extended discussion of the emergence of the Edge of Disorder law in Packard's Bugs to illustrate how robust weakly emergent patterns can provide deep insights into the behavior of certain complex systems. The same point applies throughout the study of complex systems. The global behavior of complex systems exhibits certain robust weakly emergent patterns, and those patterns play an important role in our best current scientific efforts to explain those complex systems. Thus, weak emergence passes the third test, and it is a viable conception of emergence.

This returns us to the central topic of this paper: the connection between weak emergence and computer simulations. The viability of weak emergence makes it more important to understand these connections. Although weak emergence can be defined independently of computer simulations, it still has two deep connections with computer simulations. One connection is that computer simulations provide many vivid and dramatic computational examples of weakly emergent phenomena. The other is that computer simulations are among the best ways to explain weakly emergent phenomena in natural systems. It is because of the bottom-up, micro-causal generative processes by which computer simulations produce weak emergence that we know that the conception of weak emergence is coherent and consistent and so passes the second test for viability. And it is because computer simulations are among the best ways to explain many examples of emergence in nature that we know that the conception of weak emergence passes the third test for viability.

NOTES

1. An excellent recent overview of the science of complex systems and computation is Mitchell (2009).
2. I discussed these same results in my first explanation of weak emergence (Bedau 1997).
3. Later I will adopt the terminology "macro" and "micro" for whole and part.
4. There is a gray area at the physical border of a jam, concerning whether cars at the edge of a jam are, or are not, part of the jam; this is a complication in my story, which requires a longer discussion. I spell out a similar gray area on the border between life and nonlife; see Bedau (forthcoming).
5. I also believe that nominal emergence passes all three tests and that strong emergence passes the first two tests but fails the third (Bedau 1997, 2003); however, I will not argue those points here.
6. The material in this section is adapted from Bedau (2008).
7. The different variants of weak emergence in the literature include those by Wimsatt (1986, 1997), Reuger (2000a, 2000b), Bedau (1997, 2003) and Boogerd et al. (2005). Different conceptions of weak emergence focus on different kinds of explanatory complexity, but all agree that weak emergence involves some distinctive kind of explanatory complexity.
8. An unfortunate complication in this discussion is that different philosophers use the phrase 'weak emergence' in different ways. For example, what I call "weak emergence" differs from what Achim Stephan calls "weak emergence" (Stephan 2006), and both differ from what Krisztian Balog calls "weak emergence" (personal communication).
9. There is an analogous definition that makes no reference to the incompressibility of our explanation of how macro-properties are generated, but instead refers to the incompressibility of the actual process that generates them, as follows: If P is a macro-property of some system S, then P is *weakly emergent* if and only if P is actually generated by an incompressible process from various prior micro-facts, some of which involve S. I think that the definition of weak emergence using the concept of incompressible generating processes is the more fundamental one, and it can explain the other definition using the concept of incompressible explanations. Briefly, our explanations of macro-phenomena are incompressible because the structure of our explanations mirrors the structure of the micro-causal processes that generate those structures, and those processes are themselves intrinsically incompressible.
10. For examples of compressible and incompressible explanations involving cellular automata, including the Game of Life, see Bedau (2003).
11. The explanation can apply to indeterministic systems by including complete information about the system's indeterministic micro-state changes in the information from which the explanation is sought. System boundary

conditions are handled in a similar fashion (Bedau 1997). My notion of incompressible explanation is closely connected with Chaitin's notion of random sequence (Chaitin 1975, 1988) and Wolfram's subsequent notion of incompressible computation (Wolfram 1985, 2002), as well as the notion of a dynamical system that must be simulated to discover its generic properties (Crutchfield et al. 1986).

12. I owe this suggestion to an anonymous referee.
13. See also Bedau (2003, 163). Hovda (2008) is one example of how to explicate the notion of degrees of weak emergence.
14. Rather, correcting this mistake again; the consistency of emergence and reduction is a point often emphasized by William Wimsatt (1986, 1997).
15. The computer-generated proof of the four-color theorem is one specific kind of example of a proof that one could work through in principle but not in practice.
16. I owe this objection to an anonymous referee.
17. Most of the material in this section was originally described earlier (Bedau 1997); I return to this example here because it so nicely illustrates the central connections between weak emergence and computer simulations described in this paper, and also because the Bugs models is so simple and yet the Edge of Disorder law potentially provides a profound insight into many evolving systems found in nature.
18. Elsewhere one can find further details about exactly where the edge of genetic disorder is and how mutation rates move with respect to the edge (Bedau and Packard 2003; Buchanan et al. 2004).
19. Many of the points in this section were originally made in Bedau (2008).
20. Different kinds of computational systems have been called "simulations," so I should clarify what I mean. The simulations I have in mind are those that crawl the micro-causal web (recall above) and generate global properties out of myriad local interactions. In addition to cellular automata, so-called "agent-based" models are good examples of simulations that crawl the micro-causal web. They explicitly describe how local causal processes unfold over time, and global properties are merely certain kinds of aggregations of local properties.

REFERENCES

Assad, Andrew, and Norman H. Packard (1992). "Emergence." In *Emergence: Contemporary Readings in Philosophy and Science*, eds. Mark A. Bedau and Paul Humphreys, 231–234. Cambridge, MA: MIT Press. (Originally appeared as Section 2 of "Emergent Colonization in an Artificial Ecology." In *Towards a Practice of Autonomous Systems: Proceedings of the First European Conference on Artificial Life*, eds. Francisco Varela and Paul Bourgine, 143–152. Cambridge, MA: The MIT Press, 1992).

Bar-Yam, Yaneer (1997). *Dynamics of Complex Systems*. Boulder, CO: Westview Press.

Bedau, Mark A. (1997). "Weak Emergence." *Noûs* 31 (Suppl. 11): 375–399.

Bedau, Mark A. (1998). "Philosophical Content and Method of Artificial Life. In *The Digital Phoenix: How Computers Are Changing Philosophy*, eds. Terrell W. Bynam and James H. Moor, 135–152. Oxford: Basil Blackwell.

Bedau, Mark A. (2003). "Downward Causation and Autonomy in Weak Emergence." *Principia* 6: 5–50. (Reprinted in *Emergence: Contemporary Readings* in *Philosophy and Science*, eds. Mark A. Bedau and Paul Humphreys, 155–188. Cambridge, MA: MIT Press, 2008.)

Bedau, Mark A. (2008). "Is Weak Emergence Just in the Mind?" *Minds and Machines* 18(4): 443–459.

Bedau, Mark A. (2010). "Weak Emergence and Context-Sensitive Reduction." In, *Emergence in Science and Philosophy*, eds. Antonella Corrandi and Timothy O'Conner. New York: Routledge.

Bedau, Mark A. (Forthcoming). "A Functional Account of Degrees of Minimal Chemical Life." *Synthese*.

Bedau, Mark A., and Paul Humphreys, eds. (2008). *Emergence: Contemporary Readings in Philosophy and Science*. Cambridge: MIT Press.

Bedau, Mark A., and Norman H. Packard (1992). Measurement of Evolutionary Activity, Teleology, and Life. In *Artificial Life II*, eds. Christopher Langton, Charles Taylor, J. Doyne Farmer, and Steen Rasmussen, 431–461. Redwood City, CA: Addison Wesley.

Bedau, Mark,A., and Norman H. Packard, (2003). "Evolution of Evolvability via Adaptation of Mutation Rates." *BioSystems* 69: 143–162.

Boogerd, Fred C., Frank J. Bruggeman, Robert C. Richardson, Achim Stephan, and Hans V. Westerhoff (2005). "Emergence and Its Place in Nature: A Case Study of Biochemical Networks." *Synthese* 145: 131–164.

Broad, Charlie D. (1925). *The Mind and Its Place in Nature*. London: Routledge and Kegan Paul.

Buchanan, Andrew, Mark Triant, and Mark A. Bedau (2004). "The Flexible Balance of Evolutionary Novelty and Memory in the Face of Environmental Catastrophes." In *Artificial Life IX: Proceedings of the 9th International Conference on the Synthesis and Simulation of Living Systems*, eds. Jordan Pollack, Mark A. Bedau, Phil Husbands, Takashi Ikegami, and Richard Walker, 297–302. Cambridge, MA: MIT Press.

Chaitin, Gregory J. (1975). "Randomness and Mathematical Proof." *Scientific American* 232: 47–53.

Chaitin, Gregory J. (1988). "Randomness in Arithmetic." *Scientific American* 259: 80–85.

Crutchfield, James P., J. Doyne Farmer, Norman H. Packard, and Robert S. Shaw, (1986). "Chaos." *Scientific American* 255: 46–57.

Deamer, David (2008). "Experimental Approaches to Fabricating Artificial Cells." In *Protocells: Bridging Nonliving and Living Matter*, eds. Steen Rasmussen, Mark A. Bedau, Liaohai Chen, David Deamer, David C. Krakauer, Norman H. Packard, and Peter F. Stadler, 19–38. Cambridge, MA: MIT Press.

Hempel, Carl, and Paul Oppenheim (1965). "On the Idea of Emergence." In *Aspects of Scientific Explanation and Other Essays in the Philosophy of Science*, ed. Carl Hempel, 258–264. New York: Free Press.

Holland, John H. (1975). *Adaptation in Natural and Artificial Systems: An Introductory Analysis with Applications to Biology, Control, and Artificial Intelligence*. Ann Arbor: University of Michigan Press.

Hovda, Paul (2008). "Quantifying Weak Emergence." *Minds and Machines* 18(4): 461–473.

Jiang, Yi, Brian Travis, Chad Knutson, Jinsuo Zhang, and Pawel Weronski (2008). "Numerical Methods for Protocell Simulations." In *Protocells: Bridging Nonliving and Living Matter*, eds. Steen Rasmussen, Mark A. Bedau, Liaohai Chen, David Deamer, David C. Krakauer, Norman H. Packard, and Peter F. Stadler, 407–432. Cambridge, MA: MIT Press.

Kim, Jaegwon (1999). "Making Sense of Emergence." *Philosophical Studies* 95: 3–36.

McLaughlin, Brian P. (1992). "The Rise and Fall of British Emergentism." In *Emergence or Reduction? Essays on the Prospects of Nonreductive Physicalism*, eds. Ansgar Beckerman, Hans Flohr, and Jaegwon Kim, 49–93. Berlin: Walter de Gruyter.

Mitchell, Melanie (2009). *Complexity: A Guided Tour*. New York: Oxford University Press.

Morowitz, Harold J. (2002). *The Emergence of Everything: How the World Became Complex*. New York: Oxford University Press.

Mouritsen, Ole G., and Ask F. Jakobsen (2008). "Forming the Essential Template for Life: The Physics of Lipid Self-Assembly. In *Protocells: Bridging Nonliving and Living Matter*, eds. Steen Rasmussen, Mark A. Bedau, Liaohai Chen, David Deamer, David C. Krakauer, Norman H. Packard, and Peter F. Stadler, 385–406. Cambridge, MA: MIT Press.

Nagel, Kai, and Steen Rasmussen (1996). "Particle Hopping Models and Traffic Flow Theory." *Physical Review E* 53: 4655–4672.

Packard, Norman (1989). "Intrinsic Adaptation in a Simple Model for Evolution." In *Artificial Life*, ed. Christopher G. Langton, 141–155. Redwood City, CA: Addison-Wesley.

Rueger, Alexander (2000a). "Robust Supervenience and Emergence." *Philosophy of Science* 67: 466–489.

Rueger, Alexander (2000b). "Physical Emergence, Diachronic and Synchronic." *Synthese* 124: 297–322.

Sawyer, R. Keith (2005). *Social Emergence: Societies as Complex Systems*. New York: Cambridge University Press.

Silberstein, Michael, and John McGeever (1999). "The Search for Ontological Emergence." *Philosophical Quarterly* 49:182–200.

Stephan, Achim (2006). "The Dual Role of 'Emergence' in the Philosophy of Mind and in Cognitive Science." *Synthese* 151: 485–498.

Stano, Pasquale, Giovanni Murtas, and Pier Luigi Luisi, (2008). "Semisynthetic Minimal Cells: New Advancements and Perspectives." In *Protocells: Bridging Nonliving and Living Matter*, eds. Steen Rasmussen, Mark A. Bedau, Liaohai Chen, David Deamer, David C. Krakauer, Norman. H. Packard, and Peter F. Stadler, 71–100. Cambridge, MA: MIT Press.

Sugiyama, Yuki, Minoru Fukui, Macoto Kikuchi, Katsuya Hasebe, Akihiro Nakayama, Katsuhiro Nishinari, Shin-ichi Tadaki, and Satoshi Yukawa (2008). Traffic Jams without Bottlenecks: Experimental Evidence for the Physical Mechanism of the Formation of a Jam. *New Journal of Physics* 10. doi:10.1088/1367-2630/10/3/033001.

Weinberg, Steven (1987). "Newtonianism, Reductionism, and the Art of Congressional Testimony." *Science* 330: 433–437. (Reprinted in eds. Mark A. Bedau and Paul Humphreys, 345–357, 2008).

Wimsatt, William C. (1986). "Forms of Aggregativity." In *Human Nature and Natural Knowledge*, eds. Alan Donagan, Anthony N. Perovich, Jr., and Michael V. Wedin, 259–291. Dordrecht: Reidel.

Wimsatt, William C. (1997). "Aggregativity: Reductive Heuristics for Finding Emergence." *Philosophy of Science* 64(4): 372–384.

Wolfram, Stephen (1985). "Undecidability and Intractability in Theoretical Physics." *Physical Review Letters* 54: 735–738.

Wolfram, Stephen (2002). *A New Kind of Science*. Champaign, IL: Wolfram Media.

6 Agent-Based Modeling and the Fallacies of Individualism

Brian Epstein

Agent-based modeling is starting to crack problems that have resisted treatment by analytical methods. Many of these are in the physical and biological sciences, such as the growth of viruses in organisms, flocking and migration patterns, and models of neural interaction. In the social sciences, agent-based models have had success in such areas as modeling epidemics, traffic patterns, and the dynamics of battlefields. And in recent years, the methodology has begun to be applied to economics, simulating such phenomena as energy markets and the design of auctions.[1]

In this paper, I aim to bring out some fundamental limitations and tradeoffs to agent-based modeling in the social sciences in particular. Two misconceptions about social ontology, pertaining to the relation between social macro-properties and individualistic properties, are widespread in social theory and modeling. These issues lead current models to systematically ignore factors that may be significant for modeling macro-properties. To treat the problem, I suggest that we give up on two deeply held assumptions: first, that agent-based (and other) models can provide the microfoundations for macro-properties and second, that models have to avoid ontological redundancy. Abandoning each of these is painful but may be less costly than the alternative.

AGENT-BASED MODELING IN THE SOCIAL SCIENCES

In the social sciences, as in the natural sciences, most mathematical models are "analytical" models rather than computational ones. Analytical models typically consist of systems of differential equations, giving structural relationships between variables of interest, such as the Lotka-Volterra equations, describing predator-prey dynamics, and the Susceptible Exposed Infectious Recovered (SEIR) model of epidemic propagation.[2] The most familiar agent-based models are cellular automata, with agents represented as states on a fixed geographical grid. The "Sugarscape" model, introduced by Joshua Epstein and Robert Axtell, is a well-known example.[3] With agent-based models, it is easy to construct a large population of heterogeneous agents. Sugarscape, for instance, represents a population of individuals who may be young, middle-aged, or elderly, who can "see" only nearby cells and who can see distant cells, and who have slow or fast metabolisms. Interestingly, it also includes environmental resources in the model: occupying the cells in addition to people are quantities of "sugar" and "spice," with which the agents interact.

Agent-based modeling is typically understood to be a way to provide micro-foundations for changes in macroscopic properties. Consider some macroentity, say France, represented in Figure 6.1 as a hexagon.

France has a variety of macro-properties, such as its inflation rate, unemployment, and monetary financial institution interest rates. One way to model the interrelations among these variables is to come up with macrolaws (L_M) that govern the relationships. In macroeconomics, the values of parameters might be determined by statistical estimation. A micro-foundational model attempts to decompose and eliminate the macro-properties. Instead of modeling the macro-properties themselves, it models interactions among individual agents, each of whom starts out in a particular state, with the system updating over time, as shown in Figure 6.2. At any time-step, the macro-properties can be "read off" of the micro-states.

Many analytic micro-foundational models have been developed in economics, but they tend to involve a "representative agent" framework, in which individuals are represented as if the aggregate of their choices is equivalent to the decision of an aggregate of identical individuals.[4] Agent-based modeling allows more complex initial states and transition rules to be incorporated.

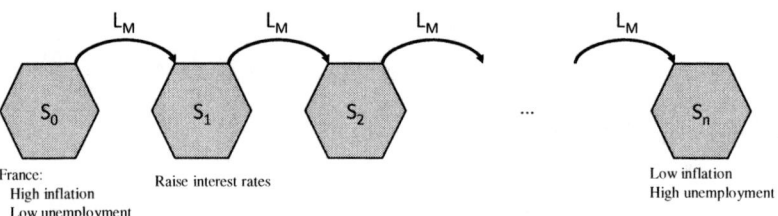

Figure 6.1 Macroentity represented as a hexagon.

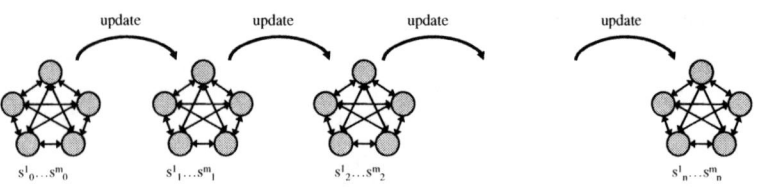

Figure 6.2 Interaction among individual agents.

Agent-Based Modeling and the Fallacies of Individualism   117

Although the best-known agent-based models are cellular automata, there are many different types, with more general mathematical characteristics. Cellular automata involve changing states of a fixed homogeneous set of cells that neighbor one another. They also typically involve a synchronous updating schedule, all the cells being updated at the same time.[5] In a more general agent-based model, one can choose to model very different sets of objects, apart from locations or people, and agents do not need to be modeled on a grid. An agent-based model, for instance, might involve a network of power plants interconnected by power lines. The network configuration itself can also be dynamic. In a model of the evolution of an epidemic, people move over time, so the network of interactions changes from one moment to the next. And different agent-based models involve different updating schedules, so that agents can update and interact asynchronously in various ways. Finally, processes in the system can be stochastic as well as deterministic.

Generically, these can be described by what might be called "graph dynamical systems," mathematical structures that include cellular automata and other structures as subcases.[6] One simple and very general class of structures, widely discussed in connection with agent-based modeling, is the sequential dynamical systems.[7] A typical agent-based model starts with a set of objects or agents, each in a certain state and with a network of connections among them. In a sequential dynamical system, this is represented as

1. A finite undirected graph G(V,E) with n nodes (or vertices), representing objects or agents, and m edges. G also has no multiedges or self loops.
2. A domain D of state values, representing the properties of each node.
3. A local transition function set F={$f_1, f_2, \ldots, f_n$}, with each f_i mapping D^{δ_i+1} into D, where δ_i is the degree of node i. (f_i takes each node and its neighbors at a state, and yields a new state for the node).
4. A total order π on the set of nodes, specifying the order in which they update their states.

A configuration K of the system is an n-vector ($k_1, k_2, \ldots, k_n$), where each k_i is an element of D. In a time-varying sequential dynamical system, the topology of the graph or the local functions can vary or evolve over time, as well as the configuration of the sequential dynamical system.[8]

Even the simplest models, as I mentioned, involve heterogeneous sets as agents. This is a significant advantage over analytic models that are limited to a single representative agent or a very small amount of heterogeneity if they are to be tractable. More generally in agent-based modeling, objects are not just simple things but have a variety of properties, behaviors, and tasks they can execute. The properties of agents are usually intrinsic properties, and the behaviors are either internally driven or can be triggered by changes in the objects with which they are connected. Connections between

118 Brian Epstein

agents are causal: one agent does something, or has a property, and that triggers a property change in another agent. Moreover, environmental factors are also just more agents, so even in simple models agents representing individual people can interact dynamically and bidirectionally with agents representing environmental properties as well as with one another.

In some sense, agent-based models can model objects at multiple scales. The widely used modeling program Swarm, for instance, allows agents to be aggregated sets of other agents. Likewise, a prominent recent book on agent-based modeling in business depicts organizations like companies as being built up out of divisions, which are built out of groups, which are built out of individual employees.[9]

Given such a compositional hierarchy, typical agent-based models take to the behavior of the high-level agents to be determined from the "bottom up," with the behaviors of the high-level agents determined by the behaviors of the components making them up. In Swarm, for instance, one might model a population of rabbits, each composed of rabbit parts, each of which is composed of cells. When a rabbit receives a message from the scheduler governing that agent, the behavior of that agent is determined by

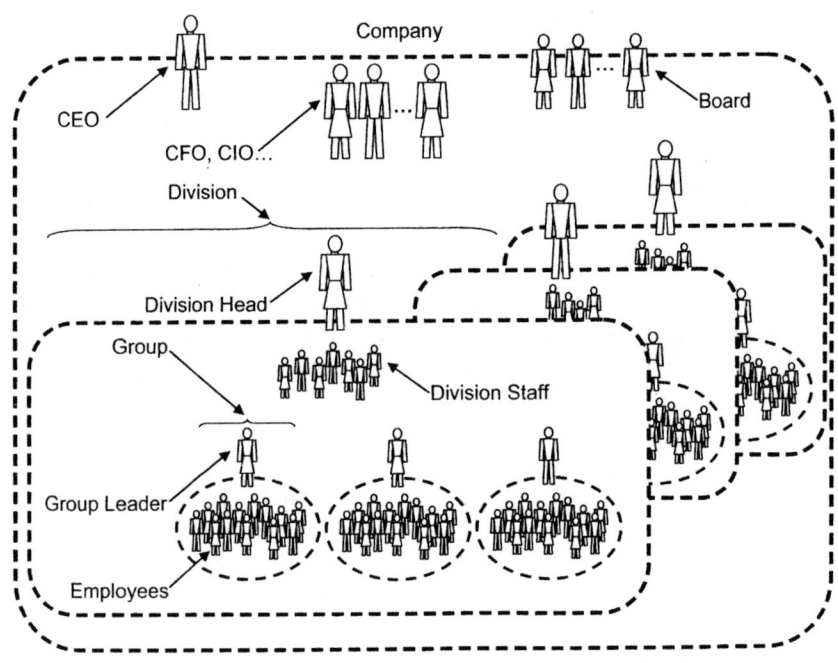

Multiscale organization with employee subagents.

Figure 6.3 Multiscale organization with employee subagents. Courtesy of Oxford University Press from *Managing Business Complexity: Discovering Strategic Solutions with Agent-Based Modeling and Simulation* by Michael North and Chares Macal (2007).

the behaviors of the swarm of the agent's constituent agents.[10] Inasmuch as an agent is constituted by a set of lower-level agents, the "causal efficacy" of the high-level agent is fully exhausted by the causal efficacy of the constituent agents.[11]

The successes in agent-based modeling, in such areas as epidemiology, power markets, and combat simulation, may lead us to be confident that we will be able to extend agent-based modeling more generally. But as soon as we move from these areas to more typical social properties, I will argue, the methodology becomes less effective. Related shortcomings apply just as much to analytic models as they do to agent-based models. But it may be mistakenly thought that agent-based modeling is immune to these problems, and the problems can be raised particularly clearly in the context of agent-based models.

TWO FALLACIES OF INDIVIDUALISM

"Ontological individualism" is typically taken as a truism in the philosophy of social science and is a background assumption of both analytical and computational models in the social sciences. Ontological individualism is a metaphysical claim about the relation of social facts or properties to individualistic facts or properties.[12] It is usually cashed out in terms of a form of the supervenience relation, to analyze the "local" claim *the social properties of any entity exhaustively depend on that entity's individualistic properties,* or the "global" claim *the social properties holding in a world depend on the individualistic properties holding of and among individual people in that world.*[13] However, ontological individualism—even understood as a global supervenience claim—does not just fail to be a truism but is false.[14]

The two fallacies I will discuss below are both fallacies about the ontology of social properties. They are what we might call "anthropocentric" fallacies, arising as a result of overestimating the extent to which social properties depend on individual people.[15] Obviously social properties are introduced by people. That does not mean, however, that these properties are fully determined by (or are exhausted by or supervene on) individualistic properties. The sources of this falsehood can be traced to two different problems, as I will discuss just below.[16] Before introducing them, let me begin with a brief comment on the relevance of ontological considerations to modeling.

Ontology in Modeling

While it may not be obvious that ontology is pertinent to modeling at all, ontological assumptions at least are tacitly built into the construction of any model. Suppose one wants to model a pot of water boiling on the stove.

A natural first step is to break the system into parts—e.g., the water, the pot, and the burner. One might then model the water with a molecular dynamics simulation, the pot with some analytic heat equations, and the burner with a Monte Carlo simulation.

Different models of some entity may treat the same entity in incompatible ways, such as modeling water as an incompressible fluid when modeling hydraulics and as a collection of particles when modeling diffusion.[17] But in a single model of physical stuff, we typically want the constituents to be mutually exclusive or nonoverlapping. In constructing a model of water boiling on a stove, the modeler thus breaks it down into parts that interact with one another, avoiding redundancy or internal conflict in the model. Once the modeler has constructed the analytic heat equations to treat the pot, she does not want to redundantly include the pot molecules in the molecular dynamics simulation that is treating the water molecules. It is usually straightforward to accomplish that.[18] All the ontological work the modeler does is to draw a spatiotemporal boundary around each object and then treat their causal interactions with one another. However, as I will discuss in a moment, typical social properties are different from typical physical properties in this respect.

To discern the role of ontology in model construction, it is important to retain a clear distinction between ontological dependence and causal dependence. The temperature of the water is causally affected by (and we might say "causally depends on") the temperature of the pot. But the water does not depend ontologically on pot molecules, whereas it does on the water molecules. Changes in the water consist in nothing more than changes in the water molecules; they are not caused by them.[19]

The Locality Fallacy

Many physical properties depend (ontologically) on spatiotemporally local features of the objects they apply to. Whether I am hot or cold depends on the temperature of the molecules of which I am materially constituted.[20] Some social properties are also local to me. My having the property *dancing an Irish jig* depends only properties spatiotemporally local to me. I need to be moving my body in a certain way, and there needs to be a floor beneath me which I am tapping in a certain way. Although the conditions for what it takes to have the property *dancing an Irish jig* were defined by certain Irishmen, the conditions for my having the property only involve my body and a small region around it.

A great many social properties, however, are not locally dependent. This point has been noted by, among others, Currie 1984, Ruben 1985, and Pettit 2003, though they did not draw out the implications for social theorizing or modeling. An example of a property that is not locally dependent, for instance, is the property *being President of the United States*. The fact that Barack Obama has that property does not depend on his intrinsic

Agent-Based Modeling and the Fallacies of Individualism   121

properties, nor does it depend on the properties of the White House, the places he has traveled, or the people he has come in contact with. Rather, it depends on a variety of properties of other people and things. For instance, it depends on certain current and historical properties of the electoral college, and the facts on which those properties depend in turn, as well as on such facts as the existence of the United States. and the U.S. government, etc., and all the factors on which those depend in turn. For modeling purposes, many of these can often be left in the background, as a practical matter. But nonlocal dependence may in many cases be important for model construction.

The "locality fallacy" is the fallacy of taking a nonlocal property to be a local property, i.e., taking some property P holding of an object x to depend (ontologically) on factors local to x, when P's holding can in fact depend on factors that are not local to x. A model can implicitly commit the locality fallacy if in modeling a nonlocal property, the only factors it takes into account are local factors and causes that impinge on the local factors. To be sure, many social properties that are in fact nonlocal can be usefully modeled, for many purposes, as if they are local properties. Similarly, it is often possible to do a nice job modeling a pot of water boiling on the burner by modeling only the burner and the water, while overlooking the pot. At the same time, if *all* models of the pot of boiling water, even very detailed ones, completely ignore the pot, we might ask whether the pot is being overlooked because we have a blind spot about pots. Likewise, local models of properties that are ontologically dependent on nonlocal factors are fine for many purposes. But if our approach to modeling both systematically and unconsciously ignores nonlocal factors, it is reasonably to ask whether this is a design flaw in our approach.

Many social properties are nonlocally dependent in very straightforward ways. Consider, for instance, an obviously extrinsic property of a social group, such as being in the National League playoffs, or being charged as a corrupt organization under the Racketeer Influenced and Corrupt Organizations (RICO) act. The factors determining the holding of such a property depends on factors that are not local to the bearer. But for such cases, it is unlikely that a modeler would fall prey to the locality fallacy. The troublesome cases are the ones where the nonlocality is not so straightforward.

Consider, for instance, the fact *the average age of the Tufts freshman class is 18½*. This obviously depends in part on certain local properties of a collection of approximately 1300 individuals. But only in part: that fact is not local to those 1300 individuals, for similar reasons that *being President* is not local to Obama. To be a member of the freshman class depends on many factors that the freshmen may not even be aware of or that may not even impinge on them causally. Suppose, to simplify, that there are only four freshman, P, Q, R, and S, aged 17, 18, 19, and 20, respectively. Evaluating *the average age of the freshman class* in the actual world, we consider the ages of P, Q, R, and S, which average to 18½. And suppose that all four of them go to a day-long lecture one day during the fall term. Over the course of the day, their individualistic properties, including their ages, remain relatively unchanged. But imagine that while they are sitting in the auditorium, the world changes radically around them: at 10 a.m., P's parents and Q's parents win the lottery and immediately withdraw their kids from school; at 11 a.m., S's parents go bankrupt and withdraw S from school as well; at 1 p.m., P's parents have second thoughts and re-enroll P; and then at 3 p.m., the board of trustees dissolves the school entirely. Over the course of the day, the individualistic properties of P, Q, R, and S remain more or less constant, but the value of the function *the average age of the freshman class* fluctuates, from 18½ at 9 a.m., jumping to 19½ at 10 a.m., dropping to 19 at 11a.m., dropping further to 18 at 1 p.m., and then becomes undefined at 3 p.m. This function fluctuates in virtue of changes in properties other than the individualistic ones of the freshmen themselves. It is not that the values of these functions do not depend on the properties of the four individuals, but rather, that they also depend on those nonlocal properties that figure into determining the holding of the property *being a freshman*.

Not every social property of an individual or group is nonlocally dependent. Consider the choices of a pair of prisoners, each given certain information and certain alternatives. Then the only factors on which the output of the "choice" function applied to the pair of prisoners depends are their local characteristics. The same is true for the audience in an auditorium in the example Thomas Schelling discusses in the introduction to Schelling 1978. To determine why an audience has spontaneously organized to sit bunched together in the seats at the back of the auditorium, as opposed to populating the better seats, the only factors that pattern depends on are again the local characteristics of the individuals in that audience. The reason is that the property *being in an auditorium*, like *being a molecule in a balloon* and unlike *being a freshman*, plausibly depends only on the characteristics of that local spatial region.

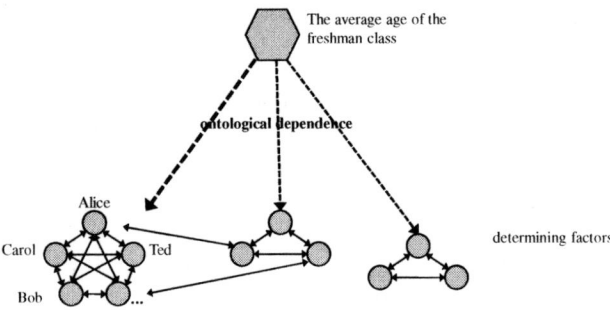

Figure 6.4 Ontological dependence and causal interactions.

In Figure 6.4, the dotted arrows represent lines of ontological dependence, and the thin solid arrows represent causal interactions. The value of the function *the average age of the freshman class* ontologically depends on a variety of individualistic properties, some of the bearer of the property, represented as the pentagon on the lower left, but also depends on properties of other members of the population apart from the bearer of the property. Some of these may causally interact with the members of the class, and some may not. Given the ontological dependence of the macro-property on the wider population, it is possible to change that macro-property through a change in a nonlocal property, even if that nonlocal property does not even causally interact with the freshmen who are the bearers of the property.

One might still have reservations about the pertinence of nonlocality to modeling social properties such as *being President*, because *being President* seems to have the following characteristic. Although it is put in place by a variety of nonlocal factors, such as being elected by electors, once that has taken place, those nonlocal factors seem to become irrelevant. Although the choices of the electors were relevant in making Obama President, he remains President even if the electors or the population as a whole now have changed their attitudes. This suggests that subsequent to the election, things like *the decisions of the President* or *the actions taken by the President* depend only on his local properties, at least until the next election.[21]

It is true that a change in the attitudes of voters across the country does not suffice to discharge Obama. But that does not mean that the nonlocal factors on which *being President* depends are irrelevant to models of the property. Even after he has been elected the determining factors may change. It is not only having been elected, that Obama's having the property *being President* depends on, other factors, such as not having been impeached and convicted, having the government in place, not having been removed in a coup, and so on. Many changes external to Obama could make the property *being President* fail to hold of him, whatever his local characteristics. It may be acceptable to relegate these to the background in a short-duration model of the U.S. presidency, but doing so may be a poor strategy for modeling different countries or time periods or models of longer durations. For other extrinsic properties, it may be even less satisfactory to leave all nonlocal factors in the background.

As with all modeling, one always must be selective about the factors to include. For certain models of *being President* it may be useful to ignore the nonlocal factors on which the property depends, just as we might ignore nonlocal factors in modeling the average age of the freshman class. But for different purposes, it may be preferable to ignore the local ones and model only the nonlocal ones. To return to the boiling water analogy: for some models it is wise to ignore the pot. But for other models, it may be the burner, or even the water, that is practical to neglect, in favor of the pot. It is not clear why there should be a systematic bias for incorporating the local factors on which social properties ontologically depend and for ignoring the nonlocal ones.

The Levels Fallacy

The levels fallacy is an equally pervasive problem in models of social properties. Roughly, it is that if we want to model entities or properties at the social level, we only need to decompose them into entities or properties at the level of individuals. This seems to stem from the same sort of reasoning that motivated the Oppenheim-Putnam picture of the sciences being divided into compositional "levels."[22] Their idea was that objects can be arranged in a compositional hierarchy of levels: (i) elementary particles, (ii) atoms, (iii) molecules, (iv) cells, (v) multicellular living things, and (vi) social groups. The sciences correspond to the study of objects at these levels. Among the conditions they imposed on a hierarchy of levels were (i) for something to be in a particular level of the sciences (except the lowest level), it needs to be fully decomposable into objects of the next lower level; i.e., that each level is a "common denominator" for the level immediately above it, and (ii) nothing on any level should have a part on any higher level.[23]

There are a number of problems with this picture. First, it is clearly not the case that objects at a given level do fully decompose into objects at the next lower level. The picture is reminiscent of the belief of early cell biologists[24] that the human body was exhaustively composed of cells. That, of course, is not the case: we are made up partly of cells, but we also are made of digestive fluids, blood plasma, bone matrix, cerebrospinal fluid, intracellular material, mucus, and so on. Bodies do not decompose into objects only at the cellular level.

Likewise, it is a common but glaring mistake to think that social objects decompose into individual people. Consider, for instance, Tartine Bakery on the corner of 18th and Guerrero in San Francisco. On a typical day, racks of croissants are being baked in the ovens, the cash register is ringing, bakers are working with flour and sugar and butter, customers are lining up out the door, credit cards are being processed, banks are being debited and credited, accountants are tallying up expenses, ownership stakes are rising in value, and so on.

The employees are critical to the operation of Tartine. Is it, however, plausible that the employees exhaust the parts of the bakery? The employees are plausibly part of the bakery: if there were no bakers, then it would arguably be an empty shell and not a bakery. But a bakery also needs ovens, and the ovens are parts of the bakery. It is further plausible that among the parts are also its butter and its flour, and even its cash registers and coins and bills exchanging hands. Historically, there have been various attempts to forcefit the dependence-base of social entities and properties into some preferred class, supposedly in the service of individualism. Most notably, behaviorists argued that social entities are exhaustively composed of the nonintentional behaviors of individual people, and psychologistic approaches took social entities to be exhaustively composed of psychological states. These,

Agent-Based Modeling and the Fallacies of Individualism 125

mercifully, have largely died off, but the assumption of an individualistic base seems to linger on. The neat hierarchical composition of corporations out of nested groups of individuals shown in Figure 6.3 is an example. That picture overlooks everything but the employees.

Equally significantly, even if the compositional pyramid were correct, that would not mean that social facts depend exhaustively on individualistic facts. Simply because a group is constituted by individuals does not imply that the group is identical to its individuals. Work on "coincidence" in recent years has argued for a gap between the constitution and the identity of ordinary objects, so that two entities may be identically constituted, and yet have different actual and modal properties.[25]

Consider the factors on which some social property, such as *having a $100 million liability for hurricane damages in New Orleans after Hurricane Katrina,* depends. A large number of factors determine whether or not such a property holds of some insurance company.[26] Among the factors may be the psychological states of employees of the company, but another factor their liability depends on is the actual hurricane damage in New Orleans. Corporations, universities, churches, governments, and so on, depend on a wide variety of objects, including contracts, liabilities, insurance policies, stocks, bonds, etc. These objects, in turn, plausibly depend on complex sets of other objects. Even the stratification of objects into levels itself is dubious: at best, the division of objects into levels is fuzzy and pragmatic.[27]

Instead, social entities and properties depend on a radically heterogeneous set of other entities and properties. They can be cross-level, they can be at multiple scales, and they do not have to be individualistic at all, as depicted in Figure 6.5.

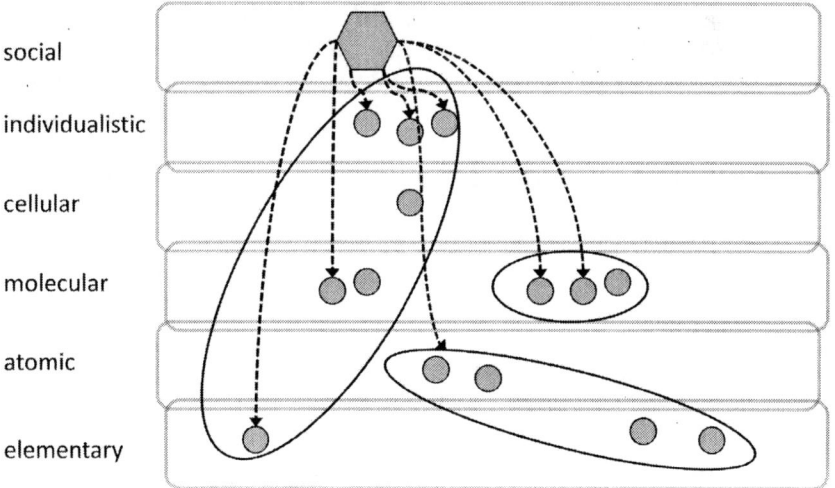

Figure 6.5 Radically heterogeneous set of other entities and properties.

In the figure, a social property such as *having a $100 MM liability for Katrina damage* may depend on certain properties of the insurance company bearing the property, represented by the large oval in the diagram. And it may depend on other objects, such as buildings, swimming pools, cats, vats of bacteria producing biofuels, particle accelerators, and on and on, represented by other ovals in the diagram. Again, these dotted arrows are arrows of ontological dependence, not arrows of causation.

AGENT-BASED MODELING AND THE FALLACIES

The locality and levels fallacies are mistakes pertaining to the relation between macroscopic facts and microscopic facts. For some facts, however, these issues may not arise. Let us call macroscopic facts *simply dependent* if they are local and depend solely on the next lower level of facts.[28] How much sense this makes depends, of course, on how much sense we can make of levels in the first place. But as we proceed, it is useful to contrast models in which it is assumed that all macro-properties to be modeled are simply dependent, from those that are not.

Agent-based models have the resources to avoid the naïve fallacies of individualism, by incorporating heterogeneous ontologies of agents. But the nonlocal and cross-level dependence of social objects and properties is often overlooked entirely, so they do not even avoid the fallacies in naïve forms. More significantly, however, there are limits to how much even sophisticated models can avoid the fallacies.

In the following sections, I argue that the problems arising from the locality and levels fallacies create a set of new problems for agent-based modeling, beyond those that arise for modeling simply dependent facts. After setting up some distinctions and terminology in the next section, in the following section I consider some issues that occasionally arise in modeling simply dependent facts and can usually be addressed though judicious choice of variables to model. I then argue in that nonlocal and cross-level dependence turn these minor issues into apparently insurmountable obstacles. Subsequently I turn to refinements of agent-based modeling, proposing ways of designing models for addressing these problems but at the same time having to make certain new compromises.

Distinguishing the Target, Base, and Causal Ontologies

In the design of a model, the modeler implicitly has a set of entities of interest, or "target" entities, in mind, although it may not be apparent from the implementation. An agent-based model targeting a macroscopic entity or property M will generally result in different choices than one targeting some other entity or property M'. Given a set of target entities, the first step in model design is to identify some set of microentities on which that target set ontologically depends. These factors we will call the "base ontology."[29] The next step, then, is to identify the other entities that causally interact with the base. These factors we will call the "causal ontology."

For a given target, we sometimes model just a fragment of the base ontology, and often (if not always) model just a fragment of the causal ontology. Given a target ontology T, I will distinguish the "modeled base ontology" M^B from the complete base ontology of T and distinguish the "modeled causal ontology" M^C from either the complete set of factors interacting with the complete base ontology, or the complete set of factors interacting with the modeled base ontology. Together, I will call $M=M^B \cup M^C$ the "modeled ontology."[30]

If we suppose that the macroentities being modeled are simply dependent, the base ontology can be entirely local to the target, and consist of entities at level n-1.[31] For a simply dependent target, the modeled base ontology can thus often be the entire base ontology. Given the base ontology, the "modeled causal ontology" is chosen on the basis of being entities in the world that are expected to interact in a significant way with those in the base ontology.

Only when the target is simply dependent is the relation between it and the modeled base ontology so straightforward. When the target is not simply dependent, the determination of the modeled base and modeled causal ontologies from the target becomes far more problematic.

The Dynamics of Ontology Under Simple Dependence

Even with the assumption that the target is simply dependent, there are some interesting problems arising from the fact that the relevant causal factors and even the relevant dependence base may change over the course of a model's evolution.

In a standard graph-dynamical system, like a sequential dynamical system, the configuration K (i.e., the state of the nodes) changes over time, but the graph itself is static. A time-varying sequential dynamical system accommodates the fact that the interactions among objects are not static, as the system evolves and across circumstances. For both, however, the modeled ontology itself is fixed at the outset.[32]

Where it is not known in advance how much causal influence some factor will have, a natural approach is simply to include it in the causal ontology. If the modeled ontology is fixed at the outset, we simply have to choose a big ontology even if we know that an object being modeled has a predictably changing causal ontology over time. For instance, to model a car traveling down a highway, where the things the car interacts with at time t are different from those it interacts with at t', a natural if inefficient treatment is to expand the ontology to include the larger set of causal features the system will have contact with over the duration of the simulation.

For systems that are well-behaved,[33] inefficiencies of fixing a causal ontology in advance are often not overwhelming. But even for simply dependent properties, changes as the system evolves may mean that new factors in unpredicted locations or at unpredicted scales may become relevant. As a system develops, the fragment of the causal ontology that is included in the model may thus need to change, in order for the model to be both reasonably tractable and reasonably accurate. In other words, we may not only need the network of interactions to be time-dynamic but the nodes as well. Rather than just a single modeled causal ontology M^C, the causal ontology may need to become a sequence $(M^C_0, M^C_1, \ldots)$.

Although this seems simple—just modify the representations of time-dynamic graphs to allow the nodes to change over time—it creates a potentially severe problem. Consider a modeler observing her model during a run, evolving over time from steps $t_0 \ldots t_i \ldots t_n$. Suppose that at time t_i, the modeler realizes that the model has evolved in such a way that some entity that was not included in $M^C_0 \cup M^C_1 \cup \ldots \cup M^C_{i-1}$ will have significant causal influence. For instance, suppose she is modeling the orbits of planets and finds that one planet is thrown out of its orbit partway through a run and enters a different system that she had assumed she could neglect. Accommodating the new elements into the causal ontology may not be as simple as just adding them into M^C_i and letting the run continue. Because, of course, over the simulation up to t_i, it might have been influenced by other entities in the domain, and given those influences it will have reciprocal effects on the other parts of the ontology, given the way the system in fact evolved. To model it properly, the run may need to be restarted from scratch, to incorporate the missing node and its influences. Even admitting small changes to the causal ontology may turn a computationally tractable model into one that must be repeatedly rolled back.[34]

Furthermore, even with a simply dependent target, it may not only be the causal ontology that needs to be dynamic, but the base ontology as well. If we are modeling the gas escaping from a chamber or the traffic at an intersection, for instance, the composition of the target changes over time. This means that the base ontology is also not static but perhaps should become a sequence $(M^B_0, M^B_1, \ldots)$.

For simply dependent targets, however, these issues tend to be easily accommodated. In many such cases, a changing base ontology does not require a change in the modeled ontology at all, since the changes in the base ontology simply involve a shift from a factor being part of the modeled causal ontology to being part of the modeled base ontology. That is, the overall ontology does not need to change if the modeled ontology M already includes the factors left out of M^B_t. This occurs if the difference between M^B_{t+1} and M^B_t is a subset of M^C_t. One of the reasons that modelers do not make a distinction between the base and the causes is that it can actually be an advantage to conflate the two, when factors can flip in and out of the base ontology over time. But as we will see, failing to distinguish the base ontology from the causal ontology is acceptable only so long as the base ontology itself is well-behaved.

The reason these considerations do not foreclose the possibility of modeling altogether is that we generally choose the entities to model judiciously, and for these entities, the world is reasonably well behaved. Simply dependent target facts do not guarantee good behavior, but inasmuch as we are dealing with systems that are basically dependent on local and single-level factors, the dynamics of causal ontology are often sufficiently predictable.

With many social properties, however, these assumptions should not be expected to hold at all. When the target is simply dependent, the dynamics of the base ontology, and even the dynamics of the causal ontology are typically small issues. But they can balloon in importance when the target is not simply dependent.

When Targets are not Simply Dependent

If the target is not simply dependent, it may be a problem to choose a reasonable modeled ontology altogether. Intuitively, the problem with nonlocal dependence is that the dependence base can become indefinitely *broad,* and the problem with cross-level dependence is that it can become indefinitely *deep.*

A non-simply dependent target may depend on factors that are spatiotemporally remote from it. To model a property such as *being President of the United States,* for instance, facts about the entire voting population may need to be included. These factors may change over time and across circumstances, meaning that the modeled causal ontology cannot be restricted to factors that are spatiotemporally local or connected to the President.

Even more significant, however, is the implication of non-simple dependence on the choice of a modeled base ontology. With changes over time and across circumstances, the choice of the modeled base ontology can have to change radically and discontinuously, if it is to be a reasonable dependence base for the target. Even for the simple property *being President of the United States,* this may be the case. This property depends on certain properties of the voting population. Yet the voting population is not static. *Being President of the United States,* for instance currently depends in part on the votes of a large group of 18- and 19-year-olds who were not in the electorate a few years ago. The dependence base is significantly different today than it was then.

If there is single-level dependence, this may not be disastrous. Recognizing that social properties such as *being President of the United States* may depend on all the members of the population, we may simply choose to model the entire population. There are not so many people in the population, that the problem becomes computationally intractable. The failure of single-level dependence, however, erases this possibility.

For example, consider some complex but important social fact that we may want to model, such as the fact *American International Group (AIG) has $2.3 trillion in credit default swap obligations*. This ontologically depends on a great variety of factors, including properties of houses, paper, contracts, other corporations, its employees, assets backing various bonds, etc. Suppose, for instance, that a third of that obligation involves being contracted to replace the payments from mortgage-backed bonds, should the issuers default. Simplifying a bit, each of the mortgages backing those bonds consists of a disjunction—either a stream of payments from the homeowner or a house that the issuer has the right to seize if the payments stop. American International Group's instantiating the property *having $2.3 trillion in credit default swap obligations* thus depends in part on the houses. To model this as an intrinsic or psychological property would grossly mischaracterize the factors determining whether the property is instantiated. Instead, the factors on which the property ontologically depends can be any of thousands of different aspects of the housing stock, other banks, and so on. All of these are eligible to be components of the modeled base ontology.

It is impossible to model the entire world in infinite detail, so the modeled base has to be a judiciously selected subset of the complete base ontology. As I noted above, if we are modeling a simply dependent property, such as the density of traffic at an intersection, then changes to the base ontology are usually easily accommodated. But when the target is not simply dependent and instead depends on highly nonlocal and cross-level factors, the choice of a modeled base becomes far more problematic. Moreover, it becomes increasingly likely that as the model runs a much more varied set of factors will become relevant. The factors that a modeler might reasonably incorporate are likely to change nonlocally and are also likely to change at different depths of texture: whereas at time t the modeler may have reasonably ignored everything at the level of housing stock, water supplies, or banking charters; any one of those might be the most relevant aspect of the base at time t+1. This problem does not arise for simply dependent properties.

Still worse, when these changes in the modeled base ontology run into the choice of a causal ontology interacting with the base, the problem explodes. Normally, for each M^B_i a corresponding causal ontology M^C_i would to be chosen, incorporating the factors that significantly causally influence the entities in M^B_i. But if M^B_i is spatially and level-wise ill-behaved, this makes the sequence of significant causal influences at least as badly behaved. And further, it increases the risk of the need for rollbacks. If even minor additions to the causal ontology potentially required rollbacks, with the present problems we might never be able to get past the first few steps of a simulation, without having to restart over and over again.

The Hard Problem: How Even to Determine the Base Ontology

These threats to the tractability of a social model are potentially game-enders. But there is a more difficult problem still. Given a total ontology M_t consisting of a base ontology M^B_t and a causal ontology M^C_t: what is M_{t+1}? The problem is that presuming that M^B_t is chosen to be a reasonable subset of the complete dependence base of the target ontology for use at time t, there is no reason to suppose that the state of M^B_t (or M^C_t) has enough information to determine or even recommend what a reasonable M^B_{t+1} might be.

Under the assumption that the target is simply dependent, the modeler could be generous about the factors at a single level to be included in M_t and confidently predict any that changes in the modeled base and modeled causal ontologies would be local or nearly local to M_t. But without simple dependence, the realization of the target in the dependence base can change any which way, both broadly and deeply. A choice of the dependence base to model at a given time, as a subset of the complete dependence base, does not determine which subset of the (potentially global) dependence base of the target should be chosen subsequently.

To see this, consider again the AIG example. Which factors in the complete base ontology are significant to determining the target can differ quite radically even in relatively nearby worlds and times. And the nearest set of circumstances to the actual one might be circumstances that involve a radically different set of nonlocal determining factors. It may be the following are three likely paths the world could take, that would potentially change the fact *AIG has $2.3 trillion in credit default swap (CDS) obligations*:

1. Houses fall into disrepair because homeowners do not have the incentive to keep them up, and hence lose value at a rapid rate, and so on . . .
2. Depositors are losing confidence in banks and withdrawing their money en masse, causing them to collapse, . . .
3. A rash of resignations among executives are leading the AIG board to choose to default on various instruments, . . .

These are three very different possible paths that the world might go in, which may change the fact in question. Note that these do not only involve causing changed states of the entities in the dependence base but may involve radically different dependence-bases themselves, as circumstances change. In path 1, houses which were once constitutive in part of the obligation may be destroyed; in path 2, the houses may remain but the counterparties to the transactions may disappear; and in path 3, contracts that were in part constitutive of the obligation may be torn up.

From a distance, we can see changes in the base coming: the housing stock, the banks, and the stress among executives are being modeled, each of which may flag a change in the modeled base. But while those factors are externally observable by modelers with the target ontology in mind, the model itself at t only includes the modeled ontology. It does not include the elements of the complete base that will be significant factors in determining the target at t+1. Nor do the factors in the modeled base determine which other factors will be significant parts of the dependence base of the target at t+1. The dependence base of the fact *AIG has $2.3 trillion in CDS obligations* at t+1 is not necessarily included in M_t, nor is it determined by M_t, regardless of how judiciously M_t is chosen.

Ignoring or minimizing changes in the base ontology may be a reasonable idealization for simply dependent targets. But when we move even to straightforward properties that are not simply dependent, these idealizations can collapse spectacularly.

REFINING AGENT-BASED MODELING

At this point, things seem bleak. The modeler does her best to choose a starting point for the simulation, based on the target entities of interest. But the ontology needs to be dynamic, in order to accommodate the changes in the realization of the targets, and every dynamic change in the ontology threatens to demand a rollback of the model to the beginning. On top of that, the modeler herself, in order to determine how the base ontology changes over time, needs to keep appealing back to the target. She can have as finely grained a model as she likes, and still the target entities cannot just be put to the side, to be read off of the model, as it evolves. It seems the only routes are either to accept the approximation of all targets as well-behaved simply dependent entities—e.g., treating AIG's liability as if it depends on features of the AIG employees alone, which may be tantamount to modeling a Ferrari by simulating its air conditioner—or else to model everything at all levels in infinite detail. If so, we might just throw up our hands.

This pessimism, I think, is unwarranted. Instead, I suggest that we give up on two deeply held assumptions I mentioned at the outset: first, that agent-based (and other) models provide the micro-foundations for macro-properties and second, that models have to avoid ontological redundancy. Abandoning each of these is painful but may be less costly than the alternative.

Proposal: Explicitly Including the Macroscopic Properties in the Ontology

Across circumstances and time, the relevant dependence base of a target may be heterogeneous and volatile. If we do not model the complete dependence base of the target (i.e., much of the entire world in infinite detail), then the determination of the dynamics of the base ontology requires appeal to the macroscopic target itself. But it is possible to include the target and other macro-properties in a model, so long as we are willing to make some sacrifices.

Consider *AIG having $2.3 trillion in CDS obligations*, given a particular modeled base ontology M^B_i for it at time t, and a causal ontology M^C_t. To determine M^B_{t+1}, we need to consider the dependence base of the target property given the state of M^B_t. For instance, if the system has evolved such that the banks are nearing default at t, then M^B_{t+1} will be chosen to involve the factors determining bank defaults. This can be done computationally—but just not on the basis of M^B_t alone. Instead, the target entity can be included in the model, not just "read off" of it, as an object or agent with causal efficacy and causal factors affecting it.[35]

In a sense, this proposal is a radical capitulation. At the outset, I mentioned that the basic idea of an agent-based model has seemed to many people to be the provision of micro-foundations for macro-properties. The explicit inclusion of macroentities in the model undercuts this aspiration. But in light of the nonlocal and cross-level dependence of social entities, I suggest that true micro-foundations—even computational and nonreductive ones—are a pipe dream. There is no reason to think that most social properties are well behaved with respect to a fixed or tractably dynamic set of microentities on which they depend. And since a model of every detail of the entire world is possible only in science fiction, the exclusion of macro-properties from models entails a reduction in their accuracy.[36]

Surreptitiously, this may already be taking place in agent-based modeling. It is likely that many agent-based models implicitly include macroscopic factors in the models, even while they seem to be microfoundational. For instance, in models of traffic patterns, parameter setting is frequently done by tweaking them until they generate the desired macroscopic properties. As far as I know, no one has investigated the question whether this is illicit when applied to agent-based models and whether it compromises micro-foundations. But there are many ways macro-properties can be hidden under the covers in a model, and it may not turn out to be such a bad thing.[37]

The Problem of Redundancy

Including macro-properties in models, however, carries with it a different and rather significant risk: the potential for redundancy in the model. In the model of the boiling water, the modeler implicitly makes sure that her base ontology is nonredundant. If she decomposes the water into individual molecules, those supplant entities that are already composed of individual molecules, such as waves and eddies and vortices.[38] If both macroelements and the microelements composing them are included, both with causal effects, there is a risk of double-counting causes. If both *the CDS obligations of*

AIG and *the housing stock on which it partly depends* are included in the modeled ontology, there are subtler but similar risks of redundancy.

This is a large problem, but not necessarily an overriding one. When properties are simply dependent, there may be no reason to take the risk of double-causation. But the exclusion of macro-properties from a model, when the base ontology is not well behaved and when we do not model the entire world, threatens how accurate the model can be anyway.

Agent-based models have always traded off computational tractability with the complete inclusion of potentially relevant factors in the modeled ontology. They have not thought to compromise nonredundancy. But nonredundancy is simply another desideratum, the partial sacrifice of which does not spell the complete failure of a model. If the threats of redundancy can be kept under control, then it may be worth it to admit some redundancy in the model so as to improve the choice of factors to be modeled over time and across circumstances, while retaining computational tractability.

Doing so also requires keeping track of the target, base, and causal ontologies separately, with the constitutive graph and the causal graphs separately represented, rather than collapsing all the nodes into a single modeled ontology, and taking all the graph edges to be causal interactions.

MODERATING THE PICTURE OF AGENT-BASED MODELING

As I mentioned at the outset, agent-based models are advantaged in many ways over analytic models, in terms of how comprehensively and realistically they can model the world. It is typical for agent-based models to incorporate a variety of environmental factors, even factors that are not necessarily local to the people in the population. The objects taken as the components of the model can be much more general than the individualistic factors employed in a typical model in analytical economics. Agent-based modeling is not strictly individualistic, so it does not have the same shortcomings of individualism. Sometimes, individualism is presumed in implementations of agent-based modeling, since individuals are regarded as the locus of social properties, and there is a failure to distinguish constitution of an entity from the factors that determine its properties. But this is not an in-built limitation of agent-based models.

Agent-based models have well-known disadvantages as well, of course, such as the difficulty of interpreting, testing, and drawing generalizations from them. The problems I have discussed here, however, are equally applicable to analytic models as they are to agent-based models. It is because agent-based modeling does seem to have an ontological advantage that it is most useful to point out the ontological problems with social modeling in this context. Inasmuch as correcting for the fallacies of individualism means changing how agent-based models are built, *a fortiori* it means rethinking analytical models as well.

To summarize some of the implications for agent-based modeling: First is the value of making the choice of target ontology explicit. The representations of the target, base, and causal ontologies should be kept separate, as should the representations of the dependence graph and the causal graph, given that all can evolve separately. Given an explicit target, the complete base ontology ought to be understood, so that a modeled base ontology can be judiciously chosen and changed as the system evolves. The base ontology, of course, can include heterogeneous properties and agents, nonlocal and cross-level. To accommodate a changing base ontology, macro-properties may need to be included explicitly in the model, not just as rollups of microagents, but having causal interactions. In modeling changes in the ontologies, as well as the causal evolution of the states of the system, some pre-modeling may be employed and/or allowances for rollback made, as the evolution of the system forces reassessment of the most significant dependent and causal factors. Finally, nonredundancy may be traded off explicitly against the goal of ensuring that the modeled ontology is well chosen over time, in idealizing the model so as to maintain its computational tractability.

APPENDIX: ON THE FAILURE OF LOCAL SUPERVENIENCE

In the paper, I describe the "locality fallacy" as the fallacy of taking a nonlocal property to be a local property. The reason this is a fallacy is that many social properties in fact ontologically depend on "nonlocal" factors. In this appendix, I discuss this dependence in more detail than above, and in particular, consider how "local supervenience" typically fails for social properties.

To begin, it is crucial to note that "local" is being used in two different ways in the present discussion. In the body of the paper, I used "local" to characterize a certain kind of property of an object. As I mention in Note 20, I mean the idea of a property being "local" to be somewhat looser than a property being "intrinsic." Intuitively, an intrinsic property of an object is a property that holds or fails to hold regardless of what other objects there are in the world. A local property of an object is one that holds or fails to hold regardless of what the world is like at a spatiotemporal distance from the object.

In speaking of supervenience claims, the term "local" is generally used differently. Supervenience relations are typically defined as holding between sets of properties—a set of "A-properties," such as mental properties or chemical properties, supervening on a set of "B-properties," such as neural properties or physical properties. In *local* supervenience claims, the dependence of A-properties on B-properties is cashed out as a comparison between pairs of objects in any possible worlds. Taking any possible pair of objects, a difference in the A-properties of the pair implies a difference in the pair's B-properties. Notice that a local supervenience claim might be true even if it is not intuitively "local" at all. For instance, it is trivially true that a set of A-properties will locally supervene on the set of A-properties itself, regardless of what the A-properties are. They could be the most extrinsic, globally dependent, holistic properties one could imagine, and that local supervenience claim would still hold.

But prevailing usage being what it is, in this appendix I will only speak of "local" as applied to supervenience claims, to avoid confusion. To refer to the "local" or the "individualistic" or the slightly-more-than-intrinsic *properties* on which social properties are often taken to depend, I will speak of "L-properties" rather than "local properties."

A common way of formalizing local supervenience is as "weak local supervenience," as defined by Jaegwon Kim:

> (WLS) A-properties *weakly locally supervene* on B-properties if and only if for any possible world w and any objects x and y in w, if x and y are B-indiscernible in w, then they are A-indiscernible in w (Kim 1984: 163).

Applying this definition to the present case, social properties weakly locally supervene on L-properties if and only if for any possible world w and any entities x and y in w, if x and y are L-indiscernible in w, then they are socially indiscernible in w. Two objects are L-indiscernible or socially indiscernible if and only if they are exactly alike with respect to every L-property or every social property, respectively.

Supervenience has come under criticism in recent years. Many philosophers have begun to doubt its utility for capturing the intuitive strength of dependence claims involving social properties. One reason is that the field has become hair-splittingly technical. There are now so many versions of supervenience (weak, strong, local, global, regional, multiple-domain, etc.) that it is unclear which interpretation of supervenience is the appropriate one to use, if any. Second, a number of philosophers have grown skeptical that supervenience is sufficient to capture the "dependence" of one set of properties on another. It is common to note that supervenience is only a modal relation, involving necessary co-variation of properties, while true dependence claims would seem to need more.

For some uses of supervenience, these are indeed problems. For our purposes, however, they are actually an advantage. If we wanted to *defend* the ontological dependence of social properties on L-properties with a supervenience claim, we would have to show two things: (i) that the supervenience claim was true and (ii) that the demonstrated supervenience claim was sufficient to capture the intuitive dependence claim. To *deny* dependence, however, the case is strongest if we can successfully deny even the weakest of the conditions for dependence to hold. To demonstrate the failure of dependence with a failure of supervenience, it does not matter that supervenience is not sufficient for dependence. What matters is that some form of

Agent-Based Modeling and the Fallacies of Individualism   137

supervenience is *necessary* for dependence. And while it is possible to deny that the strongest forms of supervenience are necessary for dependence, it can hardly be denied that the weaker forms are.

As I mentioned in the body of the paper, it is widely acknowledged that the local supervenience of social properties on L-properties fails, following Currie's point that Gordon Brown's being Prime Minister does not depend on his own L-properties. This intuitive point is correct. However, it turns out that if we stick to the most commonly used form of local supervenience, it fails on a technicality, rather than for the intuitive reasons people have taken it to. This threatens to mangle the case for local supervenience failure. By shoring up the definition of local supervenience, however, we can highlight the underlying reasons it fails. (For a related discussion of these points, see also Epstein 2009, 210–212.)

To bring out the problem, it is useful to consider a social property of a group, rather than a property of an individual person. Suppose we wish to assess whether *being a freshman class* (let us call that property F) supervenes locally on L-properties, interpreting local supervenience as (WLS) and employing a standard supervenience test. To do so, we take an actual case of a freshman class in context c_1, say the current one at Tufts. Let us call the collection of the members of the Tufts freshman class M. For the doppelganger case, suppose there is a collection of people L-indiscernible from M in some context c_2; call that sum N. In c_2, however, suppose that those individuals' parents have acted differently, failing to enroll them in college at all. The individuals in N are then not members of the Tufts freshman class, or any freshman class at all. Thus F does not hold of N. As stated, the scenario looks as though it demonstrates dependence failure. However, it does not in fact successfully make the same point.

It is important to be careful about which objects we are tracking across possible worlds—i.e., which objects x and y we are tracking to see whether setting their L-properties to be the same guarantees that they have the same social properties. Intuitively, we want to consider whether the social properties of the freshman class are fixed by its individualistic properties. So we want to compare the freshman class in c_1 with the freshman class in c_2, fixing them to have the same individualistic properties. But of course, there is no freshman class in c_2, so this cannot work. Without taking care, the supervenience case does not even get set up. This is the reason I set the case up in terms of mere collections of individuals M and N: across contexts c_1 and c_2, we fix the L-properties of the collections of individuals to be indiscernible, and then evaluate whether they have the same social properties.

But in taking care on this tracking issue, on closer inspection even properties that intuitively should locally supervene will fail to, if we use (WLS). In fact, (WLS) will fail even if we only consider objects in the local context, not even considering the doppelganger context at all. Suppose 10 of us choose to form a literary group. Let us stipulate that membership in the group is a matter only of mutual agreement among us. As with the case above, one of the entities in the local environment is the collection of the ten of us. The problem, however, is that the properties *being a literary group* and *being a freshman class* do not hold of mere collections. This is a point that is familiar in discussions of the metaphysics of "coincident objects." A statue, for instance, may have aesthetic properties that the lump of clay constituting it does not, and a person may have mental properties that the lump of tissue constituting her may not. (See Zimmerman 1995; Fine 2003; Koslicki 2004; Bennett 2004). As applied to social groups, this problem is even more apparent and yields local supervenience failure not only for properties like *being a freshman class,* that intuitively should fail, but it also yields local supervenience failure for any number of properties like *being a book group* that seem as though they should supervene locally on L-properties. What is needed is an understanding of local supervenience that has the ability to preserve the intuition that certain properties of groups *do* supervene locally on L-properties.

One way of solving the problem is to choose the properties judiciously, to show the expected failure and success of weak local supervenience using (WLS). The (WLS) is satisfactory if we ignore property F, *being a freshman class,* applied either to the Tufts freshman class (to which F holds necessarily) or to sum M (to which F does not hold), but rather consider applying to M the related property *coinciding with the membership of a freshman class*. Call this property F'. F' is a social property, one that does apply to M, and that fails to supervene on the L-properties of M. This, then, is an example of a social property (albeit an unusual one) that does not get retained in the indiscernible counterfactual case. Inasmuch as F' is a social property, the argument demonstrates the failure of local supervenience of social properties on L-properties. It seems, however, that we ought to be able to put on a stronger case than this.

A different approach is to follow Zimmerman, Bennett, and others in weakening local supervenience to be less stringent about the entities to which properties are taken to apply. The idea of "coincident-friendly local supervenience", roughly speaking, is that when we assess the social properties of any L-indiscernible pair, we do not only see if *that pair* is socially indiscernible. Rather, we look around the domain for other pairs of objects that coincide with and are L-indiscernible from the respective members of the original pair. In other words, in assessing whether property F applies to M in context c_1 and a sum N in c_2, we do not stop when we see that F does not hold of sums at all, and hence not to M or N. Rather, we look for other entities that coincide with M and N, respectively, and see if they have property F. If there is *any* pair of such entities, then F is regarded as holding for that pair.

On this slightly weaker but still plausible interpretation of local supervenience, supervenience failure does not follow just from the fact that ordinary social objects exist and have properties that do not apply to sums. Instead, it fails for more intuitive reasons, and in fact it shows how properties like F, in addition to properties like F', also fail to supervene on L-properties.

For property F, *being a freshman class,* suppose that in the doppelganger case there are no universities in c_2, so there is no freshman class there at all. (As always, N remains L-indiscernible from the actual collection of individuals M.) In the actual case, there is an entity coinciding with M and L-indiscernible from it that has property F. But in c_2, although there is an entity that is L-indiscernible from M, no entity has property F. Local supervenience thus fails. This conforms to the intuitive point. If *holding the office of President* does not supervene locally on L-properties, then neither does *being a president;* and if *coinciding with the membership of a freshman class* does not supervene locally on L-properties, then neither does *being a freshman class.* Hence the set of social properties that fail to supervene locally on L-properties does not only include the peculiar membership properties of individuals or groups but also the more intuitive properties that apply to social entities as well. This conception of supervenience also avoids forcing supervenience failure for those social properties that should locally supervene on L-properties, such as *being a literary group.*

Despite some technical hiccups, the intuitive point thus remains unchanged: social properties fail to depend ontologically on the "local" properties of the individuals or groups that bear them. Since social properties may depend on the wider population, they can change or fail to hold even if those wider factors do not interact causally with the bearer. Again, we often make the choice not to model parts of the dependence base of any given property. But it is not clear why we should systematically ignore all nonlocal factors on which social properties ontologically depend and only model the local ones.

NOTES

1. See, for instance, Tesfatsion 2003; Epstein 2005; Axtell 2006; Samuelson and Macal 2006.
2. Lotka 1925; Volterra 1926; Anderson and May 1992.
3. Epstein and Axtell 1996.
4. Typical assumptions of representative agent models, as Kirman 1992 points out, are that "all individuals should have identical homothetic utility functions (that is, ones with linear Engel curves); or that all individuals should have homothetic utility functions, not necessarily identical, but that the relative income distribution should be fixed and independent of prices" (120; cf. Lucas 1975; Woodford 2003). Some nascent analytic heterogeneous agent models are discussed in Heer and Maussner 2005. For criticism of the homogeneity of representative agents, see Kirman 1992; Hoover 2006.
5. Mortveit and Reidys 2008 limit cellular automata to synchronously updating systems, though the term 'cellular automaton' is sometimes applied more broadly to asynchronously updating ones.
6. Other less general structures that fall under generic graph-dynamical systems include Boolean networks, graph automata, concurrent transition systems, recurrent Hopfield networks, etc. Different agent-based modeling software packages (e.g., Swarm, Repast, Netlogo) provide for many of these variations.

7. Barrett and Reidys 1999; Barrett et al. 2000; Barrett et al. 2002; Barrett et al. 2006; Mortveit and Reidys 2008.
8. This is a minor variant on the presentation in Barrett et al. 2006.
9. North and Macal 2007, 25. Below I criticize this picture.
10. Minar et al. 1996.
11. This does not entail that the properties of the high-level agent are reducible to those of the constituent agents. In a similar vein, Kevin Hoover has discussed macroeconomic expectation variables that are determined by the properties of individuals and yet are not reducible to the expectations of individuals (Hoover 2009, 406). But, as Hoover stresses, that macroeconomic variable "does not possess causal capacities of the type championed by Cartwright 1989 that can be carried from one context to another. It is not cause or effect, but a summary statistic without causal efficacy." My thanks to an anonymous reviewer for pointing out this connection.
12. Historically, the ontological issues pertaining to the dependence of social properties on the properties of individual people were mixed up with issues pertaining to reductive explanation. Recent work has been careful to separate ontological individualism from "explanatory individualism." It is common for contemporary anti-reductionists to endorse ontological individualism and deny explanatory individualism (cf. Lukes 1968; Pettit 2003).
13. See Kim 1984, 1987; McLaughlin 1995; McLaughlin and Bennett 2005. In Epstein 2009, I discuss supervenience claims in connection with ontological individualism.
14. That social properties globally supervene on individualistic properties is a weak claim. In Epstein 2009, I weaken it further and argue that it nonetheless fails.
15. Elsewhere I have written about these in detail. See also Epstein 2008. Here I will characterize them intuitively rather than technically, and focus on how they are particularly destructive when combined with one another.
16. We have to tread carefully in applying troubles with individualism to agent-based models in the social sciences. As I will point out, the factors that agent-based models incorporate are often not restricted to those factors that can plausibly be understood to be individualistic. So in arguing against agent-based modeling, we cannot just apply the individualistic fallacies directly. However, a version of the same fallacies will nonetheless extend to agent-based modeling.
17. Cf. Teller 2001. I am grateful to an anonymous reviewer for drawing my attention to this reference and example.
18. This is because the objects being modeled are intrinsically individuated, i.e., their identity properties, such as *being a pot* or *being a block of ice*, depend only on intrinsic features of the object itself.
19. This distinction is sometimes overlooked because equations and functions typically do not distinguish the different uses of the equals sign or the function symbol, and hence "constitutive equations" are not visibly distinct from "dynamic" and other equations.
20. The idea of a property being "local" I mean to be somewhat looser than a property being "intrinsic." The reason I distinguish a property being local from a property being intrinsic is that individualistic properties are arguably not intrinsic properties of individual people. But they are, to some extent, local properties (e.g., that I am standing on a stone floor is plausibly an individualistic property of me; cf. Epstein 2009). For intuitive purposes, though, it is fine to use "local" and "intrinsic" interchangeably, until we want to be precise about talking about the relation between social properties and individualistic properties. Intuitively, an intrinsic property of an object is a

property that holds or fails to hold regardless of what other objects there are in the world. A local property of an object is one that holds or fails to hold regardless of what the world is like at a spatiotemporal distance from the object.
21. I am grateful to two anonymous reviewers for pressing me on this. One anonymous reviewer usefully called this a "Markov condition for ontological dependence."
22. Oppenheim and Putnam 1958.
23. See Kim 2002 for an attempt to weaken these conditions, and Epstein manuscript (forthcoming) for reasons this weakening fails.
24. Virchow 1860.
25. Fine 2003; Bennett 2004; Koslicki 2004. On the constitution of social groups, see Uzquiano 2004.
26. For present purposes, ontological dependence may be understood as supervenience, so the question as to what the property depends on can be understood as the question of the property's supervenience base. In (Epstein 2009) I present a more detailed discussion of the varieties of supervenience involved and their relation to ontological dependence in the case of social properties.
27. Wimsatt 1994, Epstein manuscript (forthcoming).
28. A way to specify this more precisely: taking facts to be property exemplifications, a pair <M,P>, where M is a set of n-level entities and P is a set of n-level properties, is *simply dependent* if and only if there is a pair <M', P'> such that: (i) M' is a set of (n-1)-level entities and P' is a set of (n-1)-level properties; (ii) The entities in M are exhaustively composed by entities in M'; (iii) P supervenes locally on P'; and (iv) For every entity m in M, m is locally individuated by the P' properties of the M' constituents of m, (i.e., identity conditions for m are local n-1-level properties). This definition involves a levels hierarchy of properties as well as a compositional hierarchy of entities.
29. The base ontology need not be unique.
30. Certainly the choices of target, base, causal, modeled-base, and modeled-causal ontologies are not explicit parts of model-construction, but it is not implausible to see them as implicit.
31. This does not require that the macroentities be reducible to microentities. One way of putting the reason simple dependence does not entail reducibility is that there may not be type-identities between the levels. Another way of putting this point is that simple dependence may hold because a global supervenience relation holds between entities at n and n-1, which may be compatible with nonreducibility. (Although some philosophers have recently argued that supervenience entails reducibility. See, for example, Kim 2005).
32. Typically, this reflects the common Galilean idealization in modeling of treating open systems as if they were closed or isolated and neglecting factors that have little causal influence (cf. Strevens 2004; Weisberg 2007)
33. The idea of a system being "well-behaved" is intuitive, but difficult to analyze. If a system is highly volatile, and develops along many radically different paths in different scenarios, interacting with radically different parts of the world in each, then it is not well behaved.
34. It is easiest to describe changes to be made to the modeled ontology as being chosen by the modeler or some external observer. But in fact this is not a necessity: an "external perspective" can be taken computationally, to modify the dynamical system being modeled. For instance, a governing process might be triggered when some entity that was expected to stay in the center of a cellular automaton runs off to the edge, and restart the simulation with a bigger grid. Or consider a model with a "synthetic" population, having agents generated by some statistical algorithm. A process might automatically detect

that people out in the suburbs, who were expected to be superfluous, actually will materially affect the end results and then generate an expanded population and restart the run.
35. Along with the nonlocal dependence of macro-properties, the causal interactions may also be "wide." cf. Yablo 1997.
36. If macro-properties do need to be included in models, then presumably micro-foundational responses to the Lucas critique (Lucas 1976) need to be revisited.
37. On this issue, see Epstein and Forber, manuscript (forthcoming). Some critics of representative agent models have pointed out that the properties of representative agents are simply macroscopically measured parameters divided by the population, which they have argued are macroscopic properties in microscopic clothing. Using macroscopic properties to set the parameters of agents may be similar. On the other hand, some philosophers have argued that the traditional injunctions against using desired results in hypothesis generation (i.e., against the use of "old data") are overstated. Which may make the use of macrodata in the setting of parameters of agents acceptable and also may allow us to count these parameters as genuinely micro-properties, even though they are determined on the basis of macro-properties.
38. This applies even to recent approaches to multiscale modeling, such as the methods described in E et al. 2007

REFERENCES

Anderson, Roy M., and Robert M. May (1992). *Infectious Diseases of Humans: Dynamics and Control.* Oxford: Oxford University Press.
Axtell, Robert (2006). *Agent-Based Computing in Economics,* http://www2.econ.iastate.edu/classes/econ308/tesfatsion/ACEIntro.RAxtell.pdf, accessed 15 September 2008.
Barrett, Chris, and Christian Reidys (1999). "Elements of a Theory of Simulation I: Sequential CA over Random Graphs." *Applied Mathematics and Computation* 98: 241–259.
Barrett, Chris, Henning Mortveit, and Christian Reidys (2000). Elements of a Theory of Simulation II: Sequential Dynamical Systems. *Applied Mathematics and Computation* 107(2–3): 121–136.
Barrett, Chris, Stephen Eubank, Madhav Marathe, Henning Mortveit, and Christian Reidys (2002). "Science and Engineering of Large Scale Socio-Technical Simulations." Proceedings of First International Conference on Grand Challenges in Simulations at San Antonio, Texas.
Barrett, Chris, Keith Bissett, Stephen Eubank, Madhav Marathe, and V.S. Anil Kumar (2006). "Modeling and Simulation of Large Biological, Information, and Socio-Technical Systems." Proceedings of the Short Course on Modeling and Simulations of Biological Networks, at AMS Lecture Notes.
Bennett, Karen (2004). "Spatio-Temporal Coincidence and the Grounding Problem." *Philosophical Studies* 118: 339–371.
Cartwright, Nancy (1989). *Nature's Capacities and Their Measurement.* Oxford: Clarendon Press.
Currie, Gregory (1984). "Individualism and Global Supervenience." *British Journal for the Philosophy of Science* 35: 345–358.
E, Weinan, Bjorn Engquist, Xiantao Li, Weiqing Ren, and Eric Vanden-Eijnden (2007). "Heterogeneous Multiscale Methods: A Review." *Communications in Computational Physics* 2(3): 367–450.

Epstein, Brian (2008). "When Local Models Fail." *Philosophy of the Social Sciences* 38: 3–24.
Epstein, Brian (2009). "Ontological Individualism Reconsidered." *Synthese* 166(1): 187–213.
Epstein, Brian (Forthcoming). "Science without Levels." Unpublished manuscript.
Epstein, Brian, and Patrick Forber (Forthcoming). "The Perils of Tweaking." Unpublished manuscript.
Epstein, Joshua M. (2005). "Remarks on the Foundations of Agent-Based Generative Social Science." Washington, DC: The Brookings Institution, Brookings CSED Working Paper #41.
Epstein, Joshua M., and Robert Axtell (1996). *Growing Artificial Societies: Social Science from the Bottom Up.* Cambridge: MIT Press.
Fine, Kit (2003). "The Nonidentity of a Material Thing and its Matter." *Mind* 112: 195–234.
Heer, Burkhard, and Alfred Maussner (2005). *Dynamic General Equilibrium Modelling.* New York: Springer.
Hoover, Kevin (2006). "A NeoWicksellian in a New Classical World: The Methodology of Michael Woodford's Interest and Prices." *Journal of the History of Economic Thought* 28 (2): 143–149.
Hoover, Kevin (2009). Microfoundations and the Ontology of Macroeconomics. In *Oxford Handbook of Philosophy of Economics*, eds. Harold Kincaid and Don Ross. Oxford: Oxford University Press, 386–409.
Kim, Jaegwon (1984). "Concepts of Supervenience." *Philosophy and Phenomenological Research* 45: 153–176.
Kim, Jaegwon (1987). "'Strong' and 'Global' Supervenience Revisited." *Philosophy and Phenomenological Research* 45: 153–176.
Kim, Jaegwon (2002). "The Layered Model: Metaphysical Considerations." *Philosophical Explorations* 5(1): 2–20.
Kim, Jaegwon (2005) *Physicalism, or Something Near Enough.* Princeton: Princeton University . Press.
Kirman, A. (1992). "Whom or What Does the Representative Agent Represent?" *Journal of Economic Perspectives* 6: 117–136.
Koslicki, Kathrin (2004). "Constitution and Similarity." *Philosophical Studies* 117: 327–364.
Lotka, Alfred J. (1925). *Elements of Physical Biology.* Baltimore: Williams and Wilkins.
Lucas, Robert E. (1975) "An Equilibrium Model of the Business Cycle." *Journal of Political Economy* 83: 1113–1144.
Lucas, Robert E. (1976). "Econometric Policy Evaluation: A Critique." In *The Phillips Curve and Labor Markets*, eds. Karl Brunner and Allan H. Meltzer. Amsterdam: North-Holland, 19–46.
Lukes, Steven (1968). "Methodological Individualism Reconsidered." *British Journal of Sociology* 19: 119–29.
McLaughlin, Brian P. (1995). "Varieties of Supervenience." In *Supervenience: New Essays*, eds. Elias Savellos and Umit Yalcin. Cambridge: Cambridge University Press, 16–59.
McLaughlin, Brian P., and Karen Bennett. (2005)."Supervenience." *Stanford Encyclopedia of Philosophy,* http://plato.stanford.edu/entries/supervenience, accessed 15 September 2008.
Minar, Nelson, Roger Burkhart, Chris Langton, and Manor Askenazi (1996). The Swarm Simulation System: A Toolkit for Building Multi-Agent Simulations, http://swarm.org/images/b/bb/MinarEtAl96.pdf, Accessed 15 September 2008.
Mortveit, Hennig, and Christian Reidys (2008). *Introduction to Sequential Dynamical Systems.* New York: Riedel.

North, Michael, and Charles Macal (2007). *Managing Business Complexity: Discovering Strategic Solutions with Agent-Based Modeling and Simulation.* New York: Oxford University Press.

Oppenheim, Paul, and Hilary Putnam (1958). "Unity of Science as a Working Hypothesis." In *Minnesota Studies in the Philosophy of Science.* Minneapolis: University of Minnesota Press, 3–36.

Pettit, Philip (2003). "Groups with Minds of Their Own." In *Socializing Metaphysics,* ed. Frederick Schmitt. Lanham, MD: Rowman and Littlefield, 167–193.

Ruben, David-Hillel (1985). *The Metaphysics of the Social World.* London: Routledge and Kegan Paul.

Samuelson, Douglas, and Charles Macal (2006). "Agent-Based Simulation Comes of Age." *Operations Research Management Science (OR/MS) Today* 33(4):34–38.

Schelling, Thomas C. (1978). *Micromotives and Macrobehavior.* New York: W. W. Norton.

Strevens, Michael (2004). "The Causal and Unification Account of Explanation Unified – Causally." *Noûs* 38: 154–76.

Teller, Paul (2001). "Twilight of the Perfect Model Model." *Erkenntnis* 55(3): 393–415.

Tesfatsion, Leigh (2003). "Agent-Based Computational Economics." Iowa State University Economics Working Paper, Ames IA.

Uzquiano, Gabriel (2004). "The Supreme Court and the Supreme Court Justices: A Metaphysical Puzzle." *Noûs* 38(1): 135–153.

Virchow, Rudolf Ludwig Karl (1860). *Cellular Pathology: As Based Upon Physiological and Pathological Histology.* Trans. F. Chance. London: John Churchill.

Volterra, Vito (1926). Variazioni e fluttuazioni del numero d'individui in specie animali conviventi. *Memoria della Regia Accademia Nazionale dei Lincei* 6(2):31–113.

Weisberg, Michael (2007). "Three Kinds of Idealization." *Journal of Philosophy* 104(12): 639–659.

Wimsatt, William (1994). "The Ontology of Complex Systems: Levels of Organization, Perspectives, and Causal Thickets." *Canadian Journal of Philosophy* Supplement 20: 207–274.

Woodford, Michael (2003). *Interest and Prices: Foundations of a Theory of Monetary Policy.* Princeton: Princeton University Press.

Yablo, Stephen (1997). "Wide Causation." *Philosophical Perspectives* 31(11): 251–281.

Zimmerman, Dean (1995). Theories of Masses and Problems of Constitution. *Philosophical Review* 104: 53–110.

7 Scientific Models, Simulation, and the Experimenter's Regress

Axel Gelfert

INTRODUCTION

In this paper, I analyze the question of whether computer simulation is, in any special way, affected by what has variously been called the "experimenter's regress" (Collins 1985) or "data-technique circles" (Culp 1995). Such a regress, it has been argued, may obtain when the only criterion scientists have for determining whether an experimental technique (or simulation) is 'working' is the production of 'correct' (i.e., expected) data. It may seem plausible to assume that techniques of computer simulation are especially prone to such regress-like situations, given that they are further removed from nature (in ways to be specified) than traditional experimentation. In public perception, too, there appears to be a gap between the trust that is placed in the experimental success of science, as opposed to its use of computer simulation methods (e.g., in predicting global climate change).

The rest of this paper is organized into eight sections. The first section summarizes the main idea of the *experimenter's regress*, as developed by Harry Collins (1985) on the basis of a case study in experimental astrophysics. The following section then addresses the question of whether computer simulation can properly be thought of as a form of 'experimenting on theories'. The section "Formulating the 'Simulationist's Regress'" identifies three clusters of scientific questions, where one might expect regress-like situations to arise in connection with computer simulation; this is followed by a distinction, in "Anatomy of a Regress", between what I call a 'software' and a 'hardware' aspect of replicability, both of which might be thought of as contributing to worries about a potential 'simulationist's regress.' I then consider the 'Response from Robustness' according to which the experimenter's regress can be dissolved whenever independent measurement techniques generate sufficiently 'robust' data. However, as I argue in "Models as Targets", in the case of simulation, this standard response is not typically available, since in actual scientific practice, computer simulation studies do not always have real systems as its targets; what is being 'simulated' are (sometimes quite abstract) mathematical models. The penultimate section, "Robustness, Invariance, and Rigorous Results" develops a richer notion of robustness that pays special attention to the invariances (and failures of invariance) of the target as well as other rigorous results that hold for the corresponding mathematical models. The final section aims to 'deflate' the Simulationist's Regress, by concluding that while data-technique circles may sometimes pose a problem for both experimentation and simulation, the practice of simulating mathematical models has sufficient internal structure and contributes new layers of assessment, such that it is no more prone to regress-like situations than traditional experimentation.

THE EXPERIMENTER'S REGRESS

The experimenter's regress takes the form of a challenge to the standard view of experiments as providing an objective way for testing theories and hypotheses and is based on the observation "that experiments, especially those on the frontiers of science, are difficult and that there is no criterion, other than the outcome, that indicates whether the difficulties have been overcome" (Collins 2005, 457). Unlike other well-known challenges to the possibility of crucial experiments (e.g., the Duhem-Quine thesis), the idea here is not that we can always save a hypothesis in the face of *disconfirming* evidence. Instead, the experimenter's regress casts doubt on the possibility of saying objectively when a given empirical finding is confirming or disconfirming evidence in the first place.

For a measurement to count as good evidence, Collins argues, we must assume that it was produced by a good instrument, but a good instrument is just one that we recognize as producing good results. This is seen as introducing a circularity: "we won't know if we have built a good detector until we have tried it and obtained the correct outcome. But we don't know what the correct outcome is until . . . and so on *ad infinitum*." (Collins 1985, 84) Controversies about the validity of empirical measurements cannot be closed by objective facts because there are no conclusive criteria "other than the outcome of the experiment itself" that scientists could apply to decide whether the experimental apparatus was working properly or not (Collins 2005, 457). In response to two of his critics, Collins identifies as the target of the experimenter's regress the popular belief that science "had eliminated disputes about standards in matters to do with observation of the natural world" (Collins 2002, 154). The experimenter's regress, in this sense, has an explicative role with respect to science: it is meant to show both "what it is that is taken to be the normal standard for experimental replication—getting the right result" and the source of that belief: namely, *"the ready availability of a standard just in those cases where experimental replicability is not needed for proof"* (Collins 2002, 155; italics original).

In support of his claims, Collins draws on a detailed case study of the scientific debate surrounding the search for gravitational waves, which are predicted by general relativity theory and which would manifest themselves

Scientific Models, Simulation, and the Experimenter's Regress

as minute disturbances of the measured gravitational constant. Collins analyzes how one scientist, Joseph Weber, built a detector—basically a massive piece of metal hooked up to extremely sensitive sensors that would detect any disturbance passing through the device—and then announced to the physics community that he had detected gravitational waves. Weber's claims were immediately contested by other scientists, some of whom built their own detectors (and could not corroborate the findings), whereas others disputed Weber's results by criticizing the validity of the procedures that Weber had used to calibrate his own instruments. To simplify somewhat, the dispute was never settled on objective grounds, but petered out slowly as Weber's track record as an experimenter came under fire and his reputation gradually eroded. In the end, Collins argues, when agreed-upon methods fail "scientists turn to a variety of criteria broadly similar to those used in making common-sense judgments in ordinary life" (Collins 2005, 458)—concerning reputation, institutional affiliation, track record, personal ties, conduct, and so forth. Whether Collins's argument does indeed extend to "the practice of disputed science" (Collins 2002, 155), in general, has been a bone of contention for some time (see Franklin 1994 for a wholesale critique); in the present paper, I shall limit myself to the question of whether *computer simulation studies* are especially prone to versions of Collins's regress argument.

SIMULATION AS 'EXPERIMENTING ON THEORIES'

Simulations as routine tools of scientific inquiry are relatively more recent than scientific models; however, like the use of scientific models, the practice of computer simulation has often been located (though not unproblematically) at an intermediate level between theory and experiment or experimental data. This is evident from the way practitioners of computer simulation themselves have, from early on, described what it is they are doing. In his study of the historical origins of 'Monte Carlo' simulations in high-energy physics, Peter Galison quotes the author of a 1972 review article on computer simulation, the computational physicist Keith Roberts, as follows:

> Computational physics combines some of the features of both theory and experiment. [. . .] It is symbolic in the sense that a program, like an algebraic formula, can handle any number of actual calculations, but each individual calculation is more nearly analogous to a single experiment or observation and provides only numerical or graphical results. (quoted in Galison 1996, 137)

In the early days of computational science, 'simulation' could be hands-on in a quite direct way. Herbert A. Simon, in his "Comments on the History of 'Simulation,'" recalls how, in an early (1956) paper, he and colleagues spoke of "hand simulation" to indicate "that the program was not yet running on the computer, but that we had hand simulated its processes and thought we could reliably predict its behavior." The irony of then manually implementing what would nowadays be an automated procedure is not lost on Simon: "Here we [were] talking of people simulating a computer rather than a computer simulating people!"[1]

At a descriptive level, computer simulationists share many of the procedures of experimentalists, among them "a shared concern with error tracking, locality, replicability, and stability" (Galison 1996, 142), as well the initial challenge of "getting the simulation to behave properly at all" (Kennefick 2000, 26). It is not at all unusual to come across, especially in the writings of early computational physicists, references to computer simulations as 'numerical experiments,' 'computer experiments,' and so forth.[2] While such phrases convey a sense of how the practice of computer simulations is experienced by those who engage in it, they are also philosophically vague. As Hans Radder notes, the phrase "computer experiments' is really an umbrella term for 'various sorts of hybrids of material intervention, computer simulation, and theoretical and mathematical modeling techniques" (Radder 2005, 271). As such, it stands in need of elaboration, and the conceptual relations between computer simulation and (traditional forms of) experimentation need to be investigated.

One bone of contention is the significance of materiality in traditional experimentation as opposed to the (allegedly) more ephemeral character of computer simulations. It has been variously suggested that computer simulation fails to meet certain criteria associated with 'proper' experimentation. Thus Francesco Guala (2005) argues that, whereas in traditional experimentation, the same material causes operate ('at a "deep," "material" level') in both the experimental and target systems, in computer simulations one can at best expect an "only abstract and formal" correspondence relation (of similarity or analogy) between the simulating and the target systems (214). However, as Wendy Parker points out, while *material* similarity may often allow us to extract information about the target system, it is only one dimension along which one can assess the power of experiments and simulations to enable inferences about the target system. As a case in point, she refers to the example of weather forecasting. Re-creating a system that is materially continuous with the target system—that is made "of the same stuff," i.e., clouds, water, mountains—is rarely feasible, nor is it very promising for the purpose of day-to-day weather forecasts (Parker 2009, 492). By running computer simulations of the underlying weather dynamics, scientists are in a much better position to make warranted predictions about tomorrow's weather. Hence, it need not be generally the case that, as some authors have suggested, traditional experiments "have greater potential to make strong inferences back to the world" (Morrison 2005, 317).

Scientific Models, Simulation, and the Experimenter's Regress

The tendency to think of computer simulations as, in some fundamental sense, more 'abstract' than traditional experimentation, may be partly due to an ambiguity in the term 'computer simulation' itself. If one understands by 'computer simulation' the successful implementation of a computational template, which is then executed to generate a set of simulated data, then each run may be considered a new simulation. Assuming that the actual device, on which the simulation is being run, works correctly, there does not seem to be much room for questions of validity (and the corresponding 'shared concerns'—of error tracking, replicability, etc.—that dominate traditional experimental practice): the correct outcome is simply whatever result is generated, and typically displayed, by the computer. This picture, of course, is too narrow, since by 'computer simulation' one does not simply mean the outcome of any particular run of an algorithm; computer simulations are deployed in concrete settings, according to the goals of the specific project at hand, in order to investigate target systems across a range of parameter values. When thus viewed in terms of their pragmatic role in scientific inquiry, simulations, too—including the devices they run on, and the programming code—are subject to evaluation, revision, and improvement. As Parker puts it, although "computer simulations *per se*"—understood in an abstract sense, as a time-ordered sequence of states—"do not qualify as experiments, computer simulation studies do," where 'computer simulation studies' refers to the investigative activity on the part of a researcher, who uses computer-generated results as input for further inquiry (Parker 2009, 495).

As this brief discussion suggests, characterizing computer simulations as 'numerical experiments' requires greater care than is typically exhibited by those practitioners of simulation studies who employ the term. Paul Humphreys may well be right when he writes that "claims that [computer simulation] methods lie 'in between' theorizing and experimentation are [. . .] best interpreted metaphorically" (Humphreys 2009, 625). However, this should be understood as a *caveat*, not as a prohibition against pursuing the parallels between scientific experimentation and computer simulation, and treating simulation studies as a form of experimental practice—though a novel one that raises its own, recognizably philosophical problems.

FORMULATING THE 'SIMULATIONIST'S REGRESS'

If one accepts that computer simulation studies are in many ways on a par with experiments, and indeed may be thought of as continuous with experimental practice, then one should expect to encounter arguments concerning their replicability which are similar to the ones that gave rise to the formulation of the experimenter's regress. Note that, at this stage, it is not necessary to endorse any sweeping claims that the experimenter's regress does *in fact* undermine the possibility of resolving experimental disputes through further empirical inquiry and rational debate. All that is required at this point is the acknowledgment that, to the extent that replicability raises serious questions for the practice of scientific experimentation, analogous questions can be raised for 'numerical experiments,' i.e., the practice of computer simulation studies.

In direct analogy with Collins's discussion, one can then define what one might call the 'simulationist's regress,' as applicable to those situations *where the best or only test of a simulation is its own disputed result*. As in the case of experiments, the regress will typically become apparent only in stubborn cases of disagreement, when mutually agreed methods of resolving disputes have failed. Before turning to specific examples of how the simulationist's regress might unfold, it is helpful to adopt a bird's-eye perspective and ask for general characteristics of situations where one might expect the regress to become salient. Collins's example of the search for gravitational waves is instructive in this respect, since it combines three important characteristics, each of which may contribute to the emergence of a regress. First, the example concerns the search for an *as yet unobserved*, and in many ways *causally isolated*, phenomenon. (Gravitational waves can neither be created, nor manipulated, nor do they have any observable effects outside the context of specific purpose-built detection devices.) Second, any theoretically predicted effect would be extremely small, thereby rendering even the *existence claim* that comes with a purported observation highly controversial. (Much of the controversy about gravitational waves concerned the question of how they would manifest themselves in any proposed experiment—i.e., what their experimental 'fingerprint' would be.) Third, even on the most optimistic astronomical theories, measurable gravitational waves will be fairly rare, thus leading to a *sparsity of empirical data*. Taken together, these three factors give rise to the regress problem described by Collins: how are experimental disputes to be resolved, if the claims in question concern the existence of causally isolated (or otherwise inaccessible) processes, for which empirical evidence is sparse?

The example of gravitational waves is special, insofar as it *combines* the three aspects distinguished above, thereby making the alleged regress especially salient. In the remainder of this section, I wish to argue, first, that less salient versions of the same problem can arise even when not all three aspects are instantiated, and second, that analogous considerations apply in the case of computer simulations. For one thing, simulations often take place under conditions of sparsity of empirical data with which to compare the simulation results—either because such data is not readily available, or because it cannot be generated at will in a controlled experiment. As Eric Winsberg points out: "[S]imulations are often performed to learn about systems for which data are sparse. As such, comparison with real data can never be the autonomous criterion by which simulation results can be judged" (Winsberg 1999, 289). Whether or not a simulation result is to be believed, is thus not generally something that can be settled by comparison with empirical data alone, at least not in cases of genuine dispute about the validity of the simulation itself.

Scientific Models, Simulation, and the Experimenter's Regress   151

In order to illustrate how regress-like worries can arise in the case of computer simulations, I want to describe briefly, without getting sidetracked by the nitty-gritty detail of a complete case study, three classes of actual scenarios, in which the best test of a simulation may be its own disputed result. Each example corresponds to one of the three aspects distinguished earlier, that is *causal inaccessibility*, *novelty of the phenomenon* (which issues in an *existence claim*), and *sparsity of* (real, empirical) *data*.

The first scenario is based on an existing case study from astrophysics (again!) and concerns simulation-based studies of the collapse of neutron stars. It is perhaps no coincidence that astrophysics is a rich source of examples, given that the target systems under investigation are often causally remote, in the sense that they do not lend themselves to manipulation or causal intervention in a controlled experiment. In his detailed case study, Daniel Kennefick (2000) reconstructs the controversy between different groups of scientists working on the astronomical analysis—and computation simulation—of how neutron stars may, given certain conditions, form binaries (i.e., systems of two neutron starts orbiting one another), which eventually collapse into one another, thereby forming a black hole. Specifically, Kennefick analyses the reaction of the astronomical community to the claims made by a pair of computational physicists, Jim Wilson and Grant Mathews, who thought they had identified, through the use of computer simulation, a scenario in which one of the neutron stars forms a black hole—through a process Kennefick dubs 'star crushing'—*before* it swallows up the other one. For various reasons, this claim contradicted what most analytically minded theorists of general relativity expected, who immediately criticized the simulation results (in sometimes a rather wholesale fashion) as erroneous and due to programming errors. Focusing his attention on the divergent methodologies (and their corresponding criteria of what constitutes a valid derivation) of relativity theorists and computational physicists, Kennefick argues that theoretical predictions are subject to a "theoretician's regress," which arises from the "near impossibility of discriminating between the results of rival calculations which fail to agree, *by the process of criticizing the calculations*" (Kennefick 2000, 33–34; italics added).

The second scenario, in which one might expect regress-like situations to emerge, can be illustrated by the simulation-aided search for the presence, or absence, of phase transitions in complex systems. Although phase transitions are among the most salient phenomena in nature—one need only think of the freezing of water, the occurrence of spontaneous magnetization in certain metals, or the phenomenon of superconductivity—physicists are hard-pressed to come up with theoretical models that successfully explain their occurrence. The study of many-body models poses complex numerical challenges, and the full set of equations describing a many-body system is almost never analytically solvable. This has led to a proliferation of simplified mathematical models—such as the Ising model or the Heisenberg model—to capture the 'essential physics' that is thought to be responsible for the phase transition.[3] In order to evaluate such models, and whether they can account for phase transitions, computer simulations are being run to determine whether there is a region in parameter space where a phase transition occurs. The occurrence of a phase transition is typically indicated by the asymptotic behavior of certain variables (such as the correlation length, or a suitably defined order parameter) as the system approaches the critical point. For most many-body models in three dimensions, it is unknown whether they are indeed capable of reproducing phase transitions; as a result, the search for models that have this capacity takes the form of a search for regions in parameter space where the relevant variables display 'critical' behavior. Hence, in the case of a purported positive result, we are faced with an *existence claim*—'that there *is* a phase transition in *this* region of parameter space, which matches the kind of phase transition in the target system'—where the best evidence for the success of a simulation is its own disputed result.

For the third kind of situation where one might expect to find a version of the simulationist's regress, I want to point to large-scale simulations arising within what is sometimes called Earth System Analysis, such as global climate models.[4] At first it might seem counterintuitive to lament a '*sparsity* of empirical data,' given that one challenge of simulating global climate change consists in integrating a diverse range of variables and complex data sets, and the relations between them. What is important, however, is not the actual *amount* of data involved, but the availability of *independent* data sets for the same variables. For example, one would typically like to compare the outcome of different simulation runs for different parameter values (e.g., certain levels of greenhouse gas concentrations), against experimental data corresponding to these different situations. However, when it comes to climate models, no independent experimental access to the target system—the Earth's climate—is possible: the only experiment that takes place is the irreversible climate change that we, as humans, are currently inflicting on the planet and its geochemical cycles. It may sometimes be possible to turn to substitutes for independent experimental 'runs': in the case of climate modeling, for example, one might simulate different episodes in the Earth's climate and compare the simulation results with historical climate records. Although it may sometimes be an option to run simulations of past, observed climate anomalies in order to test one's numerical models, one should not necessarily expect this to resolve any disputes unless there is a prior agreement about the— typically *sparse*—data that goes into any given simulation. In cases in which such agreement is lacking, one can easily imagine scenarios where the validity of the corresponding results is challenged—on the basis of worries about the admissibility of past evidence or by contesting the independence of the relevant data subsets—thereby rendering disputes about the significance of simulation results irresolvable.

ANATOMY OF A REGRESS

Having identified scenarios from different areas of research where one might expect regress-like situations to occur in computer simulation studies, let us approach the problem from a more systematic point of view, before discussing possible responses to it. Rather than ask which scientific questions or disciplines might be particularly susceptible to the simulationist's regress, I want to refocus attention on its similarities and dissimilarities with the experimenter's regress. I shall do so by developing two complementary perspectives on the practical side of computer simulations, relating in turn to the software and hardware aspects of their implementation.

Numerical techniques and computing power have become important drivers of research across many scientific disciplines. Heavily tested numerical software packages have turned computational methods into a commodity that is available to a wide range of researchers. Here is how two researchers in metabolic drug design describe the situation in their discipline: "Modern programs such as *Gepasi* (Mendes 1993) and *SCAMP* (Sauro 1993) are easy to use, run on universally available equipment, and powerful enough to handle most of the problems likely to interest the metabolic simulator" (Cornish-Bowden and Eisenthal 2000, 165–166). While over-reliance on pre-packaged computational software may lead to epistemic problems of its own, regarding how to assess the validity of the results it generated, it is important to note a crucial contrast with possible cases of the simulationist's regress—such as Kennefick's example of "star crushing." In cutting-edge science—especially where it operates under conditions of sparsity of data, or deals with existence claims and situations of causal isolation—simulations are typically purpose-built and tailored to specific theoretical scenarios and situations. Thus, in the 'star crushing' example, Wilson's and Mathews's results were attacked by a majority of relativity theorists and computational astrophysicists, not because of doubts concerning the validity of those approximations that were regarded as standard—as Kennefick notes, a number of elements in Wilsons and Mathews's simulation "resemble[d] tricks commonly used for many years in analytic GR [=general relativity]" (Kennefick 2000, 14)—but because of theoretical considerations that made the alleged crushing effect unlikely, along with a subsequent failure of other numerical groups to replicate the effect. When replication fails (or is not even attempted because the result is deemed erroneous for other reasons), the proponent of the disputed simulation result faces a similar challenge to the experimenter, in that "the experience of making a calculation work, which is part of what convinces the theorist that the result can be believed, is so particular to the specific theoretical apparatus and its operation that it is of little help in convincing those who disbelieve the result on other grounds" (Kennefick 2000, 33).

The fundamental problem, thus, is one of the *replicability* of simulations, i.e., their ability to generate stable data. Achieving replicability has both a 'software' and a 'hardware' aspect, which relate to different levels at which failures of replicability can (and sometimes do) occur. I shall begin with a discussion of the 'software' aspect, which takes its lead from the earlier observation that cutting-edge simulations—of the sort that are most likely to give rise to potentially regress-like disputes—are typically tailored to specific theoretical problems or scenarios (such as the theoretical description of collapsing neutron star binaries in terms of approximations to Einstein's field equations). In such cases, there is no agreed-upon procedure for deriving (stable and computationally tractable) computational templates from fundamental theory.[5] As Kennefick rightly points out: "In numerical work a great part of the effort is in getting the simulation to behave properly at all" (Kennefick 2000, 26). As with experiments, this stage of setting up simulations often involves tacit know-how on the part of the investigator and is rarely fully documented. Often, there may be no shorthand way of documenting how a particular simulation has been implemented, other than by restating the code and, in effect, inviting anyone who really wants to test its adequacy and stability to implement the code on another computing device. Even stating which general (theoretically justifiable) methods of approximations have been employed—e.g., perturbation theory, discretization, or Monte Carlo methods—is rarely self-explanatory, as the application of such methods to specific problems (in the form of a specially programmed piece of software) is a case-by-case affair: "Clean, abstract presentations of formal schemata disguise the fact that the vast majority of theoretical templates are practically inapplicable in any direct way even to quite simple physical systems" (Humphreys 2004, 64). The epistemic significance of this phase of 'getting a simulation to behave properly at all' lies not in furnishing *simulation results*, which may then be compared against empirical data or theoretical predictions. Rather, what goes on during this phase may, on occasion, preclude which results a simulation is capable of generating later on—once it has successfully been 'tamed,' as it were. For, this phase will be considered to have been concluded only once the simulation begins to give the kinds of results we expect it to give. But, of course, under any realistic conditions of inquiry, we would not be needing the simulation if we already knew what to expect; hence, the best or only test of whether a simulation is working 'properly' may, once again, lie in its own disputed result.

Replicability in the case of computer simulation, however, also has a 'hardware' component, which is in addition to the difficulty, discussed so far in this section, of assessing whether a particular computational template is in principle capable of successfully replicating the behavior of a given target model. The additional worry concerns the further problem of the reliability of the concrete device, on which a simulation is being run. After all, any computer is itself a material piece of hardware and assessing whether it is a 'properly functioning' device—irrespective of the stability of any computational template that might happen to be implemented on it—is subject to the experimenter's regress, or so the worry goes. Whereas the 'software' aspect of computer simulation concerns the stability, tractability, and replicability of a piece of specially programmed code on a computational device that is assumed to be functioning properly, the 'hardware' aspect concerns the replicability of the material realization of a simulation (i.e., its implementation on a concrete computing device), for various internal or external settings (i.e., for different parameter values as well as across different computers). Drawing a parallel to recent work on the philosophy of experimentation (e.g., Radder 1996), one might say that, at the 'hardware' level, replicability—while being far from theory-free—is not yet assessed under a specific theoretical interpretation of the simulation results. What is at stake is not so much the adequacy or suitability of the computational template and its programmed implementation, but the proper functioning of the computing device itself. More often than not, it is taken for granted that the physical apparatus on which a simulation is run is indeed a correctly functioning computer; however, from a foundational viewpoint that aims to identify potential sources of a regress, "the theory of the apparatus in simulation is not negligible" (Weissert 1997, 112). Even a carefully tested and designed digital computer is, by necessity, only capable of finite-precision arithmetic. This has several important implications for the replicability of simulations. For one, due to round-off errors, iterations of arithmetic operations will lead to a cumulative increase in the uncertainty of any numerical outcome. The precision of which a computing device is capable—which is in large part determined by the hardware—is especially important when dealing with regions of deterministic chaos, where minute differences can lead to very different system trajectories. This may eventually result in a loss of the true solution in the noise that is generated by the numerical procedure. As Thomas Weissert emphasizes, "This effect is not a loss of determinism in our dynamical system, but a loss of our ability to make the determination with a simulation" (Weissert 1997, 115). However, it is important not to overstate the special difficulties posed by finite-precision effects inherent in the hardware. For, it is possible to numerically estimate the propagation of finite-precision and round-off errors and their effect on the outcome of a simulation. This itself indicates that the kinds of challenges posed by the 'hardware' and 'software' aspects of replicability cannot always be neatly kept apart. Achieving replicability is a matter of eliminating sources of error and noise, and these include, among others, "computation instability, truncation error, iterative convergence error, programming mistakes, hardware malfunction, error in the mathematical form of the continuous model equations, error in the parameter values included in those equations etc." (Parker 2008, 176–177). While this list may look dispiritingly long, it also contains a grain of hope, for it suggests that, rather than being faced with a 'nesting' of independent sources of possible regress-like situations—e.g., hardware limitations, or particular features of the algorithms used in the software—it may be possible to help oneself to a range of techniques of achieving and testing replicability. Precisely because replicability is a composite phenomenon, it may be possible to assess the extent to which one 'ingredient,' relative to all others, poses a threat to replicability; one can then hope to identify conditions under which the simulation is sufficiently robust for the investigative purposes in question. This suggests a way of defusing possible worries about the dangers of a simulationist's regress—'on top of the experimenter's regress,' as it were—namely by showing that, perhaps, there is nothing special (or intrinsically less tractable) about regress-like situations in the case of simulation, as compared to experimentation. What one is dealing with, in either case, is the challenge of achieving replicability, and while this may take different forms in different experimental or simulation contexts, it is not unreasonable to expect that new contexts may also give rise to new benchmarks or mechanisms of how the stability of one's results can be ensured.

THE RESPONSE FROM ROBUSTNESS

Among the possible responses to 'data-technique circles,' of the sort represented by the experimenter's and the simulationist's regress, two general approaches can be distinguished. First, one can begin with scientific practice, by identifying strategies that scientists employ in order to build confidence in their (experimental or simulation) results. Allan Franklin, on the basis of case studies of experimentation, has drawn up a list of such strategies (Franklin 1986, 2009), the most important of which are *comparison* (with known features of the system under investigation), *calibration* (of a given method against its performance in other systems), *verification* (of one experiment by other experiments or experimenters), *retrodiction*, *consistency*, *intervention* (including variation of input parameters), and *elimination* of likely sources of error. Several of these strategies have recently attracted attention in relation to computer simulation (Weissert 1997, 122–125; Parker 2008). I shall here, however, focus on a second, more abstract approach to the question of when we are justified in placing trust in the (non-obvious or even controversial) results of scientific experimentation and simulation.[6] This response takes its cue from the familiar problem of when to trust what others tell us. Imagine that a hearer encounters a series of unexpected instances of testimony, each reporting the same unusual event. Should she believe what she is told, even when she has *prima facie* reason to think it implausible? It depends on whether the testifiers are independent or not. If the hearer has good reason to believe that the testifiers she encounters *really are* independent, and their testimonies concur in all relevant respects, then this should boost her belief in the truth of the testimony, even when the asserted fact is individually implausible. Moreover, it would be rational for her to believe the reports, given their unanimous concurrence, even when she knows each testifier to be *individually unreliable*. With independent pieces of evidence, the likelihood of error is the likelihood that *all* the evidence is wrong at the same time (and that all its ingredients are wrong *in exactly the same way*, given their concurrence)—and this becomes an ever more remote possibility the more concurring instances of evidence are involved. It is for this reason that police routinely accept the concurring, uncontradicted testimony of, say, accomplices even when they know them to be individually unreliable on most occasions—just not on this one.

Applying similar considerations to the case of 'data-technique circles,' one arrives at what one might call the *response from robustness*. This response to Collins's experimenter's regress has been aptly argued by Sylvia Culp (1995), who describes the basic idea as follows:

> When comparable data can be produced by a number of techniques and the raw data interpretations for these techniques do not draw on the same theoretical presuppositions, this remarkable agreement in the data (interpreted raw data) would seem to be an improbable coincidence unless the raw data interpretations have been constrained by something other than shared theoretical presuppositions. (Culp 1995, 448)

Culp's paper takes the form of a case study of different DNA sequencing techniques, each of which exploits different parts of the overall framework of the theory of DNA. For example, one technique might exploit the effect of certain enzymes that cut DNA at particular points in the sequence, whereas the other might be based on the individual replacement of particular nucleotides with different bases. By employing different techniques, "each of which is theory-dependent in a different way," it is possible—Culp argues—to eliminate the overall "dependence on at least some and possibly all shared theoretical presuppositions," thus giving rise to bodies of data that have been produced by *de facto* independent techniques (Culp 1995, 441). If these data sets all support the same theoretical interpretations, then we can accept these results with good reason, since the concurrence of independent techniques would otherwise be a near miracle. Culp argues not only that it is possible to eliminate theory-dependence (and with it the danger of circularity) in this fashion but additionally that 'it is the production of robust bodies of data that convinces scientists of the objectivity of raw data interpretations' (ibid.).

Following on from the empirical claim that robustness is what *actually* generates agreement among scientists, three problems with Culp's account need to be mentioned. First, there is a tension with the historical observation—impressively substantiated by Hasok Chang's recent (2004) study of the evolution of techniques of thermometry—that, in many areas of science, calibration across theoretically interdependent techniques is the norm, whereas true independence is the exception. This historical point is but the flipside of a second, more systematic consideration. What would be the alternative to calibrating a given measurement technique against existing standards and techniques that are theoretically 'of a piece,' as it were, with the technique under investigation? We would be thrown back on our prior (theoretical and pre-theoretical) expectations of what a good result would be. However, these expectations are no more independent from one another than, for example, the various measurement techniques for temperature discussed in (Chang 2004). Hence, even if all our seemingly 'independent' measurement techniques were indeed calibrated only according to our (pre-)theoretical expectations, given that these expectations do not form an independent set, the alleged independence of our techniques would once again be undermined. Finally, if we *did* have independent measurement techniques that were neither calibrated against one another, nor calibrated according to the same expectations and evidence, how could we be so sure they would all be measuring the same quantity? Or, alternatively, how could disputes about whether or not a newly proposed technique does *in fact* measure the same quantity, possibly be resolved?[7] It would appear that total independence could only be achieved at the cost of incommensurability of one sort or another.

It should be noted that thinking of experimentation merely in terms of the deployment of *measurement techniques* is guaranteed to paint an impoverished picture of the nature of experiment. It is a striking feature of many of the examples that are cited in support of the experimenter's regress—most prominently, the detection of gravitational waves—that there is not actually all that much *experimenting* going on. The experiments in question tend to be cases of passive detection; they do not involve the systematic exploitation of causal powers and their "redeployment" in other parts of nature—in short, the kind of "intervening in nature," that, in the eyes of Ian Hacking and others, is crucial to the trust we place in experiments[8] (Hacking 1983, 38). A similar sentiment appears to motivate Parker's emphasis that, if one is to think of computer simulation as a kind of *experimental activity*, one must go beyond its definition as a "time-ordered sequence of states" and introduce the broader notion of *computer simulation studies*, which is taken to also include actions on the part of the simulationist, through which he intervenes in a system (Parker 2009, 495). As I shall argue in the remainder of this paper, the notion of robustness, too, must be broadened beyond the mere insistence on independent techniques of data-generation if it is to be of use against the simulationist's regress.

MODELS AS TARGETS

Just as one finds different views on whether or not computer simulation should properly be regarded as a kind of experimentation, one also finds divergent views on the question of what the 'target' of computer simulation really is. What does simulation typically aim at? Eric Winsberg has argued that, whereas experiments "often play the role of providing crucial tests for theories, hypotheses, or models," simulations ordinarily cannot, since they assume significantly more background knowledge—including knowledge "about how to build good models of the very features of the target system that we are interested in learning about" (Winsberg 2009, 587). What matters for simulations, on this account, is their external validity: how well a simulation performs with respect to the (physical) target system and its relation to the outside world. Indeed, it is the "possession of principles deemed reliable for building models of the target systems" that, according to Winsberg, plays a criterial role in *defining simulation* in the first place: such possession of background principles is what justifies our treatment of a simulation, implemented on a digital computer, as an adequate stand-in for the target (Winsberg 2009, 588).

It is helpful, in this context, to briefly return to the earlier distinction (see Simulation as 'Experimenting on Theories') between 'computer simulation' *simpliciter* and 'computer simulation studies.' Both often get subsumed under the umbrella term 'simulation' but differ in relevant respects. When one speaks of a simulation (*simpliciter*) of a physical system P, there is a clear sense in which the term 'simulation' is employed as a success term: for something to succeed—however imperfectly—in simulating P, it is necessary that P exist and that one's simulation stand in the right sort of relation to it. Typically, this is achieved through building theoretical models of the target system P and implementing appropriate computational templates on a computing device. Clearly, then, someone who intends to simulate a physical system P must possess (or at least take himself as possessing) sufficiently reliable principles for modeling P. This, I take it, is what lies at the heart of the claim that what distinguishes experimentation from simulation is that the latter requires 'possession of principles deemed reliable for building models of the target systems.' However, if one means by 'simulation' a general kind of *theoretical activity* (as suggested by the expression 'computer simulation studies'), rather than the establishment of a *relation* between a computational procedure and a particular physical system P, then more exploratory uses of simulation might be legitimate, even in the absence of specific theoretical knowledge about any particular target system. In the remainder of this section, I want to suggest that *computer simulation studies* need not, and indeed cannot, always presuppose a good prior grasp of what makes something a good model. To be sure, in many cases it may be possible to assess the success of simulations directly in light of their external validity against experiments and observations, especially when there is agreement about which model properly represents the target systems (think of a Newtonian model of the solar system). However, when there is a lack of such agreement, it quickly becomes apparent that, as Margaret Morrison puts it, "strictly speaking the model is what is being investigated and manipulated in a simulation" (Morrison 2009, 45), not the target system. It is in this sense that one can—as scientists often do—speak of 'simulating a model.'

Simulations do not always have real physical (or biological, or otherwise materially constituted) systems as their immediate target. Instead, in many cases, what is being simulated are mathematical models, i.e., abstract objects that are not themselves (and do not purport to be) part of the causal fabric that a physical experiment would typically exploit.[9] That many of the most widely investigated—and in this sense 'successful'—models in science are, strictly speaking, uninstantiated is hardly news to anyone who is aware of the role that idealization and abstraction play in their construction. But models are often investigated not with an eye to how accurately they represent any particular target system but with an eye to whether or not they exhibit certain theoretically expected behaviors—as in the case of phase transitions mentioned above. In many scientific contexts, one is less interested in a particular model with fixed parameter values than in a class of models defined by some (continuously varying) control parameter. In such cases, it is often "the entire model class that becomes the entity under scrutiny" (Weissert 1997, 108). The *real* systems, in which the phenomenon of interest has been observed (e.g., spontaneous symmetry breaking, in the case of phase transitions), need not individually be accurately described by the mathematical models in question. In fact, the class of real systems can be so diverse that it would be quite misleading to jump to conclusions about the external validity of simulation results for any one subclass of real target systems.

Simulation, thus, is in many ways a less constrained practice of scientific investigation than causal-interventionist experimentation, making it more difficult to tell whether an unexpected result represents a feature of a 'real' target system or is a mere artifact. Galison makes a similar observation when he writes

> From the start, simulations presented a hybrid problem. On one side, the work was unattached to physical objects and appeared to be as transportable as Einstein's derivation of the A and B coefficients for quantum emission and absorption. But in practice, this was hardly the case. (Galison 1996, 140)

Interestingly, computer simulationists have sometimes taken this causal 'detachedness' of simulations to be an advantage: Galison quotes Herman Kahn, of the RAND corporation, as saying that Monte Carlo simulation methods "are more useful [. . .] than experiments, since there exists the certainty that the comparison of Monte Carlo and analytic results are based on the same physical data and assumptions" (Galison 1996, 143). However, unless more is said about why dependence on the same theoretical assumptions is unproblematic, rather than a compounding factor in any looming data-technique circles, such optimism is little more than a leap of faith. In the remainder of this paper, I wish to develop a position that acknowledges (and endorses) a certain degree of independence of simulation from experiment, while also providing the internal resources to fight off the danger of vicious data-technique circles that might arise from the fact that "the assumptions of the underlying model are, barring incompetence, built into the simulation and so the simulation is guaranteed to reproduce the results that the model says it should" (Humphreys 2004, 134). In order to do so, it is important not to assimilate questions about the trustworthiness of simulation to questions about the external validity of particular simulation results, but to recognize that simulation—in particular, in relation to mathematical models—introduces new dimensions along which its trustworthiness can be assessed.

ROBUSTNESS, INVARIANCE, AND RIGOROUS RESULTS

Computer simulation stands in need of justification, and the same confidence-building strategies that have been discussed for experimentation—calibration, verification, elimination of error, comparison with known results, etc. (cf. Franklin 1986; Parker 2008)—can also be extended to the case of simulation. In particular, as Winsberg puts it, "the first criterion that a simulation must meet is to be able to reproduce known analytical results" (Winsberg 1999, 189). Likewise, Humphreys writes, "In the case of computational devices, the calibration standards that must be reproduced are analytically derived results which serve as mathematical reference points" (Humphreys 2004, 117). The significance of such analytical results as "mathematical reference points," I want to argue, plays a special role in computer simulation, inasmuch as it goes beyond the mere reproduction of certain numerical values that are known to obtain (e.g., for certain parameter values) and beyond their uses in calibrating the computer *qua* physical device. Analytical results, in the form of what first came to be known in statistical physics as *rigorous results* (see Ruelle 1969; Gelfert 2005), add structure to mathematical models that can neither be deduced from fundamental theory nor inferred from (or, for that matter, compared with) experimental data. In addition to exact solutions—which, for many mathematical models, are hard to come by—other examples of rigorous results might include impossibility theorems, upper and lower bounds for certain variables, or mappings between different mathematical models. The latter—the existence of rigorous results that relate different classes of models—has also been shown to lead to qualitatively new ways of indirect confirmation of models (and *across* models). Thus, in one recent example from condensed matter physics, it has been argued that it is only in virtue of a mathematically rigorous mapping between two mathematical many-body models—the Hubbard model (for itinerant electrons) and the Heisenberg model (for fixed spins)—that the former was recognized as successfully representing a particular class of real, experimentally known physical systems (known as Mott insulators; see Gelfert 2009, 509–517 for a detailed study of this example).

It should come as no surprise, then, that such rigorous results and relations have been used by scientists for the calibration of computer simulation methods: "The partial independence of rigorous results from fundamental theory, and the fact that they are model-specific, makes them interesting 'benchmarks' for the numerical and analytical techniques of calculating observable quantities from the model" (Gelfert 2009, 506). However, the mere existence of a benchmark does not guarantee that a simulation method that conforms to it does, in fact, provide accurate results for real target system. There is a danger that the mere existence of rigorous benchmarks encourages belief in the general validity of a simulation method, once it reproduces them, even when the benchmarks—as can happen with rigorous results—have no connection to a model's actual performance in specific empirical contexts.

What, then, is the proper role of rigorous results and how can they help alleviate the threat of the simulationist's regress? In order to see how, it is important to realize that rigorous results have no analogue at the level of experimental data. They are genuinely new contributions at the level of mathematical models, and are often quite specific to certain classes of models. They also cannot be 'read off,' so to speak, from fundamental theory: they do not merely re-state fundamental theoretical assumptions. This partial independence from both theoretical assumptions and empirical data introduces a genuinely new layer of assessment: simulations of mathematical models can be evaluated in terms of whether or not they conform to relevant model-specific rigorous results. The qualifier 'relevant' is important, of course. Given that it may or may not be possible to give rigorous results an empirically meaningful interpretation, some rigorous results will inevitably be more important than others. For example, in one model a conserved quantity may indicate the preservation of particle number, while in another model a constant number may refer to an 'unphysical' (i.e., not even in principle realizable) limiting case. There may be entirely legitimate trade-offs—that is, situations in which an investigator might tolerate the violation of a rigorous result by a given simulation method, if there are principled reasons for regarding the rigorous result as of secondary importance. As an example, consider the so-called Mermin-Wagner theorem, according to which there is no long-range order (and hence no associated phase transition, such as spontaneous magnetization) at finite (nonzero) temperatures in one- or two-dimensional (1D or 2D) lattice systems. The Mermin-Wagner theorem has been rigorously shown to hold for a number of mathematical many-body models, including many (such as the Heisenberg, Hubbard, and Kondo-lattice models) that are currently used to describe, for example, spatially extended (three-dimensional) ferromagnetic systems. Interestingly, computer simulations of these models for low-dimensional (1D or 2D) systems sometimes do predict a phase transition to an ordered state. That is, such simulations strictly *violate* rigorous results that have been established mathematically for the underlying models. What should a simulationist do in such a situation? On the one hand, low-dimensional order (e.g., magnetic layers) are an experimentally established phenomenon, and there may be independent theoretical grounds to believe that those mathematical models that successfully describe the three-dimensional case should also extend to low-dimensional samples of the same substance. On the other hand, the simulation results directly contradict a mathematically rigorous result. Interestingly, there is no agreement among practitioners in the field about how to proceed. In the example at hand, some scientists argue that the simulation "approaches the physical reality (magnetic order in thin films does really exist) better than the Hubbard model in its exact form," whereas others insist that if a simulation "predicts the occurrence of spontaneous magnetization in one and two dimensions as well as in three dimensions [. . .], the validity of these predictions in three dimensions should clearly be investigated more fully."[10] As this example indicates, the existence of rigorous results merely creates a *presumption* against their violation. But rigorous results do introduce a new *kind* of standard, by which simulations of models can be judged in terms of their relative performance, one that is largely independent from empirical performance and theoretical expectations.

The standard response from robustness, discussed above in an earlier section, conceived of robustness explicitly as the concurrence of measurement techniques or, more generally, techniques of data-production. On such an interpretation, computer simulation would seem to have little to contribute, since simulated data is hardly independent enough to add significantly to robustness: as Humphreys rightly notes, "when simulated data rather than real data are fed into the simulation, the prospects for informing us about the world are minimal" (Humphreys 2004, 134–135). An exclusive focus on measurement techniques, however, would likewise construe the notion of robustness too narrowly. It is more fruitful, I believe, to think of robustness as, in Bill Wimsatt's terms, "a family of criteria and procedures." Among the heuristic principles that should guide robustness analysis, Wimsatt goes beyond independence of "derivation, identification, or measurement processes" to explicitly include three further principles, namely (i) the search for factors "that are *invariant* over or *identical* in the conclusions or results of these processes," (ii) the "*scope*" and "*conditions*" of such invariances, and (iii) the analysis and explanation of "any relevant *failures of invariances*" (Wimsatt 1981, 126). Of the four classes of considerations that Wimsatt identifies, the independence and diversity of measurement techniques makes up only one; the remaining three deal with issues of invariance or its failure. One way in which invariances manifest themselves in mathematical models are precisely the kind of rigorous results discussed earlier, especially where these relate to conserved quantities, symmetries (as well as symmetry breaking), scale-invariance, and so forth. However, not all rigorous results concern 'global' constraints, such as conservation laws or symmetries of a system. As the example of the Mermin-Wagner theorem shows, rigorous results also make predictions concerning the presence or absence of certain phenomena (such as spontaneous order) in specific systems. Other examples of rigorous results include mappings of different classes of models onto one another in specific limiting cases (as in the Mott insulator scenario), upper and lower bounds on various quantities (or on features such as the correlation length) in a system, critical exponents in the vicinity of phase transitions, or the topological relationships between orbits (e.g., in systems governed by chaotic dynamics). Given this rich panoply of rigorous results and relations, it is therefore eminently sensible, when one is simulating mathematical models, to test one's simulation methods against rigorous results—whether or not these can be given an empirical interpretation—and to come to a considered judgment whether possible violations, or failures of certain rigorous results, are significant or not. Rigorous results do not wear their significance on their sleeve, as it were, but they do provide an additional layer of assessment—one that is specific to the practice of simulating mathematical models and has no analogue in traditional experimental practice.

CONCLUSION: DEFLATING THE 'SIMULATIONIST'S REGRESS'

Where does this leave the initial worry that simulation studies may be subject to a *simulationist's regress* that is analogous to, but conceptually distinct from, the experimenter's regress and that may very well be more difficult to resolve in practice, given that independent data is so much harder to come by than in the case of traditional experimentation? On the account I have sketched, this worry appears somewhat exaggerated. It overstates the 'detachedness' of computer simulation from conditions of real inquiry, and it creates the false impression that the robustness of an inquiry is purely a matter of what kind of data it draws on and which techniques it employs to produce such data. Once the original notion of robustness as a family of criteria and procedures is restored, it becomes apparent that the practice of simulating mathematical models is misdescribed if one thinks of it merely in terms of the generation of simulation results and their external validity. Over and above the comparison between empirical and simulated data, there exist a range of procedures of assessing whether a simulation 'respects' the fundamental features of a model; rigorous results are one such class of examples. Their status as features of mathematical models themselves—i.e., as genuinely new contributions *at the level of models*, lacking an analogue in traditional experimentation—allows them to function as independent internal standards for the assessment of simulation methods.

Worries about data-technique circles deserve close attention and they can indeed become problematic in the scenarios discussed earlier (e.g., when a dispute arises over the existence of causally isolated novel phenomena). But simulation does not, in and of itself, have to bear a heavier burden of proof than traditional experimentation. One might naively think that simulation is necessarily one step further removed from nature and that, therefore, it is somehow doomed to take place 'in thin air'—thus making it prone to a special kind of regress, where the best or only test of a simulation is its own disputed result. But such a view ignores that simulation can help itself to a range of independent internal standards, rigorous results being one example. While simulation no doubt introduces a qualitatively new step in the multi-step process of investigating nature, it also provides new criteria and ways of assessment. It thus seems appropriate to end on a deflationary note, by stating that, while simulation makes qualitatively new demands on the researcher who must exercise his judgment in assessing its results, it is no more prone to data-technique circles than traditional experimentation.

ACKNOWLEDGMENTS

I am grateful to two anonymous referees for their constructive criticism, as well as to Paul Humphreys and Cyrille Imbert, for valuable comments on an earlier draft.

NOTES

1. Quoted in (Röller 2008, 52–53).
2. See, for example, the chapter on "The Computer Experiment" in (Hockney and Eastwood 1988) and references therein.
3. On this point, see also (Batterman 2002).
4. For a different historical example, see Peter Imhof's study of the uses of computer simulation in the Limits of Growth controversy (Imhof 2000).
5. On the notion of 'computational template,' see (Humphreys 2004, 60–76).
6. It is worth emphasizing that this general approach is complementary to, and compatible with, Franklin's identification of specific strategies (in particular, Franklin's admission of *statistical* arguments into the array of confidence-building strategies).
7. Chang makes a similar point with respect to the historically contested technique of Wedgwood pyrometry (Chang 2004, 127).
8. Indeed, as Franklin acknowledges, this is why any complete "epistemology of experiment" must include specific confidence-building strategies "*along with* Hacking's [criterion of] intervention and independent confirmation" (Franklin 2009; italics added).
9. Note that this is not meant as a *criterion* for distinguishing between simulation and experiment.
10. For sources and a brief discussion, see (Gelfert 2005, 733–737).

REFERENCES

Batterman, Robert W. (2002). "Asymptotics and the Role of Minimal Models." *British Journal for the Philosophy of Science* 53: 21–38.

Chang, Hasok (2004). *Inventing Temperature: Measurement and Scientific Progress*. Oxford: Oxford University Press.
Collins, Harry (2005). "Replication." In *Science, Technology, and Society: An Encyclopedia*, ed. Sal Restivo, 456–458. Oxford: Oxford University Press.
Collins, Harry (2002). "The Experimenter's Regress as Philosophical Sociology." *Studies in History and Philosophy of Science* 33: 149–156.
Collins, Harry (1985). *Changing Order: Replication and Induction in Scientific Practice*, London: Sage.
Cornish-Bowden, Athel, and Robert Eisenthal (2000). "Computer Simulation as a Tool for Studying Metabolism and Drug Design." In *Technological and Medical Implications of Metabolic Control Analysis*, eds. Athel Cornish-Bowden and María Luz Cárdenas, 165–172. Dordrecht: Kluwer.
Culp, Sylvia (1995). "Objectivity in Experimental Inquiry: Breaking Data-Technique Circles." *Philosophy of Science* 62: 430–450.
Franklin, Allan (1986). *The Neglect of Experiment*. Cambridge: Cambridge University Press.
Franklin, Allan (1994). "How to Avoid the Experimenters' Regress." *Studies in the History and Philosophy of Science* 25: 97–121.
Franklin, Allan (2009). "Experiment in Physics." In *The Stanford Encyclopedia of Philosophy* (Spring Edition), ed. Edward N. Zalta, http://plato.stanford.edu/archives/spr2009/entries/physics-experiment. Accessed November 1, 2009.
Galison, Peter (1996).:"Computer Simulations and the Trading Zone." In *The Disunity of Science: Boundaries, Contexts, and Power*, eds. Peter Galison and David J. Stump, 118–157. Palo Alto: Stanford University Press.
Gelfert, Axel (2005). "Mathematical Rigor in Physics: Putting Exact Results in Their Place." *Philosophy of Science* 72: 723–738.
Gelfert, Axel (2009). "Rigorous Results, Cross-Model Justification, and the Transfer of Empirical Warrant: The Case of Many-Body Models in Physics." *Synthese* 169: 497–519.
Guala, Francesco (2005). *The Methodology of Experimental Economics*. Cambridge: Cambridge University Press.
Hacking, Ian (1983). *Representing and Intervening: Introductory Topic in the Philosophy of Natural Science*. Cambridge: Cambridge University Press.
Hockney, Roger W., and James W. Eastwood (1988). *Computer Simulation Using Particles*. Boca Raton: CRC Press.
Humphreys, Paul (2004). *Extending Ourselves: Computational Science, Empiricism, and Scientific Method*, Oxford: Oxford University Press.
Humphreys, Paul (2009). "The Philosophical Novelty of Computer Simulation Methods." *Synthese* 169: 615–626.
Imhof, Peter (2000). "Computer Simulation in the Controversy over Limits of Growth." Working paper, urn:nbn:de:gbv:830-opus-1726; http://doku.b.tu-harburg.de/volltexte/2006/172/; 1–19. Hamburg: Technical University of Hamburg-Harburg.
Kennefick, Daniel (2000). "Star Crushing: Theoretical Practice and the Theoreticians' Regress." *Social Studies of Science* 30: 5–40.
Morrison, Margaret (2005). "Experiments versus Models: New Phenomena, Inference, and Surprise." *Journal of Economic Methodology* 12: 317–329.
Morrison, Margaret (2009). "Models, Measurement and Computer Simulation: The Changing Face of Experimentation." *Philosophical Studies* 143: 33–57.
Parker, Wendy S. (2008). "Franklin, Holmes, and the Epistemology of Computer Simulation." *International Studies in the Philosophy of Science* 22: 165–183.
Parker, Wendy S. (2009). "Does Matter Really Matter? Computer Simulations, Experiments, and Materiality." *Synthese* 169: 483–496.

Radder, Hans (1996). *In and About the World: Philosophical Studies of Science and Technology*, Albany: SUNY Press.
Radder, Hans (2005). "Experiment." In *The Philosophy of Science: An Encyclopedia*, eds. Sahotra Sarkar and Jessica Pfeifer, 268–275. London: Routledge.
Röller, Nils (2008). "*Scientia Media*: Simulation between Cultures." In *Simulation: Presentation Technique and Cognitive Method*, eds. Andrea Gleiniger and Georg Vrachliotis, 51–62. Basel: Birkhäuser.
Ruelle, David (1969). *Statistical Mechanics: Rigorous Results*, New York: Benjamin.
Simon, Herbert A., and Allen Newell (1956). "The logic theory machine: A complex information processing system." *IRE Transactions on Information Theory* 2: 61–79.
Weissert, Thomas P. (1997). *The Genesis of Simulation in Dynamics: Pursuing the Fermi-Pasta-Ulam Problem*. Berlin: Springer.
Wimsatt, William C. (1981). "Robustness, Reliability, and Overdetermination." In *Scientific Inquiry and the Social Sciences: A Volume in Honor of Donald T. Campbell*, eds. Marilynn B. Brewer and Barry E. Collins, 124–163. San Francisco: Jossey-Bass Publishers.
Winsberg, Eric (1999). "Sanctioning Models: The Epistemology of Simulation." *Science in Context* 12: 275–292.
Winsberg, Eric (2009). "A Tale of Two Methods." *Synthese* 169: 575–592.

8 Simulation and the Sense of Understanding

Jaakko Kuorikoski

INTRODUCTION

Computer simulation is widely taken to be the best, and sometimes the only, tool with which to study highly complex phenomena. However, in many fields, whether simulation models provide the right kind of *understanding* comparable to that provided by analytic models has been and remains a contentious issue (cf. Galison 1996; Lenhard 2006; Lehtinen and Kuorikoski 2007). Simulation models are often so complex that they, in turn, are difficult to understand, and it is thus worth questioning whether replacing an unintelligible phenomenon with an unintelligible model constitutes epistemic progress. One of the principle aims of science is the creation of scientific understanding and the ability to create understanding is, and should be, an important criterion by which computer simulation techniques are assessed.

Recent philosophical interest in the central role of models in scientific practice has highlighted the fact that creating understanding is not just a matter of providing more information about the phenomena to be understood. Cognitive limitations of humans set conditions on what can be understood, conditions that have to be met by using idealizations, abstractions, and outright falsehoods (Teller 2001); modeling is not just about providing the most detailed and accurate surrogate representation of the target phenomenon. The increase in the computational power available to scientists and the spread of standardized simulation packages has meant that computer simulation has become an integral modeling tool in almost all sciences. In principle, simulation models are not subject to the constraint of analytic tractability and hence can do away with many of the distorting idealizations and tractability assumptions of analytic models. However, the very fact that a simulation model can do away with such idealizations often makes the simulation model itself *epistemically opaque*: the relationship between the stipulated initial conditions and the simulation result cannot be modularly decomposed into sub-operations of sufficient simplicity so that the process could be "grasped" by a cognitively unaided human being (Humphreys 2004, 147–150). If a computer simulation itself becomes so complex that we understand it no better than the phenomenon being simulated —when the simulation reaches the complexity barrier, as Johannes Lenhard puts it (2006)—has our understanding about the phenomenon itself increased?

The aim of this paper is to argue that the assessment of whether or not a simulation model provides understanding is often be hampered by a conflation between *the sense of understanding* and *understanding proper*. The sense of understanding, the feeling of having grasped something, is a psychological *metacognitive* phenomenon and only a fallible indicator of true understanding. This mix-up can distort the appraisal of simulation models in two ways. On the one hand, simulations can increase our understanding without the corresponding sense of understanding. On the other hand, evocative visuals and vague intuitions about maker's knowledge can bring about a sense of understanding without actual increase in true understanding and thus result in an *illusion of understanding*. In this paper, I limit my focus on the latter possibility. In order to improve our assessments of the merits and drawbacks of simulations as tools for understanding and guard ourselves against possible illusions of understanding, we need to get clear about what it is that we are after in the first place. I propose that such clarity is gained by replacing appeals to the psychological sense of understanding (and vague intuitions about intelligibility derived from it) with the explication of what can be explained by the information provided by a given simulation model and by rethinking the proper way of conceptualizing the role of a single human mind in the collective understanding of the scientific community.

The structure of the paper is the following. The following section distinguishes between the sense of understanding and understanding proper and gives a broadly Wittgensteinian, deflationist account of the latter. The third section discusses the possible illusions of understanding, sensations of increased understanding without an accompanying increase in understanding, which may be created by simulation models. The fourth section discusses two possible ways of responding to the epistemic opacity of simulation models: the use of metamodels and the reconceptualization of the boundaries of the cognitive subject. All simulation models discussed are presumed to be "true" or giving an appropriately accurate picture of the simulated phenomenon throughout the paper; the question asked is whether a given simulation can provide understanding, not whether the simulation is correct or not or how we might come to know this.

THE CONCEPT OF UNDERSTANDING

Many simulation models are so complex that they go beyond the limits of human understanding. The limits of understanding cannot be adequately charted until there is an understanding of the concept of understanding itself. Yet the concept of understanding has thus far received little attention in the philosophy of science literature. Although the supposed means of conveying understanding—explanation—has for a long time constituted one of the most productive branches of the general philosophy of science, the product itself has been seen as something that cannot or should not be philosophically explicated (Newton-Smith 2000, 131). Even the basic ontological category of understanding seems to be unclear; should understanding be conceived as a privileged mental state, as some kind of superknowledge, as a cognitive act or even as a special method? This curious situation may partly be the result of Hempel's belief that the concept of understanding was pragmatic to the bone, a psychological by-product with no epistemic relevance, and that it therefore did not deserve any meticulous philosophical analysis (Hempel 1965, 413). No doubt another important factor has been the way in which the theory of explanation has mostly been done: repeated toy examples, historical cases, and appeals to intuitions. This methodology has left the theory of explanation largely stipulative in the sense that it has largely ignored the question of why we want explanations in the first place; explanation just *is* the laying out of the causal history of an event (Salmon 1984), unification of our overall worldview (Kitcher 1989), or whatever seems to save the most pre-theoretical intuitions and historical episodes of explanatory improvement.

Understanding has sometimes been associated specifically with the interpretation of meanings or intentional action and has even been seen as somehow distinct from or the opposite to explanation. Most strands of thinking about understanding as distinct from explanation share a crucial aspect of understanding of understanding with Hempel: *understanding itself is conceptualized as a phenomenon hidden inside the individual mind*. This mentalist conception of understanding is also presupposed in the more recent attack on philosophical theories of explanation by J. D. Trout (2002). Trout claims that philosophical theories of explanation rely on a criterion of goodness of explanation, according to which a good explanation should produce a feeling or a *sense of understanding*. According to Trout, this sense of understanding has little to do with epistemic progress proper and is usually just a result of well-known psychological retrospective and overconfidence biases. The accusation is that philosophical theories relying on such unreliable and contingent factors cannot capture anything epistemologically relevant. The same line of questioning can be applied to simulations: why should it be epistemically relevant that a model happens to elicit a certain kind of sensation?

Yet already Michael Scriven pointed out that it is absurd to identify the sense of understanding with understanding itself since the former can so easily be mistaken (Scriven 1962, 225). We can experience a sense of understanding without achieving understanding. Perhaps understanding should then be seen as a psychological state or activity that is not transparent to introspection of the individual in question? However, Wittgenstein persuasively argued that the grammar of understanding is not that of a state-concept in the first place, but similar to that of *an ability*: understanding is attributed according to whether somebody can reliably *do* something. Understanding is not about just possessing knowledge, but about what one can do with the knowledge and how reliably. Note that understanding is only *akin* to an ability in the sense that it would be equally wrong to think of it as a distinct species of (psychological) ability or skill (such as a mental faculty), something with a deeper underlying essence. Instead, understanding is a regulative or normative concept in the weak sense that it concerns whether people have the ability to do things *correctly* or in a *right* way. As such, understanding can only be attributed by the relevant community according to public criteria (Wittgenstein 1953, 143–159, 179–184, 321–324; Baker and Hacker 2005, 357–385; Ylikoski 2009; Ylikoski and Kuorikoski 2010; see also de Regt and Dieks 2005, 100–102; Elgin 2007, 39). These criteria do not (indeed cannot) primarily concern private cognitive processes, brain states or, even less, subjective feelings, but manifest performances. Feelings cannot be correct or incorrect –they just happen. Understanding is akin to an ability, not to a hidden state. It is fundamentally public, not private.

Cognitive processes (comprehension) taking place in the privacy of individual minds are a *causal* prerequisite for possible fulfillment of these criteria, but the processes themselves are not the criteria of understanding in the sense that we would have to know them in order to say whether somebody *really* understands something –just as we do not need to know the cognitive and neural processes that enable one to ride a bike in order to judge whether one can really ride a bike. In a sense, the Wittgensteinian account is deflationist, since it denies that there is a deeper essence of understanding behind the manifest abilities according to which understanding is attributed. In cognitive science, mental models of different varieties are the standard ways of causally explaining reasoning and comprehension (cf. Hegarty 2004 and Waskan 2009).[1] Yet the postulation of these models addresses a different question: how can individuals achieve what is demanded by the public standards constitutive of the concept of understanding itself[2]. This is the 'natural' cognitive relation between the behavior of the agent and the information providing the understanding that Trout correctly leaves for empirical science to investigate (Trout 2004, 203). Whatever computational and memory limitations these cognitive processes have, also limits the manifest performances that attributions of understanding are based on. But this does not remove the fact that understanding in itself is a categorically different kind of thing. The *correctness* of the internal models is judged by the external displays of understanding, not the other way around (Ylikoski 2009).

One should not, of course, deny the existence of feelings of illumination or 'getting it' often associated with increased understanding. The important point is that the phenomenological state of sense of understanding is only a fallible predictor of future performance to be judged against the public criteria. One can judge whether one really understands something only by comparing one's explanatory or inferential performance against an external public standard. What then can be said of the criteria that one needs to fulfill in order to be eligible for possession of scientific understanding of a theory or a phenomenon? Henk de Regt and Dennis Dieks (2005) stress the manner in which the criteria for scientific understanding have changed in the history of science.[3] When looking for a common factor in the attributions of understanding of *a theory*, De Regt and Dieks arrive at a conclusion that the attribution of understanding rests on the subject's ability to draw qualitative conclusions about hypothetical changes in the explanatory model without making exact calculations with additional inferential aids, such as pen and paper or a calculator (ibid., 151). According to them, the criterion of understanding of theories is unaided inferential performance.[4] Yet the ultimate goal is not just to understand theories or models, but to understand the phenomena that the theories and models are about—and understanding a model and understanding a phenomenon are different tasks. The question is how and to what extent intractable simulation models can provide understanding about the simulated phenomenon.

The central idea of inferential performance as the constitutive criterion of understanding can be further developed by linking it to James Woodward's account of scientific explanation in the following way: Woodward's theory of explanation tells us more specifically what *kinds of inferences* are constitutive of understanding, of theories as well as of phenomena that the theories are about. According to Woodward (2003), explanation consists in the exhibition of functional dependency relations between variables. Explanation is thus doubly contrastive; the functional relationship links the possible values of the *explanans* to possible values of the *explanandum*. These relationships of *explanatory relevance* ground understanding by providing answers to what-if-things-had-been-different questions concerning the consequences of counterfactual or hypothetical changes in the values of the *explanans* variable. These answers are the basis of inferential performance constitutive of understanding. In the case of causal explanations or (more broadly) explanations given on the basis of a causal model, the relevant hypothetical changes to consider are interventions, ideally surgical manipulations that affect only the *explanans* variable of interest and leave the rest of the model intact (apart from the changes caused by the change in the *explanans* variable dictated by the model, of course). A variable A is thus *causally (explanatorily) relevant* with respect to another variable B if there is an intervention on A that results in a change in the value of B. Thus the conception of understanding as inferential ability is fully compatible with the realist (or ontic) idea of understanding as knowledge of causes and mechanisms, as long as causal relations are understood as dependencies invariant under interventions. If a person can correctly solve for the effects of interventions in a model, she understands the model, and to the extent that these predictions match what would actually happen if the modeled system were intervened on, she understands the modeled system.

Woodward's account intimately links explanatory knowledge to our capacity to function in the world as goal-directed manipulators. Whereas the epistemic conception of explanation of the deductive-nomological (D-N) model was concerned only with predictive power, Woodward's theory stresses our role not only as passive observers but also as active agents. Understanding is the ability to make correct inferences on the basis of received knowledge and causal knowledge concerns the effects of manipulations and therefore licenses inferences about the effects of our actions. Understanding thus lies not only in correctness of inference but sometimes also in effectiveness of action based on the information to be understood. Although a mechanic may not be able to give a theoretical description about the physical laws governing the workings of a combustion engine, he does have an intimate understanding about its workings because he can effectively manipulate it—in order to fix an engine one has to know the functional roles of the parts and the effects of manipulating them. This idea probably underlies Fred Dretske's unarticulated intuition that one cannot understand how something works unless one can build it (Dretske 1994) and Johannes Lenhard's pragmatic conception of understanding (Lenhard 2006).

Since science is fundamentally a communal activity, extra criteria of being able to provide explanations and of being able to justify the information responsible for the understanding is appropriate for attributions of specifically scientific understanding to individuals. Scientific knowledge is often highly systematized into hierarchic sets of principles and the ability to justify correct inferences made on the basis of a limited set of theoretical principles, i.e., *theoretical understanding*, also characterizes much, but not all, of specifically 'scientific' understanding. Thus understanding as the ability to make correct inferences also accounts for the common intuitions linking understanding to integration of isolated facts into background knowledge (as in Schurz and Lambert 1994 and in Elgin 2007) and to unification more generally (Friedman 1974; Kitcher 1989). However, whereas unificationists have to adopt a stipulative stance in claiming that understanding simply is unification (Barnes 1992), understanding as inferential ability follows as a natural consequence from a well-ordered knowledge store or from a powerful set of argument patterns.

SIMULATION AND THE ILLUSION OF UNDERSTANDING

We have thus far distinguished between the psychological phenomenon of sense of understanding from the degree of understanding itself, which can in turn be operationalized (in the good old-fashioned strong sense) as the ability to make correct counterfactual inferences about the object of understanding. The sense of understanding is only an indicator of understanding proper, and it is the latter notion that is epistemically and pragmatically relevant. The sense of understanding has an important metacognitive role in providing immediate cues according to which we conduct our epistemic activities, and it is also an important psychological motivating factor for those activities (Ylikoski 2009). However, if the ability to create understanding is to be taken as an important criterion in the final *assessment* of simulation models, this assessment should not be based on vague intuitions about intelligibility, usually dependent on the sense of understanding, or on the sense of understanding directly, unless the sense of understanding turns out to be a sufficiently reliable indicator of understanding itself.

The trouble is that empirical studies seem to indicate that people systematically overestimate their abilities to explain the workings of mechanisms and natural phenomena. As Frank Keil and his associates have empirically demonstrated, when given the possibility of comparing their explanatory performance (inferential performance) against an external benchmark, people tend to downgrade their assessment of their own explanatory performance. This means that after receiving feedback about how well they actually did in explaining things, people consistently realize that they have been overconfident about their explanatory abilities. Only after learning more about the test-case do they begin to raise their self-assessment score towards the level they thought appropriate in the beginning of the test. This effect is *not* present in the self-assessment of the subject's ability to recollect facts or narratives and hence is not simply an aspect of the alleged general overconfidence bias.[5] (Rozenblit and Keil 2002; Keil 2003; Mills and Keil 2004) People are systematically overconfident specifically about their explanatory understanding. To the extent that the self-assessment of understanding is normally based on the sense of understanding, this suggests that the sense of understanding is an unreliable indicator of understanding.

The miscalibration of the sense of understanding is a general phenomenon. However, there is a possibility that the sense of understanding is especially biased in some specific contexts. In principle, the sense of understanding can be misleading in both ways: something can be understood without an accompanying sense of understanding (as in tacit learning), and there can be a sense of understanding without a corresponding increase in inferential ability. I will here focus on the latter possibility, the danger of *illusion of depth of understanding* (Ylikoski 2009), since there are a number of reasons to expect[6] that this danger might be especially severe in the context of simulation studies.

Mistaking Understanding of Sub-operations for Understanding of the Process

Even if the sense of understanding were a reliable indicator of inferential performance when applied to simple inferential tasks, the danger of miscalibration can be expected to increase when the task becomes more complex. First, simply the fact that the overall cognitive load increases probably makes self-evaluation more unreliable. Second, complex inferential tasks are something that the agent rarely encounters, thus there is less feedback and external benchmarks against which the sense of understanding could have been calibrated. In the case of simulated complex systems, both of these worries are present. An additional worry is that the sense of understanding of a simple sub-task can be mistaken as an indicator for true understanding of the more complicated task.

As Roman Frigg and Julian Reiss point out (2009), at the bottom level of the foundational transition rules on which the simulation is built on, simulations are always in principle understandable, since the changes of the states of basic elements (for example, state-transition rules for single cells in a cellular automata model) follow well-defined rules laid down in the programming of the simulation.[7] These rules themselves are usually simple and intuitive and thus relatively easy to understand, but the inferences concerning the results of interactions of these elements are well beyond our limited cognitive powers. The danger of an illusion of understanding arises from the possibility that since one does understand the behavior of the parts and consequently experiences a sense of understanding, this sense of understanding is taken to indicate that one also understands the behavior of the whole. However, it is well known that unless the system-level property is exceptionally well-behaving (for example is nearly aggregative in William Wimsatt's [2000] sense), one cannot simply infer from the properties of the parts to a property of the whole.

Model builders themselves are probably relatively immune to this illusion. In fact, the frequent talk about emergence may be seen as a placeholder for the acknowledged lack of understanding of why the whole behaves as it does even when the behavior of the parts is understood. The problem becomes more acute when knowledge and understanding are transferred and assessed across disciplinary boundaries.

Visualization Confused with Insight

Eric Winsberg (2003), Paul Humphreys (2004) and Johannes Lenhard (2006) have all emphasized the importance of visualization in rendering simulation results understandable. There is no doubt that additional visual representations of the simulation results are an efficient cognitive aid in the creation of understanding (e.g., Herbert and Bell 1997): we are exceptionally visually oriented creatures, and it makes sense to utilize also those cognitive capacities that involve sight and spatial comprehension. Visual aids often improve inferential performance. However, it should be noted that merely looking at colorful pictures or animations may not, by itself, provide grounds for making any additional counterfactual inferences, answers to more what-if-things-had-been-different questions, about the simulated phenomenon. The conflation of knowledge of dynamics with knowledge of underlying causes can be a source of illusion of understanding in the context of analytic models as well, but "seeing how the system works," preferably in vivid color, probably enhances the sense of understanding. To see this consider a model that can be represented both as an analytic model and as a simulation: the Lotka-Volterra (L-V) model.

The standard L-V model is a pair of first-order, non-linear differential equations that are used to describe the population dynamics of a predator and its prey. The two equations are

$dX/dt = rX - (aX)Y$
$dY/dt = b(aX)Y - mY$

in which the X is the prey population, and Y is the predator population, r the growth rate of the prey population, a tells us how efficiently the prey are gobbled up by the predators, m the death rate of the predators, and finally the first term in the latter equation relates the amount of pray devoured to the predator birth rate. In this form, the model has no analytic solution, but it is relatively easy to explore the dynamics numerically (even without a computer) and the dynamics are usually represented graphically. The graph shows how the population trajectories oscillate and the reason for this systemic behavior is easy to understand: over-predation leads to a fall in the prey-population, which leads to a fall in the predator population, which makes it possible for the prey-population to grow and so forth.

The essence of the LV model can also be simulated with a cellular automata model (or with other agent-based platforms).[8] With a running cellular automata model, one can see how local over-predation leads to disappearance of the pray, which leads to a decline in the predator population and so forth. Because each state of a given cell on the screen can now be given an interpretation as a "concrete," though virtual, hare, lynx, or pasture (especially if the cell states are given appropriate color codes), the simulation creates a feeling of seeing the dynamics in action, not just of an abstract representation. There is a sense of understanding of the abstract oscillatory dynamics. However, no new what-if-things-had-been-different questions (at least relating to non-spatial issues, of which the original model is silent) can be answered about the population dynamics on the basis of the cellular automata model. Hence, there is no increase in understanding proper, only an illusion of enhanced understanding.

As with the previous worry, this illusion is probably a serious issue mostly in contexts in which the visualizations are presented to a non-expert audience, who lack the knowledge of the required context-specific benchmarks and previous encounters with the kind of data that would be required for competent self-assessment of understanding. Neither is the illusion of understanding created by pictures limited to simulation results. For example, adding irrelevant brain pictures to a neuro-scientific research report makes the report more convincing for non-experts and even students of the relevant field (Skolnick Weisberg et al. 2008). In most fields, picture- and diagram-based reasoning is much less canonized and explicitly taught and debated than inferences concerning propositional content (text). Hence the reliability and competent self-assessment of reasoning with pictures might be more reliant on accumulated unarticulated experience and tacit knowledge. The extent to which visual presentations of simulation results are prone to create illusions of understanding is, of course, in the end an empirical matter, which might be worth exploring experimentally.

Manipulability of Dynamics Confused with Understanding of Mechanism

Johannes Lenhard (2006) argues that because of the epistemic opacity of simulations, traditional (D-N style) conceptions of explanation and understanding are inapplicable. In their place, Lenhard advocates "a pragmatic conception" of understanding based on our practical ability to manipulate the simulated phenomenon according to what we have learned from manipulating the simulation model. This fits well with the conception of understanding as akin to an ability advocated in the previous section, but it is nonetheless crucial to explicate just what is and what is not understood when only the results of changes in the initial conditions or parameter values are known.

Knowledge of dependencies between inputs and outputs provides understanding of why some of the end results (the outputs) are the way they are rather than something else. For example, let us suppose that some agent-based model evolves into an equilibrium E with a certain characteristic C and that this characteristic is dependent on the parameter value P. If we had found this out, for example by performing multiple runs with different parameter values, we would have (limited) understanding about why E has the property C rather than C'. However, unless we know how that dependency itself is dependent on the structural features of the model, say that if the agents obeyed a slightly different decision rule, then the dependency between P and C would be different in some predictable way, we would not have understanding of why the system behaves as it does, i.e., understanding of the mechanism. By manipulating the initial settings, we may be able to get the simulation to do what we want it to do, but this is different from understanding *why* it does so (cf. Humphreys 2004, 134).

The importance of understanding the dependence between a simulation assumption and the resulting dynamics is not always apparent on the pages of finished research articles, since it is the model output and, to a lesser extent, the credibility of the assumptions that are the main focus of model assessment (and peer review). However, the problem of tracing the way in which results emerge from the assumptions is recognized in practice. As an example, one of the most debated issues in theoretical ecology is the drivers and mechanisms of dispersal (the emigration of individuals from their natal environment). Especially pertinent explanandum has been the widespread male bias in dispersal. It has long been acknowledged that the costs of inbreeding and the risks of emigration lead to a sex-biased dispersal, but these factors do not explain why the dispersing sex should be the male. Andreas Gros et al. (2009; see also Poethke and Hovestadt 2002) provide an agent-based discrete-generation simulation of a metapopulation over habitat patches, each carrying a population of sexually reproducing organisms with four alleles in two *loci*, one determining the probability of emigration for males and the other for females. Emigrants die with a set probability. After dispersal phase, females compete for breeding spaces (set by the carrying capacity of a patch) and males for females. The number of offspring by mated females in a patch varies stochastically (environmental fluctuation). Simulation runs under different carrying capacities show that the male bias does not evolve when there is no difference in the between-patch variance in reproductive success for the sexes. The proposed interpretation for this is that since the fitness expectations of males and females depend in the simulation on different processes (the females compete over breeding territories and males over females), the programmed stochastic variability leads to bigger between-patch reproductive variability for males and hence leads to male-biased dispersion. However, this is by no means transparent and not the only possible interpretation of why the simulation is behaving as it is. Other possible culprits are the implicitly accounted for kin competition (relatives competing for limited local resources) and the way in which competition in general has been implemented in the simulation.[9]

It is important to notice that the possibility of conflating the manipulability of dynamics with understanding of a mechanism is a different concern from the obvious underdetermination problem, according to which many possible causal mechanisms could produce the same observable results (in the context of simulations the underdetermination problem is often called equifinality). The illusion of understanding arising from the manipulability of the system can arise even when the simulation does capture the right causal constituents of the simulated system. If the simulation is based on totally wrongheaded causal assumptions, the understanding it provides is simply false.

The Idea of Maker's Knowledge

Fred Dretske once asserted (1994) that in order to really understand a system, one should possess the knowledge that would, in principle, enable one to build the system. Also Joshua Epstein's motto "If you didn't grow it, you didn't explain it" (1999) could be (although perhaps slightly uncharitably) be interpreted as an expression of the same idea. As was noted above, there is a sense in which assembling a working system guarantees at least some proficiency in answering what-if-things-had-been-different questions and thus understanding: If you have coded the model assumptions, then you are bound to have some idea what would happen if the assumptions were slightly different (unless you have simply copied the program). However, this understanding should be distinguished from the common but unfounded intuition of maker's knowledge: a special privileged epistemic access a maker has into his or her creation.

One of the first things to do when verifying or validating a simulation model is to see whether the program can reproduce some results known to be true of or similar to the simulated phenomenon (benchmarking). This is not an easy task. As was already noted, the sense of satisfaction derived from getting the computer to finally do what you wanted it to do can be confused with an understanding of what is happening. This effect may be enhanced by the fact that simulations feel more concrete than analytic models. A simulation model is an artificial physical system, ecology, economy, or society with which the researcher can experiment (Peck 2008). Thus simulations seem like something that the researcher has actually built, rather than sets of stipulated assumptions and conclusions deduced from them. However, since understanding is ultimately a matter of correct inferences concerning the effects of counterfactual changes, if the creator cannot make these inferences, no amount of spilled blood, sweat, or tears can make the creator understand *why* the creation behaves as it does.

IMPROVING OUR UNDERSTANDING

Since the sense of understanding is in general an unreliable, and in simulation context possibly biased, indicator of understanding, we ought to replace it with something else as the central metacognitive criterion of model *assessment*. As was already noted, the sense of understanding plays a crucial role in motivating research and providing immediate heuristic cues by which the model building is guided. Thus it may not be possible or even desirable to try to altogether ignore the sense of understanding during the process of model building. Since what is central to understanding is the ability to make inferences about the effect of local changes in the system under study, explicating the dependencies that ground these inferences —what can be explained by what —also explicates the kind and degree of understanding. Thus the kind and degree of understanding can be explicated by stating the known relations of explanatory relevance. This should be the fundamental criterion against which models should be assessed. Explicating relationships of explanatory relevance can also be an effective strategy in *calibrating* the sense of understanding: reflecting whether one can actually answer any new w-questions concerning some puzzling aspect of a simulation or the simulated phenomenon can be used as a quick check on whether any increase in understanding has been achieved.

The trouble is that when simulations do become epistemically opaque, relations of explanatory relevance become hard or impossible to formulate for a cognitively unaided human. This means that simulation models that break the complexity barrier almost unavoidably provide only limited understanding. How should we then try to incorporate them into our overall scientific understanding of the world? First, we might try to build humanly tractable models about the simulations themselves. This strategy of *metamodeling* is an approach sometimes taken in agent-based computational economics, for example. Second, we might simply have to rethink the place of the individual human mind in our collective scientific endeavor.

Metamodels

In order for a complex simulation to be understandable for a cognitively unaided human individual, it needs to be described in such a way that inferences concerning hypothetical local changes become feasible. Thus one way of improving understanding about simulations is to build additional explanatory representations, *metamodels*, about them. When the interaction of system parts becomes intractable, one way to gain some inferential power concerning possible paths of development for the system as a whole is to build a new representation of the simulation at a higher level of abstraction. As an example, Cosma Shalizi and Christopher Moore (2003) claim that the way to create understanding of complex bottom-up simulations is to throw away micro-information in such a way, that the information left describes macro-states with Markovian dynamics, i.e., the behavior of, e.g., a cellular automata system (CA) is described in such a level, that the future state of the system is independent of its past when conditioned on the present state (Shalizi and Moore 2003, 10). The idea of understanding as inferential ability presented above explains why the Markov property is conducive to the creation of understanding: Markovian dynamics make it possible to draw inferences concerning the effects of temporally localized changes in the process without the need to repeatedly run the simulation with alternative settings. In order to gain some inferential power concerning possible paths of development of an otherwise intractable model, we need a new representation of the CA at a higher level of description, the Markovian stochastic process. This stochastic process can be considered as a metamodel of the original CA, which, in turn, models the primary system to be understood.

The building of "empirical" metamodels[10] using standard statistical techniques on the data generated by the simulation is standard practice in some fields using micro-simulation, such as agent-based computational economics (ACE; Kleijnen and Sargent 2000). For example, Bunn and Oliveira (2001) construct an ACE model of a wholesale electricity market to explore the possible effects of the New Electricity Trading

Arrangements (NETA) introduced in the United Kingdom in March 2001. Their model incorporates strategically interacting market participants (electricity generators and energy purchasers for end-use customers); a system operator; interactions between a bilateral market, a balancing mechanism, and a settlement process; determination of day-ahead markups on previous day price offers by means of reinforcement learning; and daily dynamic constraints. The result is an enormously complicated repeated stochastic game. In order to make some sense of their simulation results, they use their simulated data to fit a number of simple econometric models describing the characteristics of market equilibria under NETA as functions of both market structure and agent characteristics. These simple econometric models are then used to make inferences concerning the consequences of counterfactual changes in variables. Thus the econometric model provides understanding of the simulation, and, to the extent that the simulation captures the important dependencies in the actual market, also provides understanding of the modeled electricity market. As with the cellular automata models, the mere possibility of recreating some macro-phenomenon of interest from micro-foundations is not enough for achieving understanding of the phenomenon. What is required is a representation that enables answering to what-if-things-had-been-different questions concerning hypothetical changes in the values of variables or parameters.

Extending the Understanding Subject

There is no a priori assurance that all the pragmatically and theoretically important systems could be captured in such representations that would allow a cognitive agent with human limitations to reliably make counterfactual inferences about them.[11] Adding new layers of representation, models of models, may not even always be the most sensible thing to do. So far this essay has been about understanding as possessed by individual human beings. However, limiting the proper place of inferential activity to the mind of a lone heroic theorist might simply be misguided, since a great part of our cognitive practices are in any case best seen as distributed outside our minds and bodies. In a very important sense, we can understand the world better not because we have become smarter, but because we have cumulatively made our environment smarter. The question of the proper unit of cognition is especially pertinent in the case of scientific understanding, which is both massively distributed (within research groups and across the scientific community) and massively extended.

For Paul Humphreys (2004, 2009), the most important philosophical question that computational science, in general, and simulation, in particular, pose is whether we should rethink the very anthropocentric enlightenment conception of epistemology, i.e., should we let go of the presupposition that the individual human mind is the primary or default cognitive/epistemological subject. Of course, the idea of extended cognition is now generally accepted, but there is a specific point about the sense of understanding that is worth making here: the sense of understanding has probably been a major intuitive motivation for unreflectively presuming that it is the individual mind that should be taken as the seat of knowledge and understanding. For Descartes, the sense of understanding even acted as a foundational epistemic principle: the things that we feel we understand most clearly should be taken as the most secure basis of all knowledge. Our epistemic activities are to a large degree motivated by the psychological sense of satisfaction accompanying understanding, but it is still a mistake to confuse this sensation for the ultimate goal itself.

The deflationist and anti-mentalistic conception of understanding advocated here thus supports the idea that epistemological anthropocentrism should be, if not discarded, at least weakened. If the sense of understanding itself has no epistemic value, then we cannot use it to argue that attributions of understanding should be limited to conscious individual minds. In fact, unless there is some epistemic reason to think otherwise, the question of whether an extended cognitive system of a mind and a computer can be said to understand something even when the unaided mind is incapable of doing it becomes mostly definitional. In such a case, the only epistemically relevant facts of the matter are that the extended system can reliably answer a range of what-if-things-had-been-different questions about the simulated phenomenon, can successfully infer and explain, but the constrained cognitive system (the human) cannot. Insofar as the human-computer pair is reliably integrated to the appropriate scientific community (for example, the computer and the code can be subjected to effective error control that is independent of particular simulation results), whether the extended system "really" understands or not becomes a non-issue. The human may not understand the simulation, but the human-computer pair may understand the simulated phenomenon. In fact, we already attribute understanding to extended cognitive systems, since even traditional analytic modeling, which essentially involves the use of pen and paper, should also be seen as extended cognition (Giere 2002; Kuorikoski and Lehtinen 2009).

The extent that understanding is allowed to seep outside our skulls varies across scientific disciplines and probably reflects important and deeply ingrained differences in methodological presuppositions and epistemic situation. Physicists do not seem to be particularly worried about offloading cognition to machines, probably because of their relative confidence in their basic theory (from which the simulation assumptions are derived), familiarity with technology and heavy computation and, most importantly, because they have no choice. Economists have been less sanguine, probably because the underlying theory is not that strong and the investigated systems are heterogeneous and constantly changing thus creating the need for general and robust models. Many theoretical economists also conceptualize their research as economic "thinking" and thus uphold the romantic image of the lone heroic theoretician, unraveling the secrets of society within his or her mind (the contentious history of the epistemic importance of introspection probably also plays a role).

CONCLUSIONS

Explanation is an epistemic activity and thus limited by our cognitive capacities and simulations should therefore be, and in fact often are, assessed by how well they enhance our explanatory understanding. The trouble is that this assessment is often done according to vague and implicit intuitions and relying on the psychological sense of understanding. But the sense of understanding is only an imperfect indicator of understanding proper, and there are reasons to expect that it is systematically biased in simulation context. Conflating understanding of the component operations with understanding of the process, visualization with insight, manipulability with knowledge of mechanism and ability to build something with knowledge of principles of its operation, can all contribute to illusions of understanding—a sense of understanding without accompanying increase in inferential performance. Distinguishing between the sense of understanding and understanding is also crucial when thinking about what to do when our models unavoidably become too complex for our limited minds: whether we try to model our models or whether we should rethink the position of the unaided individual human mind as the privileged seat of knowledge and understanding.

ACKNOWLEDGMENTS

This paper draws heavily on joint work with Petri Ylikoski. I would also like to thank the anonymous referees, the audience at Models and Simulations 3(MS3) and the Philosophy of Science Group at the University of Helsinki for their valuable comments. This research was supported by the Academy of Finland.

NOTES

1. It appears that these mental representations represent dependency relations directly or "intrinsically," i.e., they are not propositionally structured but are analogous to actual model systems that are manipulated mentally in order to produce predictions (Hegarty 2004).
2. Jonathan Waskan seems to claim that these mental models *make* phenomena intelligible (Waskan 2008), which, of course, would just lead to a regress: in virtue of what do these mental models make phenomena intelligible? This mentalistic way of conceptualizing understanding also leads easily to confusing sense of understanding with understanding.

3. It is certainly necessary to allow certain flexibility in the criteria of understanding so as not to render all past scientific explanations unsatisfactory or downright unintelligible if they do not meet the standards of today. However, certainly understanding per se is such an integral part of our conceptual scheme that a complete lack of continuity in historical applications of this concept would be hard to accept. There should also be a place left for unabashedly normative use of the concept from our current perspective; we do make judgments as to whether some particular phenomenon was understood at some given time, and the grounds for these judgments should have continuity with arguments given in some other era. After all, explanatory progress in science is at least partly a matter of empirical discovery, not just of conceptual change.
4. It is not clear why de Regt and Dieks single out the qualitative predictions concerning consequences of changes in the explanatory model, since quantitative mathematical manipulations surely are a form or a dimension of understanding as well, usually one regarded as highly scientific (although one can, of course, stipulate that intuitive understanding must be arrived at by intuitive means). In fact, the degree of externalization of inferences and the amount of explicit calculation that would still count as constitutive of understanding vary across scientific disciplines. For example, economists are notorious for insisting on proficiency in mathematical model manipulation and for especially valuing analytical solutions done with only pen and paper as inferential aids (Lehtinen and Kuorikoski 2007).
5. Trout's hypothesis that the sense of understanding is a result of the general overconfidence bias seems therefore to be empirically falsified.
6. It is important to admit that the following worries are, in the end, empirical psychological hypotheses.
7. Also, in principle, every simulation run is just a long deduction (Epstein 2005) and hence understandable by a cognitive agent with sufficient working memory (and god-like patience).
8. One can browse the web for numerous examples of LV cellular automata applets. The idea for this example originated from Michael Weisberg's presentation in Models and Simulations 2 in Tilburg, in which he demonstrated a cellular automata L-V model. There is now a probabilistic agent-based L-V model on Weisberg's web page: www.phil.upenn.edu/~weisberg/Models/probabilistic-predation-final4.0.html.
9. I am grateful to Woutar Vahl for suggesting this example.
10. Providing understanding of the simulated phenomena is not the only aim of metamodeling. Of independent interest may be the brute behavior of the output, calibration, or sensitivity/robustness analysis. Consequently, these different aims entail different metamodeling strategies (Kleijnen and Sargent 2000). Notice that the term metamodel is sometimes also used to refer to a kind of meta-theory, according to which the simulation should be constructed and carried out.
11. Herbert Simon famously argued that most evolved or designed complex systems are likely to be modular and hierarchical (Simon 1988). However, see also Kashtan and Alon 2005 for some crucial limitations to Simon's argument.

REFERENCES

Baker, Gordon P., and Hacker, Peter M. S. (2005). *Wittgenstein: Understanding and Meaning, Part I: Essays.* 2nd revised ed. Oxford: Blackwell.

Barnes, Eric (1992). "Explanatory Unification and Scientific Understanding." *PSA: Proceedings of the Biennial Meeting of the Philosophy of Science Association*, Vol. 1, 3–12.
Bunn, Derek W., and Fernando S. Oliveira (2001). "Agent-Based Simulation –An Application to the New Electricity Trading Arrangements of England and Wales." *IEEE Transactions on Evolutionary Computation* 5(5): 493–503.
De Regt, Henk W., and Dieks, Dennis (2005). "A Contextual Approach to Scientific Understanding." *Synthese* 144:1: 137–170.
Dretske, Fred (1994). "If You Can't Make One, You Don't Know How It Works." In *Midwest Studies in Philosophy*, Vol. 19, ed. Peter French, Theodore Uehling, and Howard Wettstein, 468–482. Notre Dame (IN): Notre Dame University Press.
Elgin, Catherine (2007). "Understanding and the Facts." *Philosophical Studies* 132: 33–42.
Epstein, Joshua (1999). "Agent-based Computational Models and Generative Social Science." *Complexity* 4 (5): 41–60.
Epstein, Joshua (2005). "Remarks on the Foundations of Agent-Based Generative Social Science." Center on Social and Economic Dynamics (CSED) Working Paper No. 41. The Brookings Institution, Washington DC.
Friedman, Michael (1974). "Explanation and Scientific Understanding." *Journal of Philosophy* 71: 5–19.
Frigg, Roman, and Julian Reiss (2009). "The Philosophy of Simulation: Hot New Issues or Same Old Stew." *Synthese* 169: 593–613.
Galison, Peter (1996). "Computer Simulations and the Trading Zone." In *The Disunity of Science: Boundaries, Contexts, and Power*, eds. Peter Galison and David J. Stump, 118–157. Stanford: Stanford University Press.
Giere, Ronald (2002). "Models as Parts of Distributive Cognitive Systems." In *Model Based Reasoning: Science, Technology, Values*, eds. Lorenzo Magnani and Nancy Nersessian, 227–241. Dordrecht: Kluwer.
Gros, Andreas, Hans J. Poethke, and Thomas Hovestadt (2009). "Sex-Specific Spatio-Temporal Variability in Reproductive Success Promotes the Evolution of Sex-Biased Dispersal." *Theoretical Population Biology* 76: 13–18.
Hegarty, Mary (2004). "Mechanical Reasoning by Mental Simulation." *Trends in Cognitive Sciences* 8(6): 280–285.
Hempel, Carl (1965). *Aspects of Scientific Explanation and Other Essays in the Philosophy of Science*. New York: Free Press.
Herbert, Ric D., and Rodney D. Bell. 1997. "Visualisation in the Simulation and Control of Economic Models." *Computational Economics* 10: 107–118.
Humphreys, Paul (1993). "Greater Unification Equals Greater Understanding?" *Analysis* 53(3): 183–188.
Humphreys, Paul (2004). *Extending Ourselves: Computational Science, Empiricism, and Scientific Method*. Oxford: Oxford University Press.
Humphreys, Paul (2009). "The Philosophical Novelty of Computer Simulation Methods." *Synthese* 169: 615–626.
Kashtan, Nadav, and Uri Alon (2005). "Spontaneous Evolution of Modularity and Network Motifs." *Proceedings of the National Academy of Sciences* 102(39): 13773–13778.
Keil, Frank C. (2003). "Folkscience: Coarse Interpretations of a Complex Reality." *Trends in Cognitive Sciences* 7(8): 368–373.
Kitcher, Philip (1989). "Explanatory Unification and the Causal Structure of the World" In *Scientific Explanation*, eds. Philip Kitcher and Wesley Salmon, 410–505. Minneapolis: University of Minnesota Press.
Kleijnen, Jack P. C., and Robert G. Sargent (2000). "A Methodology for the Fitting and Validation of Metamodels in Simulation." *European Journal of Operational Research* 120: 14–29.

Kuorikoski, Jaakko, and Aki Lehtinen (2009). "Incredible Worlds, Credible Results." *Erkenntnis* 70: 119–131.
Lehtinen, Aki, and Jaakko Kuorikoski (2007). "Computing the Perfect Model: Why Do Economists Shun Simulation?" *Philosophy of Science* 74: 304–329.
Lenhard, Johannes (2006). "Surprised by a Nanowire: Simulation, Control, and Understanding." *Philosophy of Science* 2006: 605–616.
Mills, Candice M., and Frank C. Keil (2004). "Knowing the Limits of One's Understanding: The Development of an Awareness of an Illusion of Explanatory Depth." *Journal of Experimental Child Psychology* 87: 1–32.
Newton-Smith, William H. (2000). "Explanation" In *A Companion to the Philosophy of Science*, ed. W. H. Newton-Smith, 127–133. Oxford: Blackwell.
Peck, Steven L. (2008). "The Hermeneutics of Ecological Simulation." *Biology and Philosophy* 23: 383–402.
Poethke, Hans, and Thomas Hovestadt (2002). "Evolution of Density- and Patch-Size-Dependent Dispersal Rates." *Proceedings of the Royal Society B: Biological Sciences* 269: 637–645.
Rozenblit, Leonid, and Frank Keil (2002). "The Misunderstood Limits of Folk Science: An Illusion of Explanatory Depth." *Cognitive Science* 26: 521–562
Salmon, Wesley (1984). *Scientific Explanation and the Causal Structure of the World*. Princeton, NJ: Princeton University Press.
Schurz, Gerhard, and Karel Lambert (1994). "Outline of a Theory of Scientific Understanding." *Synthese* 101: 65–120.
Scriven, Michael (1962). "Explanation, Predictions, and Laws." In *Minnesota Studies in the Philosophy of Science*, Vol. 3, ed. Herbert Feigl and Grover Maxwel, 170–230. Minneapolis: University of Minnesota Press.
Shalizi, Cosma Rohilla, and Christopher Moore (2003). "What Is a Macrostate? Subjective Observations and Objective Dynamics." *PhilSci Archive*, http://philsci-archive.pitt.edu/archive/00001119/. Access date: 26.10.2010.
Simon, Herbert (1988). "The Architecture of Complexity." In *The Sciences of the Artificial*, 2nd ed., 192–227. Cambridge MA: The MIT Press. Originally published in *Proceedings of the American Philosophical Society* 1962.
Skolnick Weisberg, Deena, Frank C. Keil, Joshua Goodstein, Elizabeth Rawson, and Jeremy R. Gray (2008). "The Seductive Allure of Neuroscience Explanations." *Journal of Cognitive Neuroscience* 20: 470–477.
Teller, Paul (2001). "Twilight of the Perfect Model Model." *Erkenntnis* 55: 393–415.
Trout, J. D. (2002). "Scientific Explanation and the Sense of Understanding." *Philosophy of Science* 69: 212–233.
Trout, J. D. (2004). "Paying the Price for a Theory of Explanation: De Regt's Discussion on Trout." *Philosophy of Science* 72: 198–208.
Waskan, Jonathan (2008). "Knowledge of Counterfactual Interventions through Cognitive Models of Mechanisms." *International Studies in the Philosophy of Science* 22: 259–275.
Wimsatt, William (2000). "Emergence as Non-Aggregativity and the Biases of Reductionism." *Foundations of Science* 5: 269–297.
Winsberg, Eric (2003). "Simulated Experiments: Methodology for a Virtual World." *Philosophy of Science* 70: 105–125.
Wittgenstein, Ludwig (1953). *Philosophical Investigations*. Trans. Gertrude E. M. Anscombe. Oxford: Blackwell.
Woodward, James (2003). *Making Things Happen*. Oxford and New York: Oxford University Press.

Ylikoski, Petri (2009). "Illusions in Scientific Understanding" In *Scientific Understanding: Philosophical Perspectives*, eds. Hank De Regt, Sabine Leonelli, and Kai Eigner, 100–119. Pittsburgh: Pittsburgh University Press.

Ylikoski, Petri, and Jaakko Kuorikoski (2010). "Dissecting Explanatory Power." *Philosophical Studies* 148: 201–219.

9 Models and Simulations in Brain Experiments

Patrick Suppes

I begin by discussing the question of how we can observe the activity of the brain experimentally, and what the problems are of interpreting this activity. This is only a preliminary discussion to set the scene for the two experiments and simulation discussed later. The first experiment shows how we can identify the representation of a stimulus in the brain. In simple experiments we can do this with a high degree of success. In the second experiment, I analyze how we can identify invariance of constituents of language. To keep the discussion simple, I will only consider the identification of phonemes. All of the same methods can be used for words or sentences as larger constituents. In the final part, I consider, at a more general level, a simulation of some behavioral experiments in terms of what we hope are promising ideas about how the brain is actually computing in a physical way. The brain associations are conditioning connections, familiar from many years of behavioral studies.

If we accept the broad and essentially correct generalization that, as far as we know, human brains are among the most complicated objects in the universe, we can scarcely expect to have a simple theory of observation. There is a long history of study of the anatomy of the brain, and in fact already in *Principles of Psychology* (James 1890), there is an excellent chapter on what had been learned in the 19th century, really the first century in which the brain was studied in a way that we would consider modern in a contemporary scientific sense. Not surprisingly, as we examine the smaller units of the brain, such as neurons, we come to see that neurons themselves are incredibly complicated, consisting as they do of billions of atoms organized in a structural hierarchy of chemical and electrical activity. In the past there have been large controversies over whether signaling via synapses and axons between neurons is mainly chemical or electrical. In general terms, the conclusion broadly accepted now is that both kinds of signaling occur. In the study of the brain, there is a sharp and important distinction between cell neuroscience and system neuroscience. In cell neuroscience, which has the longest and most extensive record of scientific results, the concentration, as the name implies, is on the nature of individual cells or neurons. In contrast, system neuroscience is concerned with the organized activity of the brain at the level at which we think cognitively and act behaviorally. A good example is the production and comprehension of language. Cognitive perception, quite apart from language, has also received an enormous amount of attention and study, as have feelings, emotions, and affective responses of pleasure and pain.

There is much study of signaling between neurons, and the subject is about as complicated as any that one can think of. There has been less study of how the brain is making computations from a systems standpoint, but it is these system computations that are particularly relevant to everyday thinking and talking, as well as to psychological and philosophical viewpoints towards mental activity.

Moreover, in terms of system activity, the electric and magnetic activity of brains is much more central than chemical activity. That does not mean that chemical activity is not important in the background. For example, there are many reasons to think that language is managed almost entirely in production and comprehension by electromagnetic, not chemical, activity, but at the same time, the chemical signaling of neurons undoubtedly plays a part in the permanent storage of both declarative, episodic and motor memory, in the brain. Moreover, chemical processes are important even in the observation of the brain by what are, to some extent, indirect methods in nuclear magnetic resonance. We think we are observing magnetic resonance activity in nuclear magnetic resonance, but what we may see are surrogates for that activity: rates of use of energy, particularly oxygenation processes in the blood. But even then, fMRI, functional magnetic resonance imaging, is much better for locating activity in the brain than direct recording of that activity itself. The two principle methods for observing electromagnetic activity are first electroencephalography (EEG), which is the method for observing electric fields in the brain, and MEG, magnetic encephalography, for observing magnetic fields. Although I am stating this separation so baldly, this is not meant to imply that the ordinary physical conception of the electromagnetic field as one unified physical activity is not held to, but rather that we have, as our ordinary experience with electricity and magnetism, separate ways of observing electrical phenomena and magnetic phenomena. There is also some use of positron emission tomography (PET), but I shall not say anything about PET or fMRI here, because I am interested in the kind of experiments that are observing the electromagnetic activity directly. For this purpose EEG and MEG have the same excellent time resolution, down to at least the order of a millisecond. Notice this unit of one millisecond (ms). By modern standards of digital computer processes, the electromagnetic processes in the brain are mostly pretty slow and certainly cannot hold a candle to the speed of modern supercomputers or even simple handheld devices such as cell phones. Figure 9.1 shows a typical distribution of what is called, for reasons I will not explain, the 10–20 system of EEG sensors located on the scalp of a person. Of course,

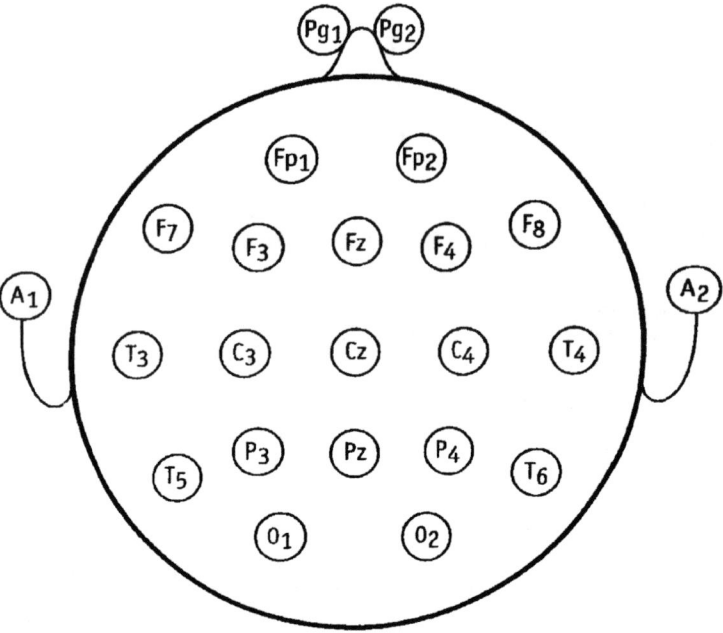

Figure 9.1 The 10–20 system of EEG sensors.

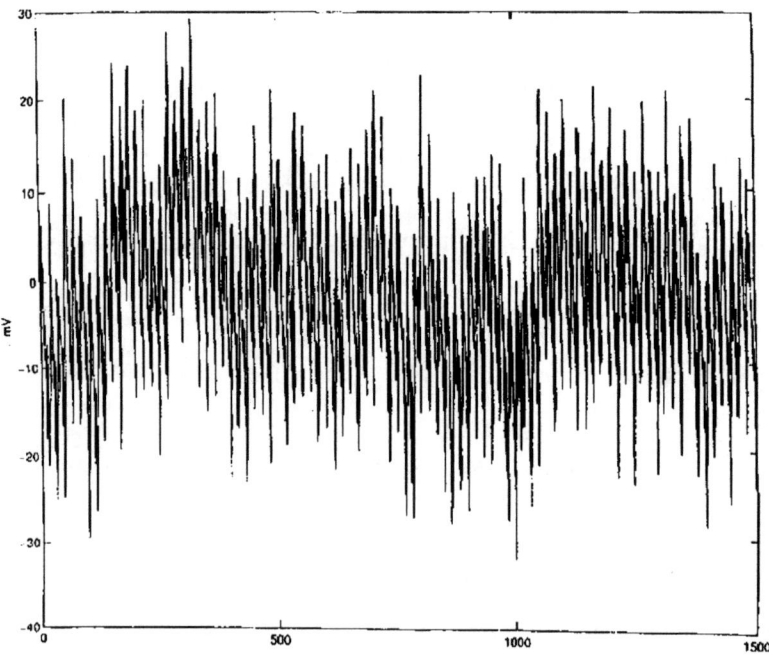

Figure 9.2 The EEG recording of one sensor to hearing the spoken word 'circle'.

in using a circle, we are distorting the actual shape of heads, for which computational adjustments are commonly made.

Figure 9.2 shows the recording of a subject listening to the word "circle." This is the raw EEG data, so to speak, with time measured in ms on the horizontal axis, and microvolts on the vertical axis. In a general way, Figure 9.1 does not look different from a spectrogram of the sound itself. But, as would be expected, it is much more difficult to find the words in the brainwaves. I show Figure 2 just to give an idea of the initial data that we begin with in analyzing, for example, data concerning language in the brain. The recording I have shown you is for a single sensor; with the equipment we are now using, we have 128 sensors. What we are now seeing is 128 images that resemble this one, each of them, or at least many of them, containing relevant information about what is being processed as representation of some external stimulus. (Of course what is being processed is not always from an external stimulus; this is a point I will return to later.) I turn now to the first experiment, and in this context I will also discuss methods of classifying or recognizing EEG signals in the brain. It needs to be clearly understood that by speaking of EEG signals, I mean that there is a brain wave that is a neural response to some presented external stimulus, for example, a spoken word. It could also be the perception of a picture, of an object, of a process, or any one of many different things. Our first task, one I will focus on here primarily, is identifying that brain wave in the brain that represents that external stimulus.

EXPERIMENT 1: A RED STOP SIGN AND THE SPOKEN WORD "GO" AS STIMULI

This experiment was generated by the desire to do our own study of what is called "brain computer interaction" (BCI). The general aim of BCI studies of the brain is to develop signals that the brain can generate, first on the basis of an external stimulus, and second, if things go well, on the basis of an internal intention to signal, and to be able to recognize those signals, so they can be used for communication by someone. An example would be an Amyotrophic Lateral Sclerosis (ALS) patient, whose peripheral nervous system is almost entirely frozen, and so the patient is no longer able to speak but is able to think and have definite appetites, needs, and the like, that the patient would like to be able to communicate. For BCI experiments, one hopes to have a signal generated from the brain that can be easily recognized. This first step is to show that two stimuli, in our case one from the visual modality and the other from the auditory modality, produce EEG brain responses that we can classify with considerable accuracy. Notice how trivial this experiment is from a behavior standpoint. We show a person, on the screen of a computer being used for the experiment, a drawing of a red stop sign on one randomized trial, and then on other such trials the screen may be blank and the person in the experiment hears the spoken word "Go." Now this would be too simple and too trivial an experiment to run if we were only interested in the subject telling us what he or she had seen or heard. Not so with the brain. We are still at a rather elementary and primitive state of understanding well its processing. This experiment and its analysis, as yet unpublished, was a collaborative effort with Marcos Perreau-Guimaraes, Claudio Carvalhaes, Logan Grosenick, and Lene Harbott.

There were 11 subjects in the experiment; each one had 1200 trials. I omit many details of the experiment itself and our methods of analysis. In fact the choice of method of analysis itself is a large and complicated question. We begin with a classical statistical method, linear discrimination classification (LDC), but with the addition of machine-learning methods I will not describe here (Suppes et al. 2009). As a second analysis, we add to the LDC the Surface-Laplacian (SL). The SL is composed of spatial derivatives of the scalp potential and is useful in localizing the EEG signals. Finally, we can improve matters still further by adding to the LDC and SL analysis independent component analysis (ICA; Bell and Sejnowski 1995).

I now summarize in Table 9.1 the results shown for each of the 11 subjects in the experiment. There were three kinds of trials. First, all 11 subjects were given trials in which they were simply asked to pay attention to the stimulus presented (STIM), the red stop sign or the spoken word "Go." Second, for seven of the subjects, after each such trial they were asked, as the task of the next trial, to imagine which stimulus had just been presented (IMAG-SAME). Third, for the remaining four subjects, they were asked to imagine the opposite of the stimulus just presented (IMAG-DIFF). In Table 9.1 the percentage of correct classification we achieve for three different computational models is shown: the direct analysis of the unfiltered data, the SL, and the SL plus ICA. The statistical measure of significance of each of the three conditions for each subject is shown.

This significance measure is the probability (p-value) of getting a result by chance (the null hypothesis): the lower the p-value, the higher the significance. The majority are better, i.e., smaller than 0.01, the classical standard of significance in experimental psychology.

In the second column on the left of Table 9.1 is shown the recognition rate for the trials on which one of the two stimuli was presented. For example, for S1, we had a recognition rate of 98.5%. When the same image on the succeeding trial was imagined by S1, we were able to recognize the imagined image correctly 83.0% of the time. We show the p-value for the null hypothesis, which is that what was recognized, was recognized just by chance. In the second group of columns, we show the results of applying the SL. We get some improvements, as you can see, by comparing the imagining results for the SL versus the results for the unfiltered data. When we go on to add ICA, that is, independent component analysis, together with the Laplacian, in the third group of columns, then there is an almost uniform improvement in results. Notice that we did not even show the results for the

Table 9.1 Results for the "Stop-Go" Experiment

Subject	Unfiltered data			Surface Laplacian			Laplacian+ICA	
	STIM	IMAG-SAME		STIM	IMAG-SAME		STIM	IMAG-SAME
S1	99.5	83.0	<0.0001	99.3	81.2	<0.0001		83.7	<0.0001
S2	97.8	60.3	0.0462	99.7	57.7	0.1225		65.8	0.0067
S3	97.8	65.5	0.0067	99.2	66.7	0.0031		69.5	0.0013
S4	99.7	67.8	0.0031	99.5	67.8	0.0031		68.7	0.0013
S5	96.3	64.7	0.0137	97.5	67.5	0.0031		69.7	0.0013
S6	95.2	54.2	0.2595	96.7	59.5	0.0775		60.2	0.0462
S7	97.3	60.3	0.0462	98.8	69.8	0.0013		71.5	0.0005
		IMAGE-DIFF			IMAGE-DIFF			IMAGE-DIFF	
S8	97.2	73.2	0.0002	98.2	76.3	<0.0001		75.3	<0.0001
S9	97.8	58.2	0.1225	99.3	60.3	0.0462		60.8	0.0462
S10	98.7	75.2	<0.0001	99.3	79.5	<0.0001		80.8	<0.0001
S11	97.8	68.3	0.0013	99.0	68.7	0.0013		67.8	0.0031

Abbreviations in the table: STIM = stimulus,
imagine the same stimulus of the last trial = IMAG-SAME,
imagine the stimulus different from that of the last trial = IMAG-DIFF.

stimulus after adding ICA, because the results were already so good. The corresponding results for imaging the opposite of the presented stimulus are shown in the bottom part of Table 9.1 for Subjects S8–S11. These imagined images are, of course, of great interest conceptually.

The general summary of the results here is that the results of the classification, in response to an image generated by the two stimuli are as good as one could expect even for a very strong BCI requirement. As would be expected, the results for images generated internally by the subject are not as easy to classify, but many of the results are significant, and in the case of two of the subjects, even quite significant for the hardest task, generating an image that corresponds to the opposite of the stimulus just seen or heard. I am referring here to the results seen for subject S8 of 75.3% and S10 of 80.8% correct. In the long run, these are what are important in understanding better how the brain imagines images.

EXPERIMENT 2: LANGUAGE AND THE BRAIN

I describe briefly here one of our most detailed and, in many ways, successful experiments on language in the brain. In a broad sense, I and the members of the group I work with on such brain research tend to think of ourselves as cryptologists of brain language. When speech is either produced or listened to, representations in essentially real time need to take place in the brain, and the question is this: Do we look on this as another way of encrypting language, in our case usually English, and can we as cryptologists decode it? That is a way of putting it that makes a lot of sense, although the kind of cryptology here is of quite a special kind. One of the things that we would expect to find, in broad terms, is some kind of structural isomorphism between the structure of speech as it is heard ordinarily and as it is represented in the brain. One question might be why should there be a structural isomorphism? I think that some reflection on the problem makes it pretty obvious that such an isomorphism is needed to make it easy to understand each other's speech. Moreover, there is some reason for thinking that such an isomorphism will have aspects that are tighter than, for example, the translation of English into Chinese. There are many aspects of spoken and written language that have, from a broad perspective across languages, different orderings of the principle constituents of sentences –verbs, nouns, adjectives, adverbs, and so forth –and in making a translation from one to the other, this order is disturbed. One of the problems of translating, say, complicated English sentences into Chinese is how to determine the appropriate word order in Chinese. Well, there is reason to think that we do not need to worry so much about word order in the brain representation of a sentence, whether it be in English or in Chinese. There are many reasons to argue for a roughly real-time one-to-one correspondence with the sounds and syllables of each particular language. If word W occurs in English before W' in a sentence, then in most cases I would expect this also to be true of the brain representation of that sentence, and so it goes for other languages. This is also a reason to believe that the preoccupation with translation into first-order logic is not a good route to understanding a natural language, because the structure of the syntax of first-order logic is so radically different from that of natural languages.

On the other hand, to expect a really mathematically rigorous structural isomorphism that we can discover at the present time is probably too demanding. Psychologists over the past century and a half have a weaker substitute that also exists in mathematics. This is a principle notion of similarity. For example, weaker than congruence of triangles in geometry is similarity. In Euclidean geometry, this means simply eliminating considerations of size and only considering shape. So two Euclidean triangles are similar if and only if they have the same shape; their sizes may be entirely different. Recognizing that a weaker notion was used, psychologists began studying the notion of similarity as a natural psychological notion, essential to the analysis of perception in the 19th century. Such a weaker concept of similarity, or its companion, dissimilarity, is appropriate for our current, relatively weak understanding of the representation of brain language. The experimental results presented here are part of a larger study involving words as well, but here I show a small introductory piece focused on four initial consonants, that is, the four phonemes p, t, b, and g. The larger study of similarity, including the theory of semiorders with thresholds, has been published in (Suppes et al. 2009).

Table 9.2 Data of Perceptual and Brain Confusion Matrices with Perceptual Data on the Left for Each Entry and Brain Data on the Right

	p^1	t^1	b^1	g^1
p^2	51/36	53/38	2/12	2/14
t^2	64/34	57/50	2/3	1/13
b^2	4/14	2/31	60/42	18/13
g^2	3/10	1/33	20/15	29/42

Fortunately, we have a beautiful study from the 1950s by George Miller and a colleague (Miller and Nicely 1955) of these very phonemes, as part of a larger set studied perceptually. Table 9.2 shows the confusion matrix for the perceptual data they obtained. The prototypes are shown on the rows and the responses on the top, so the diagonal gives you a sense of how successfully each phoneme was perceptually recognized, and the off-diagonal elements show the extent of confusion. The corresponding brain data from our experiments, which I will not describe in detail but which produced exactly the same kind of confusion matrix, are also shown in Table 9.2.

Notice that in both confusion matrices, p is most often confused with t (53 times), as evidence of the perceptual (38 times) and brain-data similarity. The same thing is true of t being confused with p (row 2, 64/34). Actually in the brain data you may notice that t is a kind of universal attractor, because in the off-diagonal elements for both b and g, t is the most frequently selected.

Here are the steps of analyzing these confusion matrices, both for perception and the brain. Note that in coupling these two together, I am doing so deliberately, because what I want to compare is the common similarities between the perceptual and the brain data, the kind of generalized structural isomorphism we would hope to find, now exemplified by similarity rather than by some simpler and stronger concept of structuralized isomorphism or congruence. So here are the steps of analysis:

1. Normalize each row of a confusion matrix to probabilities summing to 1. In other words, we have a conditional density thereby for each row of each of the matrices. To be explicit, these conditional probabilities have the following form:
2. The conditional probability

$p(o_j^+ | o_i^-)$ = Prob(Prototype j | Test sample i)

= the measure of similarity of prototype j and test sample i

For example,

$p(t^+ | t^-) = .50$

$p(t^+ | b) = .03$

So, to summarize, the three steps are, we form the conditional probability densities.

$$p(o_j^+ | o_i^-) \geq 0 \text{ and } \sum_{j=1}^{N} p(o_j^+ | o_i^-) = 1$$

Then we estimate from confusion matrix.

$$\hat{p}(o_j^+ | o_i^-) = \frac{m_{ij}}{\sum_j m_{ij}}$$

And third, the ordinal similarity differences are defined in terms of inequalities of these similarities.

$$o_j^+|o_i^- > o_{j'}^+|o_{i'}^- \text{ iff } p(o_j^+|o_i^-) > p(o_{j'}^+|o_{i'}^-)$$

To continue this development, to move towards a qualititative notion of similarity or, as we would say in measurement theory, an ordinal sense of similarity, let R be irreflexive on A. Then R is an *interval order* on A iff for every a, b, c, and d in A, if aRb and cRd, then aRd or cRb. R is *strongly transitive* on A iff for every a, b, c, and d in A, if aRb and bRc then either dRc or aRd. R is a *semiorder* on A iff R is irreflexive, strongly transitive, and an interval order on A.

I note that the algebraic notion of a semiorder corresponds to the concept of threshold introduced in the 19th century in the early psychological work of Fechner and others. We can then state a representation for finite semiorders—meaning that set A is finite—which has the following simple form:

Any finite semiorder $(A,<)$ has a numerical representation ϕ with a positive threshold such that for a and b in A

$\phi(a) + 0.02 < \phi(b)$ iff $a < b$,

which in our earlier notation reads as follows:

$$o_j^+|o_i^- \approx o_{j'}^+|o_{i'}^- \text{ iff } \left| p(o_j^+|o_i^-) - p(o_{j'}^+|o_{i'}^-) \right| \leq 0.02$$

$$o_j^+|o_i^- > o_{j'}^+|o_{i'}^- \text{ iff } p(o_j^+|o_i^-) > p(o_{j'}^+|o_{i'}^-) + 0.02$$

Table 9.3 Conditional Probability Matrices for Miller-Nicely Perception Experiment and the Brain Experiment, with Perception Probabilities on the Left and Brain Data on the Right

	p	t	b	g
p	0.47/0.36	0.49/0.38	0.02/0.12	0.02/0.14
t	0.52/0.34	0.46/0.50	0.02/0.03	0.01/0.13
b	0.05/0.14	0.02/0.31	0.71/0.42	0.21/0.13
g	0.06/0.10	0.02/0.33	0.38/0.15	0.55/0.42

In Table 9.3, we show the conditional probability matrix for the Miller-Nicely Confusion Matrix. Each row sums to 1, as a discrete probability density function. The rows are for the four test samples – phonemes on the left and brain representation of phonemes on the right. The columns on the left are for the behavioral classification responses and on the right for the prototypes of the brain representations.

Now we want to mention here that the steps of analysis of the EEG data, that is, the EEG recordings from which we produce the brain-data confusion matrix, goes through steps of analysis very similar to what was discussed for the first experiment, so I will just give a quick overview of the Fourier analysis used in the present case. To read the full details, I will refer you to the paper on these experiments (Suppes et al. 2009). In a series of eight steps:

1. Average over trials: half for prototypes, other half for test samples.
2. Mirror and smooth with a Gaussian function.
3. Use fast Fourier transforms (FFTs) applied to prototypes and test samples.
4. Filter the results with a fourth-order Butterworth bandpass filter having parameters (L, H).
5. Inverse FFT.
6. Classify by least-squares criterion, for observations in a given temporal interval (s, e).
7. Repeat (4)–(6) with new parameters (L′, H′, s′, e′) from a set of values selected on the basis of past experience until the set is exhausted.
8. Select best recognition performance and corresponding parameters.

Now we can compare the semiorder graphs based upon the qualitative orderings, as shown in Figure 9.3.

On the left is shown the Miller-Nicely data and on the right brain data from a single subject. Of course, completely different subjects were used for the brain experiment and in the Miller-Nicely one.

Now to get the invariant partial order of similarity between the Miller-Nicely perceptual data and our brain data, we intersect the two semiorders to obtain Figure 9.4.

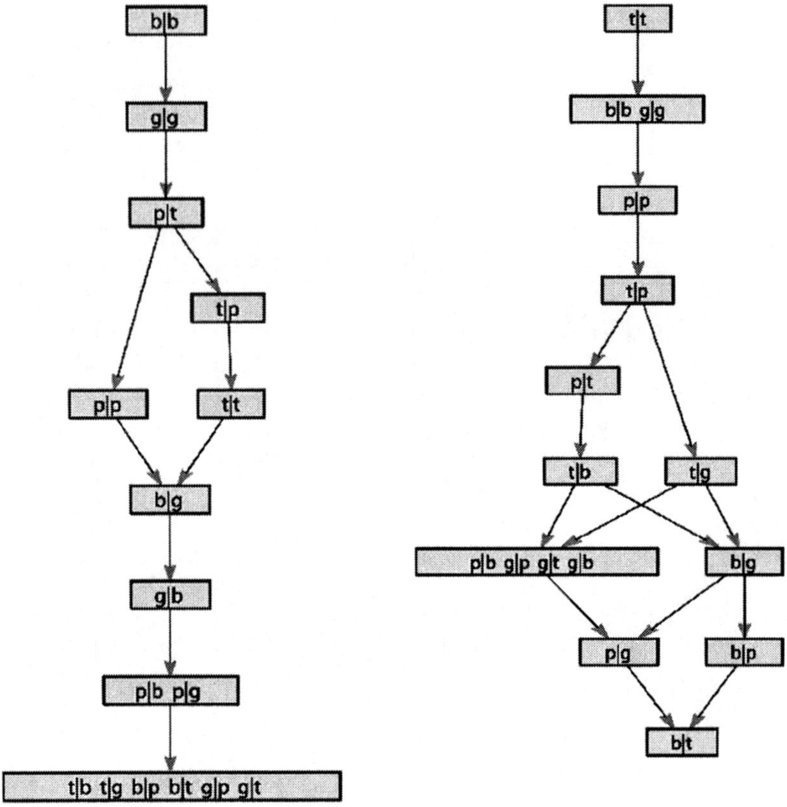

198 *Patrick Suppes*

Figure 9.3 Miller-Nicely data on the left and brain data on the right.

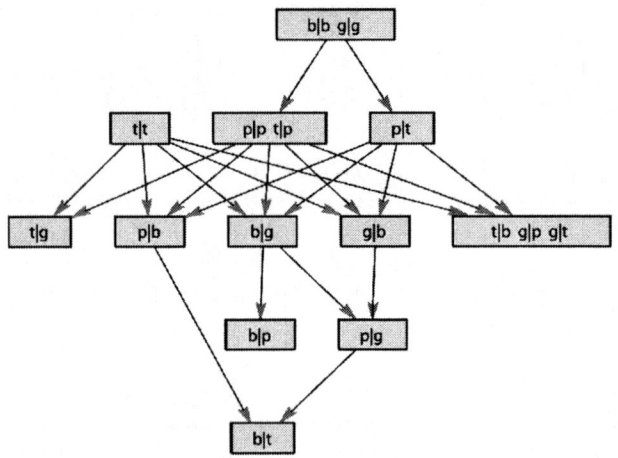

Figure 9.4 Invariant partial order that is the intersection of the perceptual and brain-data semiorders.

Models and Simulations in Brain Experiments 199

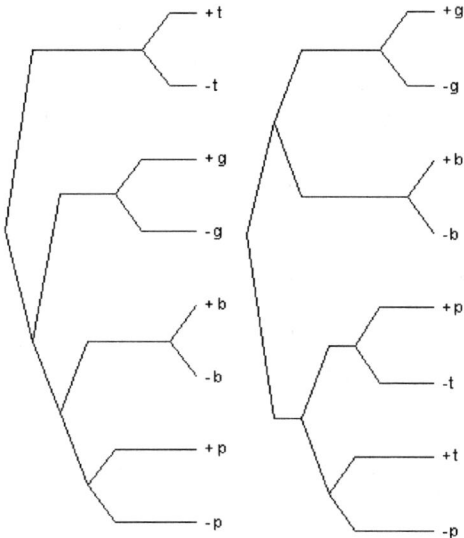

Figure 9.5 Trees derived from the two semiorders, one perceptual and the other brain data, with the perceptual (Miller and Nicely 1955) results on the left and the brain results on the right.

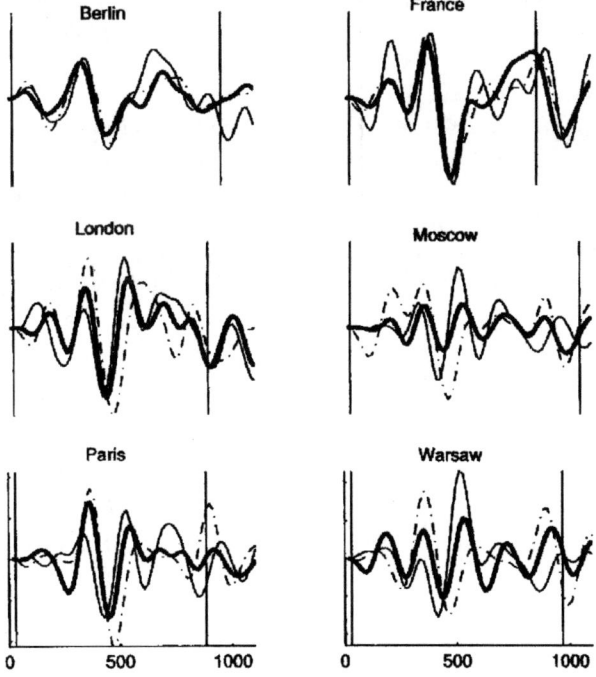

Figure 9.6 Prototypes and test samples of brain waves of six names of capitals or countries.

Notice three features of this intersection: robust discrimination, both in perception and in the brain, of unvoiced *p* versus voiced *b*; second, the strong attraction between *p* and *t* already noted in the confusion matrices; and third, the strong similarity for prototypes and test samples for each phoneme, both in perception and in the brain data.

In Figure 9.5, we show the trees that we can construct from the semiorders for the Miller-Nicely experiment and for our brain experiment, with the Miller-Nicely data shown on the left and our brain data shown on the right.

This is for the same subject mentioned earlier, as far as the brain data are concerned. Notice that the brain representation has more expected similarities than does the perceptual representation. I will not describe the construction of these trees from the semiorders; the details of this are spelled out in the paper on this experiment already referred to.

Just to give you a sense of the other part of this large experiment, I show in Figure 9.6 an analysis of prototypes and test samples for six of the nouns that were used, in sentences such as "Berlin is the capital of Germany" or "Moscow is North of London." The prototypes are in the solid black, and the thin black lines are the average test samples. Agreement is good but could certainly be better. Ideally we would like it so.

EXPERIMENT 3: SIMULATION OF STIMULUS-RESPONSE MODELS BY WEAKLY-COUPLED PHASE OSCILLATORS

The simulation I discuss is a detailed, rather technical piece of work with two physicists, José Acacio de Barros and Gary Oas. The basic idea is simple and easy to state. Among behavioral theories of learning and performance, stimulus-response theory (SR) is notable for the simplicity and formal clarity of its concepts, especially as developed by William K. Estes and his collaborators, beginning with a seminal paper by (Estes 1950).

The formal clarity of the basic concepts of stimulus sampling, responses, reinforcement of responses, and conditioning of responses to stimuli permit mathematically rigorous derivations of response probabilities for a great variety of experiments. For axiomatic analysis of the foundations, see (Estes and Suppes 1959a, 1959b), and a reduced version of the latter in (Estes and Suppes 1974), and for a rather recent review of many associated foundational topics, (Suppes 2002, Chapter 8). For a good survey of the experimental range of SR theory, see (Neimark and Estes 1967).

Both the strength and the weakness of SR theory depend on the abstractness of its concepts, as can be seen in the following informal statement of the basic axioms.

Response Axioms

- If the sampled stimulus is conditioned to some response, then that response is made.

- If the sampled stimulus is not conditioned to any response, then the probability of any response is a constant guessing probability that is independent of the trial number and of any preceding pattern of events.

Conditioning and Reinforcement Axioms

- On every trial each stimulus is conditioned to at most one response.
- The probability is c of any sampled stimulus becoming conditioned to the reinforced response if it is not already so conditioned, and this probability is independent of the particular response, trial number, or any preceding pattern of events.
- The conditioning of all stimuli remains the same if no response is reinforced, and the conditioning of unsampled stimuli does not change.

The intuitive content of each axiom seems simple and understandable at some general intuitive level. Such general explanations have an ancient pedigree.

Since Plato and Aristotle, philosophers have puzzled over these ideas, one of the better developments being Aristotle's theory of sensible and intelligible forms developed in *De Anima*, and elaborated further in the Arabic and Latin Commentaries for well more than 1000 years—Aquinas' famous commentary was written about 1500 years later. The structural isomorphism implicit in these ideas is underdeveloped in modern cognitive science but is a natural topic to consider in thinking about how the brain works. Here the metaphors of traditional philosophy of perception are not very helpful. For example, the idea of there being something simple like direct perception seems naïve when confronted by the known complexities of processing a perception of any one of the five senses.

Before turning to the brain, a different but equally enlightening perspective on SR theory can be found in the theory of computation. It is a fundamental insight of the 20th century that the general theory of computation can be based on remarkably simple devices. This was above all the fundamental discovery of Alan Turing. Following in this tradition, in (Suppes 1969), I proved such a theorem on computation for SR theory by showing that for finite automata structurally isomorphic SR models can be constructed, and I extended this in (Suppes 2002, Chapter 8) to universal register machines whose computational power is equivalent to that of universal Turing machines.

Still, from a neuroscience perspective, these computational results are, like Aristotle's concept of intelligible form, or SR theory's concepts of response and reinforcement, also too abstract. The theorems do not show what particular isomorphic structures are physically realized in the brain. The theorems do show how simple the fundamental process can be in universal Turing machines, such a register machine, or also generalizations of conditioning to associative learning models or networks, but when left at this level of abstraction, they are no closer to actual brain activity than Aristotle's cognitive theory.

Before turning to the brain, there is one other comment that should be made about SR theory. The physical description of responses and reinforcements is totally omitted in the theory, but the experimental practice has been to select responses and reinforcements that have easily recognizable, although rather general, ordinary physical descriptions that require no use of physics. In this discussion so far, I have omitted the concept of stimulus. In many if not all of the experimental applications of SR theory, the stimuli are not observable. Moreover, as the theory is ordinarily spelled out in detail, simplifying statistical assumptions are made so that only the number of stimuli presented in an experiment, or sampled on a given trial, is relevant computationally, not, the actual stimuli themselves. But, it is not necessary to develop these ideas here. What is important is to recognize that in general in SR theory there is a difference in the assumed observability of a response, as compared to a stimulus.

Fortunately, there are now many ways of trying to build more physically concrete simulations of how the brain computes. Or, put in more general, but for me, essentially equivalent terms, how the brain works. The broad answer seems to be that oscillations of many kinds are found in the brain, and in many other biological structures. From oscillations to oscillators as mechanisms for producing them is a natural step. It is then the synchronization of oscillators that, in this physical model, corresponds to the state of conditioning in SR theory or association, in the more general framework of associative learning.

But with physical embodiment, especially in the brain images of stimuli and responses, many details obviously need to be spelled out. For example, how is conditioning represented in the brain? As already remarked, in SR theory there is a clear behavioral difference between the direct observation of responses and reinforcements as the most obvious observables, on the one hand, and the conditioning parameter c, on the other. Clearly, this parameter is in no sense directly observable. So, even if we expect to find brain images of some kind of responses and reinforcements, we certainly have no such direct expectation about finding anything like a brain image of the conditioning parameter c, which is, I emphasize, by far the most important parameter of SR theory in the formulation given above. The point is that we do not expect to find a direct representation of this parameter in the brain. Yet, we need to include it in some essential way in the oscillator model, which we do with a mathematical equation written in terms of oscillator parameters.

The simulation of SR theory set up by de Barros, Oas, and me made the physics explicit, including the effort to give biologically based estimates of the many physical parameters that must be given values for simulation computations. In this simulation, each stimulus is represented by an oscillator, as is each response and reinforcement. Oscillators generally have three parameters, a fundamental frequency ω, a phase φ, and an amplitude A. Synchronization of oscillators can occur by resonance and thus transfer of energy, as in the case of marching soldiers on a bridge leading to its collapse. For purposes of brain computations, we consider only phase synchronization of oscillators and ignore amplitude. So the simulation uses weakly coupled phase oscillators. The weak coupling constants are the learning mechanisms needed for oscillators to become synchronized. In the simulator all these matters are described mathematically by the kind of differential equations familiar in physics. I emphasize that equations describing all the important details of real biological oscillators are still not feasible or are very difficult. But the intuitive physical ideas of a qualitative kind are more robust. We can all understand oscillators. Their synchronization is involved in an endless variety of biological phenomena from the synchronized chirping of thousands of crickets to the circadian rhythms of sleep and wakefulness characteristic of most mammals, including humans. Our simple, rather artificial simulation of SR theory is just meant to show how we can think about brain computations in a concrete, physical way.

In the N-stimulus oscillator model, restricted to two responses, each sampled stimulus is represented by one oscillator s_j (j=1, ... ,N), the two possible responses by oscillators r_1 and r_2, and reinforcements of r_1 and r_2 by oscillators e_1 and e_2 (to avoid confusion, whenever possible we will use lowercase letters for oscillators, e.g., response oscillator r_1, and uppercase for SR concepts, e.g., response R_1). At the beginning of an experiment, we assume all the oscillators mentioned are connected by at least weak random couplings. When a stimulus S_n is sampled with probability 1/N, we assume its corresponding phase oscillator s_j activates, i.e., the neurons of this oscillator start to fire synchronously. This activation is followed by the spreading of activation to the two response oscillators. After any given time interval, the dynamics leads to changes in the phase relations between the stimulus and the response oscillators. Depending on the couplings, the stimulus and response oscillators may phase lock with fixed phase relations. Two oscillators are phase locked if they keep a constant phase relation between each other. In this way, depending on the couplings, after a finite amount of time one response may become in phase with the stimulus while the other dephases. Shortly after a response, during reinforcement, we assume that a neural oscillator representing the reinforcement disrupts the dynamics of the stimulus and response oscillators, forcing them to phase lock to the reinforcement oscillator with a phase difference of up to π and to change their couplings accordingly. The new reinforced couplings may result in a dynamics that yields new phase relationships between oscillators, thus changing the response to a stimulus. This sequence of neural events closely follows the sequence of behavioral events.

We now turn to the mathematical description of these qualitative ideas. We assume that stimulus and response neural oscillators have the same natural frequency ω_0 such that their phases are $\varphi(t) = \omega_0 t +$ constant, when they are not interacting with other oscillators. We also assume that they are weakly coupled to each other with symmetric couplings. The assumption of coupling between response oscillators is a detailed feature that has no direct correspondence in the SR model. Let $t_{s,n}$ be the time at which the stimulus oscillator s_j is activated on trial n and Δt_r the mean amount of time it takes for a response oscillator to phase lock. The assumption of the same frequency for stimulus and response oscillators is not necessary, as oscillators with different natural frequency can entrain (Kuramoto 1984), but it simplifies our analysis, as we focus on phase locking. In a real biological system, however, we should expect different neural oscillators to have different frequencies.

At the beginning of a trial, a stimulus is sampled, and once its corresponding oscillator, s_j, along with r_1 and r_2, are activated, the phase of each oscillator resets according to a normal distribution centered on zero. After the stimulus is sampled, the three active oscillators evolve for the time interval Δt_r according to the following set of differential equations, known as the Kuramoto equations (Hoppensteadt and Izhikevich 1996; Kuramoto 1984).

$$\frac{d\varphi_{s_j}}{dt} = \omega_0 - k_{s_j,r_1} \sin\left(\varphi_{s_j} - \varphi_{r_1}\right) - k_{s_j,r_2} \sin\left(\varphi_{s_j} - \varphi_{r_2}\right)$$

$$\frac{d\varphi_{r_1}}{dt} = \omega_0 - k_{s_j,r_1} \sin\left(\varphi_{r_1} - \varphi_{s_j}\right) - k_{r_1,r_2} \sin\left(\varphi_{r_1} - \varphi_{r_2}\right),$$

$$\frac{d\varphi_{r_2}}{dt} = \omega_0 - k_{s_j,r_2} \sin\left(\varphi_{r_2} - \varphi_{s_j}\right) - k_{r_1,r_2} \sin\left(\varphi_{r_2} - \varphi_{r_1}\right),$$

where $\varphi_{s_j}, \varphi_{r_1}$, and φ_{r_2} are their phases, ω_0 their natural frequency, and $k_{s_1,r_1}, \ldots, k_{s_N,r_2}, k_{r_1,r_2}$ are the 2N+1 coupling strengths between oscillators Elaboration of further details is beyond the scope of this general account of this simulation, and can be found in (Suppes et al., forthcoming). The next important step to take is to introduce further differential equations to describe the change of the coupling strengths under reinforcement –the principal method of learning in this model. Numerical solutions of the equations are then computed to compare with behavioral data. The quantitative results, which I do not present here, are encouraging if not perfect.

CONCLUDING SPECULATION

In the current primitive state of our understanding of how language operates in the brain, the variety of experiments that may be relevant to advancing our understanding is very large. The first two experiments described above show some of the possibilities, but their limitations in terms of providing a general approach are evident. If we think of ourselves as cryptologists of the brain trying to decode the brain representation of language, we are just at the beginning of making serious progress. The simulation that constitutes the third example approaches the problem of how the brain works at a more general level. But if the oscillator representation of SR theory can be brought to a more finished stage, earlier results on language computation in SR theory, for example, (Suppes 1969a,b), could provide direct access to language computation in the brain. I do want to emphasize, however, that we do have some distance to go to reach such a goal.

Neuroscientists of this century will not be content until they understand thoroughly how the human brain physically produces and comprehends language. Each hour of every day, billions of words are being spoken and understood with little difficulty. Yet these speakers and their listeners can be found in remarkably different physical environments, from quiet rooms to shouting marketplaces. That speech can be understood so well and so easily in such a robust variety of physical circumstances is scientifically important and justifies optimism about the possibility of our coming to a deep understanding of the processes of production and comprehension.

The great robustness of the parameters of speech to noise and many other kinds of physical interference strongly suggests we will find a robust solution, even if the details turn out to be awesomely complex. For example, we may be able to model the various stages of processing spoken language from the cochlea to the cortex using a variety of neural networks that depend on so many physical parameters that an intuitive understanding of exactly how they work will not be possible. As a wartime meteorologist of many years past, I find the resemblance between the physical complexity of the brain and the world's weather system a cautionary warning. Like the weather, the brain, too, may be a physical system we come to understand much about. But the possibility of having a detailed microscopic grasp is blocked by the kind of chaos seen in the highly simplified convection model of the atmosphere (see Lorenz equations 1963), which describe a three-parameter family of three ordinary differential equations. The subtle nature of these equations is thoroughly explored in (Sparrow 1982). It is not unreasonable to think that the physical theory of how spoken language moves through the various stages of the auditory system on to the auditory nerve fibers, and finally reaching the cortex, is not any less complicated than the weather.

REFERENCES

Bell, Anthony J., and Terrence J. Sejnowski (1995). "An Information Separation Approach to Blind Separation and Blind Deconvolution." *Neural Computation* 7(6): 1129–1159.

Estes, William K. (1950). "Toward a Statistical Theory of Learning." *Psychological Review* 57: 94–107.
Estes, William K., and Patrick Suppes (1959a). "Foundations of Linear Models." In *Studies in Mathematical Learning Theory*, eds. Robert R. Bush and William K. Estes, 137–179. Stanford, CA: Stanford University Press.
Estes, William K., and Patrick Suppes (1959b). "Foundations of Statistical Learning Theory, II: The Stimulus Sample Model." Technical Report No. 26. Institute of Mathematical Studies in the Social Sciences. Stanford, CA: Stanford University Press.
Estes, William K., and Patrick Suppes (1974). "Foundations of Stimulus Sampling Theory." In *Contemporary Developments in Mathematical Psychology, Vol. 1: Learning, Memory, Thinking*, eds. David H. Krantz, Richard C. Atkinson, R. Duncan Luce, and Patrick Suppes, 163–183. San Francisco: Freeman.
Hoppensteadt, Frank C., and Eugene M. Izhikevich (1996a). "Synaptic Organizations and Dynamical Properties of Weakly Connected Neural Oscillators." *Biological Cybernetics* 75:117–127.
Hoppensteadt, Frank C., and Eugene M. Izhikevich (1996b). "Synaptic Organizations and Dynamical Properties of Weakly Connected Neural Oscillators II. Learning Phase Information." *Biological Cybernetics* 75:129–135.
James, William (1890). *Principles of Psychology*. New York: Holt.
Kuramoto, Yoshiki (1984). *Chemical Oscillations, Waves, and Turbulence*. Mineola, New York: Dover Publications, Inc.
Lorenz, Edward N., (1963). "Deterministic Non-Periodic Flows." *Journal of Atmospheric Sciences* 20:130–141.
Miller, George A., and Patricia E. Nicely (1955). "An Analysis of Perceptual Confusions among Some English Consonants." *Journal of the Acoustical Society of America* 27: 338–352.
Neimark, Edith D., and William K. Estes (1967). *Stimulus Sampling Theory*. San Francisco: Holden-Day.
Sparrow, Colin (1982). *The Lorenz Equations: Bifurcations, Chaos, and Strange Attractors*. New York: Springer-Verlag.
Suppes, Patrick (1969). "Stimulus Response Theory of Finite Automata." *Journal of Mathematical Psychology* 6:3: 327–355.
Suppes, Patrick (2002). *Representation and Invariance of Scientific Structures*. Stanford, CA: CSLI Publications.
Suppes, Patrick, José Acacio de Barros, and Gary Oas (Forthcoming). "Neural Phase Oscillator Representations of Behavioral Stimulus-Response Models." Journal of Mathematical Psychology.
Suppes, Patrick, Marcos Perreau-Guimaraes, and Dik Kin Wong (2009). "Partial Orders of Similarity Differences Invariant between EEG-Recorded Brain and Perceptual Representations of Language. Neural Computation 21: 3228–3269.

Part II
Representation

10 Representing with Physical Models

Ronald N. Giere[1]

INTRODUCTION

Every student of scientific practices knows that physical models have long been used to represent a great many things.[2] By and large, however, philosophers of science have taken for granted the representational powers of physical models. They have focused on more ubiquitous and seemingly more important theoretical models, particularly those found in mathematical physics. In this paper, I focus on physical models, comparing them first with theoretical models and finally with recently popular computational models. My aim is to show that the representational aspects of models used in science are fundamentally the same across these three categories of models.[3]

REPRESENTING WITH THEORETICAL MODELS[4]

My paradigm of a theoretical model has been a model in classical physics, something described by a few differential equations. Canonical examples include the simple harmonic oscillator and a two-body gravitational system. The many questions that can be asked about such models include some that are ontological and others that are functional. What kind of thing are theoretical models? For what can they be used?[5]

I have long thought of theoretical models as being "abstract entities" in the minimal sense of merely not being concrete (Giere 1988, 1999, 2006). A concrete simple harmonic oscillator (which neglects friction) is a physical impossibility. Because they are described using mathematics, many have assumed that such models have the status of mathematical entities; however, that is to be understood (Thompson-Jones 2010). I think this is a mistake for the simple reason that theoretical models are interpreted. There is a world of difference between the number five, a mass of five grams, and a length of five centimeters. It is simplest, I think, to regard theoretical models as having the same ontological status as imaginary entities (Giere 2009). Whatever status that might be, it cannot be too mysterious since creating imaginary objects is something that comes naturally to evolved, language-using creatures like us. A three-year-old child can do it. Thus, one can imagine a flying horse, though such a being is presumably physically impossible. There are even paintings and tapestries showing flying horses, just as there are drawings of simple harmonic oscillators. But there are no photographs of such.

One of the primary functions of theoretical models, if not the primary function, is to play a role in representing aspects of the world. The question is what role they play and how they succeed in fulfilling this role. A standard account has it that theoretical models play the role of representing things in the world in the way that words and sentences are thought to represent things in the world. In the simplest case, words refer to objects or properties, and sentences are true or false according to whether the designated objects exhibit the indicated properties. This simple picture does not work for theoretical models.[6]

The problem is that the model is an imaginary object that cannot exist in the real world, while the thing being represented is a concrete object in the real world. The relationships of reference and truth don't apply between two objects, one imaginary and the other concrete. And isomorphism, being a relationship between logical/mathematical structures, is not defined for such a pair. Any possible counterpart to isomorphism would presumably be some kind of perfect matching between the two objects, but that cannot be since one is physically impossible and the other actual. We are left with less well-defined relationships such as similarity, resemblance, likeness, or fit. For ease of expression, I will invoke similarity, though any of the others would do as well.

One problem is that nothing is just similar to anything else. Or, alternatively, everything is similar to everything else in some respect or other. What can be better defined is a notion of "similarity with respect to . . . " And for any given respect, one can have degrees of similarity, whether qualitative or even quantitative. So the relationship we are after could be called "selective similarity." But what does the selecting? The imaginary and concrete objects by themselves do not pick out any particular feature with respect to which they may be said to be similar. Another problem, as is often noted (Suárez 2003), is that (selective) similarity is a symmetrical relationship, whereas representation is asymmetric.

These difficulties can all be resolved by introducing a third element, an intentional (purposeful) agent that selects the relevant features and desired degrees of similarity, and breaks the symmetry by indicating that the imaginary object is being used to represent the concrete object. On this view, representing with theoretical models turns out to be a matter of an intentional agent using a model to represent a concrete system. Now let us see whether we reach the same conclusion regarding representing with physical models.

REPRESENTING WITH PHYSICAL MODELS

Watson and Crick's physical model of deoxyribonucleic acid (DNA) became the iconic physical model of the 20th century.[7] Since this model was manifestly not an imaginary object, but a concrete object, the ontology of physical models is different from that of theoretical models. Of course, one could imagine a physical model, which would then be an imaginary object, but unlike theoretical models, it could be fully realized as a concrete physical object.[8] What matters for representation, however, is not ontology but function. As for theoretical models, the question about physical models is what role they play in representation and how they succeed in fulfilling this role. So does an examination of the representational function of physical models lead us to conclusions similar to those reached for theoretical models?

Again following standard theories of linguistic representation, it is tempting to think that the Watson and Crick model represents a DNA molecule, or, to put it more tendentiously, that there is a representational relationship between the two objects in and of themselves. One cannot now deny that the model resembles or is similar to actual DNA molecules, that is, that there is a (strong) similarity relationship between the two objects. But again, similarity is neither necessary nor sufficient for representation.

That similarity is not strictly necessary; even the case of physical models is shown trivially by the fact that one can arbitrarily designate physical objects to represent other physical objects. At a pub, Crick might place two pencils in parallel, pointed in opposite directions, and say these represent the sugar phosphate backbone of a DNA molecule. There is some residual similarity here. But if he then places matching salt and pepper shakers next to each other between the pencils and declares that the salt shaker represents the base adenine and the pepper thymine, this designation is completely arbitrary. Nevertheless, I can think of no case where an established type of physical model is employed in science without being similar (in particular respects, of course) to the kind of object for which it is standardly used, as in this iconic DNA example. In fact, it turns out to be one of the virtues of focusing on physical models that they highlight the role of similarity in scientific representation, thus providing a strong contrast with linguistic representation where there is rarely any similarity between word and object.

From now on I will presume there is some relevant similarity between a physical model and its target. But, as has now often been remarked, similarity is also not sufficient for representation since similarity is a symmetrical relationship, while representation is clearly asymmetrical. So we are again forced to consider what it takes, in addition to similarity, to establish a representational relationship between one physical object and another.

In the case of physical models, then, it is again fairly plausible to introduce the intentions of some agent to use the model in a representational context in order to differentiate between a model merely resembling some other object and representing that object. I have called this (somewhat tendentiously, I admit) "The Intentional Conception of Scientific Representation" (Giere 2010).[9] At its most basic, the idea is that models as physical objects by themselves don't do anything, including representing. It takes an agent to use the model to represent something. Note that this characterization of scientific representation cannot be a definition as it invokes a generalized notion of representation. I doubt that a reductive definition of so basic an activity as representing is possible.[10] But more can be said.

It requires no argument to realize that representing using physical models is selective. Not all features of a model will have counterparts in the target system, and the target will have many features not represented. The Watson and Crick model had some pieces made of tin that obviously has no counterpart in a DNA molecule. We can now summarize how representing using physical models works. It works mainly by selective similarity, where the selecting is done by the agent employing the model.[11] And selection is context sensitive. In some contexts it might be important to emphasize the base pairings; in others the reverse direction of the backbone.

REPRESENTING WITH COMPUTATIONAL MODELS

Philosophers of science are only gradually becoming aware of the overwhelming changes in the way science is practiced due to the employment of computers in science (Humphreys 2004). The vast majority of models now used in the sciences exist primarily as computational models. Here I will examine just one example, chosen partly because it nicely fits with a discussion of physical models.

The Swiss Institute of Bioinformatics at the Biozentrum in Basel has developed a server with information that can be used to construct models of proteins (SWISS-MODEL). They have also developed a PdbViewer (DeepView) that automatically converts information about proteins into images that can be manipulated on screen.[12] One can rotate and zoom in and out to view aspects of a model. The viewer will even produce side-by-side images that can be seen in dramatic three dimensionality without special glasses (this takes a little practice). How are we to understand these models?

Behind the images of protein molecules, so to speak, there is a large database and an elaborate program, which also incorporate theoretical structures. It is such things that tend to attract the interest of those who think about computational models. Indeed, scientists and philosophers often identify models themselves with equations or code. These are what one works with when constructing a model. But this identification is a category mistake. It is like identifying a description with the thing described or a definition with what is defined. In scientific practice this conceptual mistake has few consequences, but it is important to be clear about the distinction for the purpose of constructing a higher-level understanding of the practice of computational modeling.

What then is the model? Like theoretical models, it is an abstract object we can identify simply as whatever abstract object is defined by the equations and code. It is because of this easy identification that distinguishing the description of the model from the model itself has little practical importance. Since computational models are theoretical models, representing with computational models should be like representing with theoretical models. It may be said to work because intentional agents invoke selective similarities between the model and those aspects of the world represented.

In working with computational models, the users invoke various partial realizations of the model, including especially the visual images generated using the overall model. These are particularly important since the images function as the primary interface between the model and model users. One can think of them as two-dimensional projections of aspects of the abstract model. More controversially, I suggest we think of these images themselves as another form of physical model, easier to manipulate, perhaps, but not different in principle.

As an intermediate step to grasping this suggestion, consider the physical models of atoms and molecules that are used in teaching chemistry. The first such "ball and stick" models were produced in the mid-19th century. Modern versions can now be purchased on the open market.[13] With these basic models, one can construct more complex chemical models roughly as one builds Lego structures (or objects with "tinker toys" for those of a certain age). The images produced by DeepView, then, can be thought of as pictures of model molecules constructed with these basic physical models. Just as one can turn the physical model over in one's hands, one can rotate the image on the screen to view it from different angles. In fact, the operative features of the human visual system employed are basically the same when viewing a computer-generated image and viewing a real physical model. That goes a long way toward explaining why the visual interface is so effective.

The reasons we had for thinking that the operative relationship between physical models and their targets is selective similarity seem to apply equally well to the derived images. There are many features of proteins not represented in the computational model and even fewer represented in the constructed images. What about the role of intentional agency in making the computer images represent something in the world? Here again the case seems even stronger than for other physical models. To some extent, most physical models have a life of their own. They may, for example, sit for long periods on a shelf without being employed in any representational task. This is far less true of computer images, which are constantly recreated by those using them in representational contexts. And, in active research, they are continually, and deliberately, modified to improve their representational virtues. For computer images, the users are never very far away.[14]

CONCLUSION

It seems possible to develop a unified understanding of representing using three major types of models found in the sciences: theoretical, physical, and computational. In all three cases, representing involves an intentional agent selectively invoking similarities between a model and an aspect of the world. Here I have been particularly concerned to bring physical models into the fold. That has the unexpected consequence of suggesting a new way of understanding aspects of the use of computational models.

NOTES

1. I am grateful to an anonymous reviewer for extensive comments on a previous draft that led to this paper being reoriented as well as substantially revised.
2. This practice, of course, was more common before the advent of computational modeling.
3. I was moved to think more about physical models by two recent volumes: (de Chandarevian and Hopwood 2004) and (Sterrett 2006). Neither, however, emphasizes the representational role of models.
4. This section is a capsule version of (Giere 2006, Chapter 4).
5. Explicit interest in ontological questions about models is relatively recent beginning with Cartwright (1999), followed by the essays in an issue of *Synthese*, edited by Gabriele Contessa, *Synthese* 172(2), 2010. The papers in (Morgan and Morrison 1999) describe various functions of models other than representational ones.
6. In fact, according to recent accounts of usage-based linguistics such as (Tomasello 2003), this account does not work for language either. I have elaborated this point in (Giere 2010).
7. Its eventual iconic status was obviously not initially recognized since the original was soon cannibalized for parts. A reconstructed version, including a few original pieces, now sits in the Science Museum in London.
8. This would be what Peter Godfrey-Smith (2006) has called an "imagined concrete thing."
9. This designation is meant to provide a direct contrast with Suárez' (2004) "Inferential Conception of Scientific Representation." This is not the place to engage in explicit criticism of this latter conception.
10. This judgment regarding the possibility of there being a reductive definition of representation is shared by Suárez (2003) and van Fraassen (2008, 7).
11. This view is shared by van Fraassen (2008).
12. One can get into the system at the following address: www.expasy.org/spdbv. Be sure to check out the link to a picture gallery of dramatic models.
13. For example, see www.indigo.com.
14. I realize, of course, that structural chemistry provides a very favorable case for thinking of images generated from a computational model as themselves models. Clearly not all outputs (e.g., a set of measurements) are models, but once one produces an image that can be manipulated, it becomes plausible to think of that image as itself a model.

REFERENCES

Cartwright, Nancy D. (1999). *The Dappled World: A Study of the Boundaries of Science*. Cambridge: Cambridge University Press.
de Chandarevian, Soraya, and Nick Hopwood, eds. (2004). *Models: The Third Dimension of Science*. Stanford, CA: Stanford University Press.
Giere, Ronald N. (1988). *Explaining Science: A Cognitive Approach*. Chicago: University of Chicago Press.
Giere, Ronald N. (1999). *Science without Laws*. Chicago: University of Chicago Press.
Giere, Ronald N. (2006). *Scientific Perspectivism*. Chicago: University of Chicago Press.
Giere, Ronald N. (2009). "Why Scientific Models Should Not be Regarded as Works of Fiction." In Fictions in Science: Philosophical Essays on Modeling and Idealization, ed. Mauricio Suárez, 248–258. New York: Routledge.
Giere, Ronald N. (2010). "An Agent-Based Conception of Models and Scientific Representation." *Synthese* 172(2): 269–281.
Godfrey-Smith, Peter (2006). "The Strategy of Model-Based Science." *Biology and Philosophy* 21(5): 725–740.
Humphreys, Paul (2004). *Extending Ourselves: Computational Science, Empiricism, and Scientific Method*. New York: Oxford University Press.
Morgan, Mary S., and Margaret Morrison, eds. (1999). *Models as Mediators: Perspectives on Natural and Social Science*. Cambridge: Cambridge University Press.
Sterrett, Susan G. (2006). *Wittgenstein Flies a Kite: A Story of Models of Wings and Models of the World*. New York: PI Press.
Suárez, Mauricio (2003). "Scientific Representation: Against Similarity and Isomorphism." *International Studies in the Philosophy of Science* 17(3): 225–244.
Suárez, Mauricio (2004). "An Inferential Conception of Scientific Representation." *Philosophy of Science* 71(5): 767–789.
Tomasello, Michael (2003). *Constructing a Language: A Usage-Based Theory of Language Acquisition*. Cambridge, MA: Harvard University Press.
Thomson-Jones, Martin (2010). "Missing Systems and the Face Value Practice." *Synthese* 172(2): 283–299.
van Fraassen, Bas C. (2008). *Scientific Representation: Paradoxes of Perspective*. Oxford: Oxford University Press.

11 The Truth of False Idealizations in Modeling[1]

Uskali Mäki

INTRODUCTION

Almost by definition, models violate the whole truth. Models also appear to involve false elements that violate nothing-but-the-truth. At the same time, good models are often expected somehow to yield true information about some real features of the world—not just true predictions but also true representations of important dependencies and mechanisms in some target domains. I have pursued the idea of models possibly being true more directly elsewhere (e.g., Mäki 1992, 1994, 2009b, 2009c, 2011).

This paper mainly discusses a complementary way of reconciling truth and falsehood, focusing on the apparent violations of nothing-but-the-truth by various unrealistic assumptions used in building and applying a model. Theoretical models in science characteristically involve idealizing assumptions that appear to be plain false: mass point, infinite velocity, perfect information, zero transaction costs, etc. These apparent falsehoods are in need of justification. I will offer a strategy of justification that construes many of them not only as functional for the pursuit of truth but also as only apparently false themselves. I will also distinguish that strategy from a nearby procedure that cannot deliver as much.

My suggestions are inspired by Alan Musgrave (1981), who designed an argument as a criticism of Milton Friedman's (1953) famous defense of the false assumptions of profit maximization and perfect competition in economic models. Friedman had claimed that a model can be just fine even though its assumptions are false, provided its predictive performance is satisfactory. Musgrave explained that once the import of those assumptions is correctly understood, they can be required to be true. Musgrave's argument was a clear advancement in the ongoing debate over Friedman's claims, but it had its own flaws that had to be removed by some revisions, elaborations, and amendments (Mäki 1994, 2000, 2004). In what follows, I will reframe and elaborate the arguments further. I will also briefly discuss related proposals by Frank Hindriks (2005, 2006), Catherine Elgin (2004), and Michael Weisberg (2007).

I will frame the discussion in terms of what I call the *functional decomposition approach* to examining modeling and the issue of truth in relation to models (Mäki 2004, 2009b, 2009c). The point of the approach is to decompose a model and the representations it is embedded in so as to determine what functions are served by the various components. Importantly for the task at hand, this strategy enables (re)locating the relevant truth bearers in modeling the world.

MODELS AND TRUTH: THE FUNCTIONAL DECOMPOSITION PERSPECTIVE

Here is a succinct formulation of my present account of models as representations (it has been discussed in more detail in Mäki 2009b, 2009c, 2011):

[ModRep]
Agent A uses object M (the model)
as a representative of
(actual or possible) target system R
for purpose P;
addressing audience E;
at least potentially prompting genuine issues of resemblance between M and R to arise;
describing M and drawing inferences about M and R in terms of one or more model descriptions D;
and applies commentary C to identify the above elements and to align them with one another.

[ModRep] should help us to locate, characterize, and resolve the issues around idealizations in modeling. The source of the puzzlement and debate that motivates the present paper derives from certain features of these idealizations: they often appear as statements that are outrageously false if taken as claims about any real target system. Reading them as such claims creates the puzzle. How to justify such outright falsehoods?

The first step is to localize those components and to understand the functions they serve in a given act of representation. Rather than reading the idealizing assumptions as claims about real target systems, they can be taken to play an important role as part of model descriptions D. They contribute to the description of an imagined model world by saying that in such a world, some factor is absent, constant, or has no effect on what is the case or what happens in the model. This is the first obvious sense in which an idealizing assumption can be only apparently false: it appears false if taken as a claim about some (or any) real-world target. When viewed as contributing to model description, its perceived falsity is seen to be based on a localization failure. However, the arguments for apparent falsehood to be presented are not exhausted by this simple step; their structure is more complex.

What I call *model commentary* plays an indispensable role in rectifying the localization failure and in justifying idealizations in terms of their functions. Commentary C identifies the key components of an act of representation and the functions they serve, coordinating them with one another. A commentary is needed because no model is itself able to specify how it relates or is supposed to relate (or fail to relate) to its (or any) target or targets. Importantly, a model commentary can be asked to spell out the purposes, audiences, and the respective desirable epistemic and pragmatic virtues to be pursued by a given modeling exercise. An adequate commentary helps remove misunderstandings that otherwise easily arise, as has been the case in the debates around idealizing assumptions. Showing how this happens is one task of this paper.

Many models are (or should be) accompanied by a commentary that spells out *isolation* as a major purpose of the model. A model is viewed as an imagined system in which some dependences and mechanisms are isolated from the interference and involvement of things not included in the system. Idealizing assumptions are vehicles of exclusion that contribute to the inclusion of selected items and relationships in the model. (Mäki 1992, 1994, 2011) By assuming that air pressure and the influence of other forces are nil, we isolate the impact of the Earth's gravity on a falling body. By assuming that there are no exports or imports, we close a model economy from the influence of foreign trade.

Whether or not idealizing assumptions are interpreted as statements about some real targets, it is unsurprising that issues of resemblance between a model and some real systems arise. The age-old debate in economics over the various "unrealistic assumptions" exemplifies this perfectly. Much of the time, this debate is carried out in terms of truth and falsity. Some say economic models are bad models because their assumptions are false. Others say they are just like all other models, always false since they contain false assumptions. I say false assumptions per se are no obstacle to a model being true.

Any component in *[ModRep]* can serve as a truth maker, so we can have truths about them all. But not all of them can naturally serve as truth bearers. The obvious truth bearers can be found in model descriptions D (about M) and model commentaries C (about anything else in *[ModRep]*). My controversial suggestion has been that parts of M itself (about target R) can function as truth bearers (e.g., Mäki 2009b, 2011)—but this is not the main subject of the present paper.

Pragmatics play a prominent role in *[ModRep]* and also in meeting the challenge of justifying apparently false assumptions as true. Although I don't take *truth* itself to be pragmatic, I take *what is relevantly true* to be pragmatically conditioned. The pragmatic context of purposes and audiences plays a key role in the pursuit of truth by model representation. For each purpose and audience combination $P_i \& E_j$, the pursued or intended truths about the target (or perhaps about something else) may be different. It is the task of model commentary C to identify the relevant $P_i \& E_j$ combinations and the respective truth bearers in the model. I call these relevant truth bearers the *truth nominees* of the model (Mäki 2004). The model commentary nominates some component parts of model representations as worthy of consideration for their truth-value. I will discuss more on this in a moment.

Suppose (parts of) the relevant audience (for example, students) raises an issue with the "unrealistic assumptions" that are used for describing the model, asking for their justification. The story we tell in response may be a long one, and as part of it, we may agree to consider those assumptions as truth nominees. But whatever it is that we will highlight in our commentary depends on the pragmatic context of modeling, including the purposes served and audiences addressed.

The larger pragmatic context of modeling includes disciplinary cultures and traditions. They shape the connections between the components of *[ModRep]*. For example, parts of economics appear to be done without much systematic attention to real target systems and issues of resemblance between model M and target R. The main attention is given to inferences among model descriptions D in the examination of the properties of imagined model worlds. Disciplinary practices of this sort tend to treat models as *substitute systems*—studied for their own sake—rather than as *surrogate systems*—studied in order to indirectly learn about real systems (Mäki 2009b, 2009c, 2011; cf. Sugden 2009). Yet, when pressed sufficiently hard, modelers in such cultures may sometimes articulate (or at least accept) interpretations of the kind suggested in the subsequent sections. Disciplinary cultures are not carved in stone. Nor are they completely uniform, there is space for some individual variation.

Treating a model as a substitute system with no issues of resemblance arising may also be due to the properties of the model (in relation to any possible target) rather than its cultural context. Some models just may not seem to have any targets (they have not yet found any or have lost the ones they were thought to have) and are therefore treated as substitute systems only.

TRUTH RE-NOMINATION

The notions of truth nomination and truth nominee (Mäki 2004) play a key role in the argument that false idealizations may be true. A *truth nominee* is any item that is, or is supposed to be, considered for its truth-value; it is a candidate for truth. *Truth nomination* is an act that assigns an item for the role of truth nominee. *Truth re-nomination* is the proposal to treat some *P2* instead of *P1* as the relevant truth nominee. The argument here requires that *P2* and *P1* be connected. The connection can be conveyed in two ways. One is by *reformulation or paraphrase*: a sentence is transformed into another related sentence by some alteration of contents. Another is by *meta-claim*: a claim is being made about a sentence. In both cases, the original sentence is replaced by some other sentence as the relevant truth nominee. *P1* is paraphrased as *P2* so as to nominate *P2* instead of *P1* as the relevant truth nominee. Or claim *P2* is made of *P1*, and *P2* is nominated for truth. In both cases, it may be the case that phrased as *P1*, the item is false, whereas phrase *P2* is true.

It is the task of model commentary to propose one or both of those two kinds of truth re-nomination so as to deal with, and perhaps resolve, the issue of apparent falsehood. The issue arises since a model description is typically put in terms of apparently false idealizing assumptions—that is, false if taken to be about the real world. But one should not be misled to thinking that this reveals their intended or proper import. Conclusions about the latter can only be based on the model commentary. Truth nomination and truth re-nomination are among the accomplishments of model commentary. Neither a model itself nor its descriptions are able to reveal the relevant truth nominees. Appearances may deceive also in modeling. The task of model commentary is to check whether they do and so to get the facts right.

So the trick accomplished by a model commentary is to relocate truth nominees within *[ModRep]*. For truth re-nomination to be justified, one must take the re-nominee as the intended or otherwise relevant truth nominee. What may first appear to be false might not be relevantly so after all, since it is not a relevant truth nominee, and its proper paraphrase or meta-claim might be true. Re-nomination is needed for revealing what real truth nominee lies behind an apparently false idealizing assumption. Among other things, this implies that one should not criticize what appears as a false idealization without understanding what assertion is intended when using it; nor should one rush to conclude that a model is false just because it seems to involve false idealizations.

Note that this strategy is greatly facilitated by the idealizing assumptions being explicitly formulated as part of model descriptions. In an early paper, I proposed distinguishing two means for theoretical isolation, one based on *silent omission*, the other based on *explicitly formulated idealization* (Mäki 1992). They both serve to exclude some items from the model world, either by leaving them out without mentioning them, or by explicitly assuming that they are constant, absent, at normal states, etc. Omissions are not included in model descriptions, while idealizations are. The latter are natural subjects for paraphrase or meta-claim. Because of their explicitness, they are easier to recognize, and because of their apparent falsity, they invite closer scrutiny and justification. Explicit formulation of idealizing assumptions with associated commentaries about their functions delivers information that will be easily missed if only silent omission is applied. However, this division is not fixed and permanent: what is implicit can be made explicit.

Idealizations such as *the economy is closed* (exports and imports are zero) and *transaction costs are zero* (no costs of acquiring information, enforcing contracts etc) are typically nowadays made explicit. Other exclusions, such as the *absence of fairness issues* or the *absence of hysteresis effects* in all or some markets, are more often based on just silent omission. But they, too, can be made explicit for closer examination. Consider an idealizing assumption saying that

[C] The economy is closed.

This is a much-used assumption in describing macroeconomic models. If one considers this as a factual assertion about the real world, it turns out that it is seldom exactly true; in lots of cases, it is very far from the truth. The key move in the present strategy is to refrain from considering [C] and other such assumptions as factual assertions but rather to turn them into other sentences that are used to make factual assertions about the world or to make claims about them and their role in inquiry. At the end, truth-values are not supposed to be ascribed to idealizations like [C] but rather to their paraphrases or meta-claims that are treated as the relevant truth bearers or truth nominees. It is the task of model commentary to point out how a sentence like [C] is to be interpreted in a given context.

Musgrave (1981) suggested that one and the same sentence such as [C] (his example was "The government has a balanced budget") can be turned into various kinds of assumption that have a chance of being true. His typology included three such types: negligibility assumption, domain assumption, and heuristic assumption. In my revision (Mäki 2000), I renamed and redescribed these and suggested that there are many other types as well. I also suggested that there are limits to this strategy. I now suggest that there are two different procedures of re-nomination (those of paraphrase and meta-claim) and that their powers in justifying the original idealizing sentence are different.

RE-NOMINATION BY PARAPHRASE: NEGLIGIBILITY AND APPLICABILITY

The first two kinds of assumption—negligibility and applicability assumption—are based on paraphrasing an apparently false idealizing assumption and thereby giving the truth re-nominee a chance of being true.

Negligibility

In general terms, a negligibility assumption is the hypothesis that some factor F that might be expected to affect the phenomenon under investigation actually has an effect upon it small enough or otherwise irrelevant to be neglected relative to a given purpose and audience (Mäki 2000, 322). This is a very important and fundamental type of paraphrase that turns an idealizing assumption into a claim about the negligibility of a factor, given the modeler's purposes and audiences. Thus, [C] could be paraphrased as:

> [NC] Foreign trade has negligible effects on the phenomenon of interest—effects that are negligibly small or otherwise irrelevant, given the purposes and audiences of the modeler.

The crucial payoff of this move is that while [C], if considered as a factual assertion, would be false, paraphrased as [NC] it may well be true. So conceived, [C] would be a claim about the economy having a property (of being closed), while [NC] is a claim about the negligibility of the economy not having that property (that is, of being open). These are claims about two very different things, thus it is not surprising that their truth-values have little to do with one another: one may be utterly false while the other may be true. The same applies to other assumptions such as *no transaction costs* and *no fairness considerations*. They may be irrelevantly false, while claims about the negligibility of transaction costs and fairness may be relevantly true or false, depending on the case.

Assumptions like [C] are properly read as parts of model description. They describe an imagined system such as a simple closed economy and are not claims about any real system directly. So they are not to be taken as candidates for truth about real-world economies at all; hence, they cannot fail as such candidates either. Given that they are not treated as such candidates, as worthy of truth nomination, it is not sensible to ascribe them truth-values: because they are not taken to make any claims about any real systems, they are not to be regarded as true or as false about such systems.

On the other hand, formulations such as [NC] provide the relevant truth nominees, and it is these assertions that are to be examined for their truth-value, and eventually to be ascribed truth-values (with whatever degree of assurance). Such ascriptions, to be successful, require factual inquiries into the world. One hopes that claims about negligibility such as [NC] are true.

If a claim such as [NC] is a relevant truth bearer, what is its truth maker? It is the fact of negligibility. So what is negligibility? In general, negligibility is a function of two sets of facts—ontic facts about the world and pragmatic facts about modeling the world. Two versions can be distinguished.

In the *first version*, negligibility is a function of the modeler's wish to highlight some fragment or feature in the causal structure of the world. A theoretical model isolates some important dependence relation or causal mechanism, while excluding a number of other factors using idealizing assumptions that appear to state that these other factors are absent, have zero value, etc. Such idealizations are camouflaged negligibility assumptions that claim that the non-absence, nonzeroness, etc of those other factors is negligible because the factors are irrelevant to the main message of the model. Only the isolated factors are viewed as relevant to the purpose of illuminating some important structural feature of the world.

The *second version* of negligibility focuses on the connection between empirical data and the implications of a model. Just as in the first version, we may be interested in a relationship between the causal efficacy of a factor and the purposes and audiences of a modeler. But here what matters is how the causal efficacy of a factor (such as foreign trade) manifests itself empirically and how this relates to the predictive output of the model. The purposes of the modeler in meeting the expectations or desires of some audience may include predicting the domestic inflation rate at a certain conventional level of accuracy appreciated by policy makers for certain policy purposes. What is negligible on this version may be very different from what is negligible on the first version.

The first version is related to the notion of *minimal model* and Weisberg's associated notion of "minimalist idealization: "Minimalist idealization is the practice of constructing and studying theoretical models that include only the core causal factors which give rise to a phenomenon" (2007, 642). I would put the idea a bit differently, with due stress on the idea that idealizations are devices of exclusion rather than inclusion: "Minimalist idealization is the practice of constructing and studying theoretical models that exclude all but the core factors by way of idealizing assumptions . . . " (see Mäki 1992).

There is another issue. Weisberg's definition only cites a causal criterion of inclusion in a model while ignoring the pragmatics of modeling. However, in actual scientific modeling practice negligibility assumptions reflect varying explanatory interests in relation to the causal structure of the world. For each different explanatory question, different facet of that causal structure—different "core factors"—may be isolated by the model, while other factors are assumed to be negligible for the purpose of answering those specific questions. And each explanatory question is a function of the varying epistemic interests of the modeler and his audiences. Negligibility is pragmatic through and through. Therefore, I would not say, as Weisberg does, that "minimalist idealization is not at all pragmatic" (2007, 645).

So in both cases the negligibility of a factor is jointly determined by an objective real-world fact and some pragmatic characteristics of scientific practice in studying the real world. And the truth-value of a negligibility assumption is determined by nothing else but just that: negligibility.

This implies that the truth-value of a negligibility assumption may vary from situation to situation. This is particularly evident in regard to the second version. *With a fixed purpose and audience*—such as predicting the rate of inflation at a certain level of accuracy expected by central bankers in some particular situation—it is more likely that [NC] is true of the U.S. economy than about the Finnish economy. This is because the U.S. economy is "more closed" and so causally less dependent on foreign trade. *With a fixed target domain*—the U.S. economy—[NC] is more likely to be true the less stringent the model users and consumers are about the desired predictive accuracy. So the truth or falsity of a negligibility assumption depends on both the ontology of causation and the pragmatics of inquiry.

There is another way of thinking of the contextual dependency of negligibility assumptions' truth-values. This makes those truth-values dependent just on the contingent facts of causation. On this approach, a *particular* purpose and audience would be built into each and every negligibility assumption; in other words, for each particular purpose and audience, there would be a corresponding particular negligibility assumption. Thus, [NC] would express a potentially limitless family of negligibility assumptions, each variant based on a particular purpose and audience combination. One implication of this way of framing things would be to make the truth-values of negligibility assumptions dependent on facts of real-world causation only.

By now it should be pretty clear how the proposal deals with truth, but some further illumination may still be useful. So let me briefly compare the notion of negligibility and Elgin's (2004) notion of *true enough* as well as the associated strategies of tolerating falsity. The notions and the strategies are closely related but not identical. I would suggest *true enough* can be taken to mean *negligibly false*. Elgin rightly recognizes that deviation from the truth is an ever-present situation in our epistemic endeavors, and she attempts to accommodate this in terms of the notion of true enough. There is a slight difference between her and my concerns and solutions. Elgin's concern seems to be with identifying the grounds for *accepting what is no more than true enough*. In passing, she comes close to my proposal: "to accept a claim is not to take it to be true, but to take it that the claim's divergence from truth, if any, is negligible" (119). Then she puts forth her dominant formulation: "We accept a claim when we consider it true enough" (119). This way of looking at things implies that no truth re-nomination needs to take place for acceptance. We simply accept falsehoods because they are true enough for some purpose.

Given how I have framed my argument, my concern and the suggested strategy is a little different. For my strategy, true enough is not good enough. I am proposing that if we "accept a claim when we consider it true enough" then we can as well turn this to the idea that we *accept a claim when we consider it true about negligibility*. This is the idea of accepting as true what is apparently false. Thanks to truth re-nomination, we can accept claims about negligible falsehood as true (rather than true enough, approximately true, close to the truth, etc). We don't only "take it" but we *claim* "that the claim's divergence from truth, if any, is negligible," while giving the claim about divergence from truth a chance of being true. That's how we can *replace what is just true enough by what is nothing short of true*.

Applicability

An applicability assumption states that a model applies to domains in which some factor F is absent or only has negligible effects on the phenomenon of interest (Mäki 2000, 323–324). The model user may be prompted to interpret and paraphrase an idealizing assumption ($F=0$) as an applicability assumption after having discovered that in some domains of application, factor F is systematically non-negligible. The model commentary then suggests that those recalcitrant domains be excluded from its proper domain of applicability and that the model only applies to cases in which F is absent or negligible.

A paraphrase of [C] as an applicability assumption amounts to turning it into a claim about a property of a model in relation to a relevant domain of phenomena. Such a claim would be a claim about the applicability of the model:

> [AC] Model M is only applicable to economies that have no foreign trade—or to economies whose actual foreign trade is negligible.

Similarly, the *no fairness* or *no hysteresis* assumptions can be turned into applicability assumptions by claiming that the model involving such assumptions only applies to domains in which fairness or hysteresis make no difference or make only a negligible difference. One obvious candidate domain to which fairness-less models do *not* apply is labor markets. This is so with respect to both minimal and predictive models.

Again, what appeared to be a false idealization—such as [C]—is not treated as making any factual claims whatsoever, hence is not nominated for truth. It is rather paraphrased as a claim about applicability, a property that is jointly determined by the properties of the model in question and the properties of some domain of phenomena (and, in case negligibility is invoked, the purposes and audiences of modeling). Considered as such a claim, [AC] is worthy of truth nomination: it may be true or it may be false. Again, ascribing a truth-value to claims such as [AC] requires factual inquiries into the real world. One hopes such claims to be true.

Applicability assumptions play a role in a research strategy in which the modeler has access to a pool of models and makes informed selections from that pool depending on the domain to which—and the purpose for which—the selected model is supposed to be applied. In a "Letter to Roy Harrod, 4 July 1938", John M. Keynes famously said, "Economics is a science of thinking in terms of models joined to the art of choosing models which are relevant to the contemporary world" (296). On an obvious reading of this statement, it is reasonable to suggest that in an ideal case "the art of choosing models" should be based on employing true applicability assumptions. On the other hand, in the actual practice of economic inquiry such assumptions are seldom explicitly formulated and systematically tested, while implicit judgment plays a major role.

RE-NOMINATION BY META-CLAIM: EARLY-STEP, TRACTABILITY, AND MORE

Negligibility and applicability assumptions as truth re-nominees emerge as paraphrases of the original idealizing assumptions. By contrast, the following re-nominations are based on making meta-claims about the original idealizations.

Early-step

A common response to the charge or discovery that some factor F is not negligible and thus should not be excluded from the model by idealization is to include the factor by relaxing the idealizing assumption. These are assumptions first made, then relaxed. To qualify as an *early-step* assumption, it is to be replaced by a *later-step* assumption at some later (logical or temporal) stage of the modeling exercise. This requires a process of de-idealization. In many such cases, a false negligibility assumption is replaced by a presumably (more) true negligibility assumption.

Many philosophers and practicing scientists appear to hold the view that models involving such early-step idealizations are unavoidably false and that they can be made truer by relaxing such assumptions and replacing them by more realistic assumptions. This is a way of enriching the picture given by the model about the world, making it look more like the world itself, in all its richness and complexity. I don't subscribe to this view of how to approach *the* truth in modeling. I think there are many different kinds of truth to be captured with models, and this is just one of many ways of capturing some of them. Some others can be captured with (minimal) models involving unrelaxed early-step assumptions. But this is a view I have defended elsewhere. Here the focus is on those early-step assumptions themselves.

Model commentaries often identify some idealizations as early-step assumptions, but what exactly is their identity? One option is to say that an early-step assumption is an idealization like [C] plus a comment that it is temporary only and will be relaxed. Another option is to add a reason why it will be relaxed, such as its negligibility paraphrase being false. But there are other possible reasons as well, and each such reason would give us a distinct type of early-step assumption. So let us consider the first (neutral) option and see what to make of a meta-claim about [C] that looks like this:

> [ESC] [C] is an element of an early-step version of model M and is relaxed in its later-step versions

We may say that while [C] is an *early-step assumption*, [ESC] is an *early-step ascription*. [ESC] ascribes an early-step status to [C], or in other words identifies [C] as an early-step assumption. The truth maker of the early-step ascription [ESC] is the fact of [C] being made (in early-step versions of a model) and being relaxed (in later-step versions). This is a fact of a feature of scientific practice, namely a sequence of assumptions. A typical—but not the only possible—reason for the obtaining of this fact is the perception that the factor identified and excluded by the early-step assumption is not negligible, implying the denial of [NC] and the acceptance of the claim, "foreign trade makes a non-negligible difference for the phenomenon of interest."

How to ascertain whether there is a respective truth maker that obtains? I can think of a number of different kinds of cases. The easy cases are those that refer to *the past or the present*: both the early-step and the later-step have already occurred. This divides into two kinds of situation. The diachronic situation involves the history of a research field: as a matter of historical fact, an idealizing assumption was first made, and then at some later time it was relaxed. The synchronic situation involves no similar historical time dimension. The structure of a textbook is an example: in early chapters, an assumption is made, only to be relaxed in later chapters of the book. (Of course, this often involves a temporal order of reading and learning, but this is not a matter of historical sequence in assumptions.)

Accessing and establishing instances of these sorts is relatively easy. In the diachronic case, it may require nothing but tracing the actual historical development of a model in the relevant literature. In the synchronic case, it may be a matter of checking a journal article or textbook that begins with a closed economy model and at a later step relaxes the assumption and ends up with an open economy model. That the required truth maker is in place seems to be easy to establish.

The more difficult cases are those that refer to *the future*: at the time of examining the idealizing assumption, the later step has not yet occurred. In one kind of future-oriented case, the orientation is descriptive, taking on the form of expectation, anticipation, and prediction: the early-step assumption is expected or anticipated to be relaxed in some later-generation models yet to be built. Depending on whether later steps will be taken, the descriptive attitude is true or false, but one can establish whether the relevant truth maker obtains only at a later time.

In other kinds of future-oriented cases, the orientation is rather a matter of attitudes such as intention, prescription, hope, or even promise: it is intended, prescribed, hoped, or promised that the assumption in question indeed will be just an early-step assumption and will be relaxed at a later step. The modeler herself may have such a purpose, while prescription and hope are attitudes that can also be held by the audiences of modeling, deeming it as desirable that later-step models be built such that the early-step assumption will be relaxed. Promise might be given by the modeler when confronted with a sceptical or critical audience and attempting to dispel worries about an excessively unrealistic model.

Intentions, prescriptions, promises, and hopes are things that are different from predictions or other descriptive claims in that they cannot as obviously be taken as truth-valued, thus not as straightforwardly worthy of truth nomination. We can talk about their realisticness in other senses, though, such as the ease and likelihood of their realization.

The truth of early-step ascriptions such as [ESC] does little to justify an idealizing assumption such as [C] in a model in ways that would rely on judgments of causal efficacy or difference making. In this respect, meta-claims of this kind are different from negligibility paraphrases like [NC].

Tractability

Frank Hindriks (2005, 2006) has usefully analyzed the role of tractability considerations in scientific practice. He has proposed that there is a separate class of assumptions, that of tractability assumptions. But more needs to be said about the identity and separateness of such assumptions. Consider this meta-claim about [C]:

> [TractC] The use of [C] in describing and examining model M makes a problem (more) tractable.

Let us say that while [C] is a *tractability assumption* (or a tractability-enhancing assumption), [TractC] is a *tractability ascription*. [TractC] ascribes the function of enhancing tractability to [C], or in other words identifies [C] as a tractability (or tractability-enhancing) assumption.

On this formulation, a tractability ascription is a claim about tractability being enhanced by an idealization. The truth makers of tractability ascriptions are facts of tractability, and facts of tractability are facts of scientific practice, relating problems and various constraints on problem solving with one another. Again, a tractability ascription may be true even though the respective tractability-enhancing idealization were false about real target systems.

Tractability considerations are very important in modeling, and many idealizations are dependent on and motivated by them. However, I am unwilling to go along with Hindriks's (2006) radical suggestion to reinterpret early-step assumptions in terms of tractability. Musgrave (1981) had characterized his category of "heuristic assumptions" in a way that stresses their early-step status but does not distinguish this from their tractability-enhancing properties. My (2000) reconstruction focused only on the former characteristic and identified the category of "early-step assumptions." Hindriks's suggestion is closer to Musgrave's in not distinguishing the two characteristics but differs from it in stressing tractability. I think early-step and tractability are two different categories that should be kept separate.

Weisberg's (2007) characterization of "Galilean idealization" is close to Hindriks's suggestion. Weisberg lumps together the two characteristics of enhancing tractability and having an early-step status. He says that the purpose of Galilean idealizations is to make theories computationally tractable and also that "advances in computational power and mathematical technique should lead the Galilean idealizer to de-idealize" (641). These are the key features of Galilean idealization.

> "The practice is largely pragmatic; theorists idealize for reasons of computational tractability. The practice is also nonpermanent. Galilean idealization takes place with the expectation [sic] of future de-idealization and more accurate representation." (642)

I think lumping these two things together in one category is not advisable. As I noted above, the early-step status typically depends on negligibility considerations. The relaxation of an early-step assumption is then motivated and prompted by perceptions of non-negligibility. This must now be linked to the issue of how the early-step status and tractability are related. There seems to be no necessary connection of the kind suggested by Hindriks and Weisberg.

Suppose an idealizing assumption is initially included in a model description for tractability reasons. There is no reason to relax the assumption—and thereby to confirm its early-step status—just because it enhances tractability. A genuine reason for its relaxation would be the belief that the factor that the assumption removes from the model is non-negligible. But this fact of non-negligibility does not explain or define the tractability-enhancing characteristics of the assumption.

An even more directly effective argument against linking tractability and the urge to relax too tightly together is to point out that it is also possible to make the discovery that the factor initially removed from the model due to tractability considerations is negligible. In such a case, there may be no reason to relax the assumption. Indeed, in such a conceivable case, the idealizing assumption would have the status as a negligibility assumption and as a tractability assumption at the same time—but not as an early-step assumption.

That an idealizing assumption initially enhanced tractability is not necessarily connected to its later relaxation: it is likely to be relaxed for reasons other than its tractability-enhancing properties. These include perceived non-negligibility and an attempt to check for negligibility (as in robustness analysis). For example, a need for greater predictive accuracy or explanatory depth may motivate relaxing an idealization regardless of what happens to tractability along the way (provided the consequences lie within some minimal boundaries of tractability). And thinking of why early-step assumptions are made in the first place, there are considerations other than those of tractability that may motivate making an idealizing assumption that is later relaxed, as we will next see.

And More . . .

We may also generalize on the idea of early-step assumption by disconnecting it from judgments of negligibility that appeal to real-world causation. We may choose to say that for something to count as an early-step assumption is neutral with respect to the reasons for this being so. Assumptions are made and relaxed, and the reasons for both may be various. Consider this meta-claim about [C]:

> [PedC] The use of [C] serves useful pedagogical purposes.

This is a claim about *pedagogical function*. It is a product of a model commentary that aligns a model description with the audience of economics students and the purpose of learning. If considered as a truth nominee, [PedC] is a true claim as witnessed by the numerous highly praised macroeconomic textbooks and lecture courses that start with expositions of closed economy models. Students are presented the simple case that will help them learn how a selected set of variables interact undisturbed by foreign trade effects. The pedagogical advantages seem obvious. This pedagogical strategy is not based on perceptions of whether the causal effects of foreign trade are negligible in the real world.

Later chapters in those textbooks (or later classes) typically relax [C] and present students with more complex open economy models with foreign trade effects included. Thanks to the pedagogical strategy, the more complex cases are now easier for students to learn. Pedagogical goals dictate the early-step strategy. On the other hand, it is obvious that causal negligibility considerations play a role in the background: there would be no need to relax [C] in the textbooks were it not the case that in many real-world domains and for many purposes, foreign trade makes a difference. But this fact is not among the truth makers of meta-claim [PedC].

Then consider a meta-claim in terms of *academic entry conditions* as part of a model commentary that coordinates model descriptions with the purpose of fortifying one's publication record and the expectations of audiences such as journal editors and referees:

> [EntryC] The use of [C] is a prerequisite of getting one's paper accepted in journal J_1, while its relaxation is required for success in journal J_2.

It does not matter whether this particular claim is actually true (with any existing two journals), but many others of this general kind surely are: they are true in virtue of facts of academic fashions, specializations, gate-keeping practices, etc. These facts may motivate strategies that rely on building early-step and later-step versions of a model (submit a closed economy version to journal J_1 and an open economy version to journal J_2). No reference might be needed to considerations of tractability or causal negligibility.

There are many other possible meta-claims about idealizing assumptions, such as

[AesC] The use of [C] yields aesthetically pleasing models.

Such an appeal to *aesthetic values* can be popular among economists and other scientists. Putting aside difficulties of measuring aesthetic pleasure as part of the relevant truth maker, meta-claims of this kind may be true even though idealizations such as [C] were false about real systems. In this respect they are no different from paraphrases [NC] and [AC]. But there are other differences, on which a few final words are next.

TRUTH RE-NOMINATION AND JUSTIFICATION

Milton Friedman (1953) had sought to justify false assumptions of a model by arguing that their falsity does not matter provided the model yields sufficiently accurate predictions. This has mostly been interpreted as an instrumentalist defense of unrealistic models (but see my deviant-realist reading, e.g., in Mäki 2009a). Following Musgrave's general idea, the argument outlined in the present paper suggests that one should care about the factual truth of assumptions and that this can be seen once it is understood that the idealizing assumptions can be used for making a variety of different kinds of claim. But this argument should not overlook some important asymmetries between these kinds.

These asymmetries emerge partly in consequence of the difference between the two kinds of truth re-nomination, those of paraphrase and meta-claim. An important difference lies in the truth makers of these two classes of re-nominees as we have illustrated them. Both have a chance of being true but do so in virtue of different kinds of fact. On the one hand, paraphrases in terms of negligibility and applicability make reference to, and may be true partly in virtue of, properties of the real target system of a model. Thanks to this, they can be used for justifying the original idealization. On the other hand, the meta-claims listed above are about the roles of idealizing assumptions in modeling practice so are true or false in virtue of pragmatic facts only—facts about disciplinary practices of inquiry and education such as early-step, tractability, academic entry, and pedagogic value. These claims may be very informative about important facts of academic practice so may be justified as such factual claims. But for them to be adequate for justifying the original idealizing assumptions, they have to be put in a larger context that contains a collectivity of modelers, a multiplicity of models, and their dynamic evolution in short, medium, and long term—as well as the various epistemic values that guide modeling activities.

Another perspective to the asymmetry is provided by the distinction (in Models and Truth: The Functional Decomposition Perspective above)

between *substitute modeling*—modeling unconstrained by, or without contact with, real targets—and *surrogate modeling*—modeling as an indirect way of accessing real targets. Re-nomination in terms of early-step status, tractability, aesthetics, pedagogy, and academic entry may support nothing but substitute modeling if left unsupported by other arguments that do appeal to real-world facts. Negligibility and applicability assumptions do make appeal to the properties of real targets and are therefore able to support surrogate modeling.

It is for such reasons that only in case of paraphrase in terms of negligibility and applicability can we say that the original idealization is only apparently false: the relevant truth-value is that of the truth re-nominee. This cannot be said about the meta-claims above. Their truth-value does not replace that of the original idealization so does not remove the worry about false idealizations.

NOTES

1. Earlier versions have been presented at Trends and Tensions in Intellectual Integration (TINT), Helsinki, February 23, 2009; at Models and Simulations 3 (MS3), Charlottesville, March 6–8, 2009; and at the conference on Modeling the World, Helsinki May 28–30, 2009.
Thanks go to the respective audiences and to an anonymous referee for helpful comments.

REFERENCES

Elgin, Catherine (2004). "True Enough." *Philosophical Issues* 14: 113–131.
Friedman, Milton (1953). *Essays in Positive Economics*. Chicago: University of Chicago Press.
Hindriks, Frank (2005). "Unobservability, Tractability and the Battle of Assumptions." *Journal of Economic Methodology* 12: 383–406.
Hindriks, Frank (2006). "Tractability Assumptions and the Musgrave-Mäki Typology." *Journal of Economic Methodology* 13: 401–423.
Keynes, John Maynard (1973). "Letter to Roy Harrod, 4 July 1938." In *The Collected Writings of John Maynard Keynes*, ed. Donald E. Moggridge. Vol. 14, pp. 296–297. Cambridge: Macmillan.
Mäki, Uskali (1992). "On the Method of Isolation in Economics." *Poznan Studies in the Philosophy of the Sciences and the Humanities* 26: 319–354.
Mäki, Uskali (1994). "Isolation, Idealization and Truth in Economics." *Poznan Studies in the Philosophy of the Sciences and the Humanities* 38: 147–168.
Mäki, Uskali (2000). "Kinds of Assumptions and Their Truth: Shaking an Untwisted F-Twist." *Kyklos* 53: 303–322.
Mäki, Uskali (2004). "Some Truths about Truth for Economists, Their Critics and Clients." In *Economic Policy-Making under Uncertainty: The Role of Truth and Accountability in Policy Advice*, ed. Peter Mooslechner, Helene Schuberth, and Martin Schürz, 9–39. Cheltenham, England: Edward Elgar.
Mäki, Uskali (2009a). "Unrealistic Assumptions and Unnecessary Confusions: Rereading and Rewriting F53 as a Realist Statement." In *The Methodology of*

Positive Economics. Reflections on the Milton Friedman Legacy, ed. Uskali Mäki, 90–116. Cambridge: Cambridge University Press.

Mäki, Uskali (2009b). "Models and Truth. The Functional Decomposition Approach." In *EPSA Epistemology and Methodology of Science. Launch of the European Philosophy of Science Association*, ed. Mauricio Suárez, Mauro Dorato, and Miklós Rédei, 177–187. Dordrecht: Springer 2009.

Mäki, Uskali (2009c). "MISSing the World: Models as Isolations and Credible Surrogate Systems." *Erkenntnis* 70: 29–43.

Mäki, Uskali (2011). "Models and the Locus of Their Truth." *Synthese*, 180, 47–63.

Musgrave, Alan (1981). "'Unreal Assumptions' in Economic Theory: The F-Twist Untwisted." *Kyklos* 34: 377–387.

Sugden, Robert (2009). "Credible Worlds, Capacities and Mechanisms." *Erkenntnis* 70: 3–27.

Weisberg, Michael (2007). "Three Kinds of Idealization." *Journal of Philosophy* 104: 639–659.

12 Idealized Models as Inferentially Veridical Representations
A Conceptual Framework
Juha Saatsi[1]

INTRODUCTION

Idealized models misrepresent the world, but typically they do so only partially. That is, idealized models typically latch onto reality in significant respects. How should we analyze the way in which idealized models can provide partially veridical representations of their target systems? Or take some idealization—a respect in which an idealized model misrepresents the world—and ask: what is its function, and how is it related to the truth-latching aspects of the model? Here I provide some answers to these questions, applicable to some idealizations.

My analysis is incited by my scientific realist inclinations. Scientific realists are in the business of defending the claim that (strictly speaking) incorrect theories and models, which are predictively successful (in appropriate measure), typically latch onto reality in a way that is explanatory of their success. The challenge most often raised against the realist concerns theories and models of past science, inconsistent with today's best-justified beliefs. But idealized models present a well-known challenge to realism as well (Cartwright 1983; Morrison 2000). The conceptual framework developed here is the first step in an argument to the conclusion that predictively successful idealized models often latch onto reality in those respects that are explanatory of their success, even if they are heavily idealized in other respects. This first step focuses on analyzing a way in which idealized models can provide partially veridical representations of unobservable aspects of the world.[2]

Much work has gone into analyzing 'partial' or 'approximate' truth in connection with linguistic representations. There are highly sophisticated accounts of the logic and semantics of verisimilitude, for example, and one can approach idealization from that perspective. An alternative approach, with a venerable tradition in philosophy of science, is to construe scientific modeling as incorporating an element of *non-linguistic representation* (e.g., Giere 1988). In the spirit of this tradition, there is incentive to analyze the partially veridical nature of scientific models directly in terms of the non-linguistic representation relation. This is the approach that I adopt,

Idealized Models As Inferentially Veridical Representations

following a variant of the semantic view of theories that is representative of this essentially non-linguistic tradition. 'Semantic view' labels a broadly model-based view of scientific theorizing. By virtue of focusing on model systems as essentially non-linguistic entities, the semantic view (arguably) liberates philosophy of science from unnecessarily tight involvement with logic and propositions, providing us with new conceptual and technical resources to analyze the nature of models, approximations, and idealizations (among other things).

There are various versions of the semantic view that come in various degrees of formalization. My perspective is most closely aligned with the version espoused by Giere (1988) and Teller (2001). At the heart of this view, I take it, is the idea that models function by allowing modelers to attribute properties to their target systems. A *model description* specifies all the properties of a *model system* and also the representation relation between the model system and the target. Although language is involved in specifying a model system and also the relation between a model system and its target, the latter relation—the relation that is critical for analyzing how models can partially latch onto reality—is itself non-linguistic. For Giere and Teller, this relationship is a symmetric one of *similarity*. I do not subscribe to this specific assumption, nor do I wish to make any universal assumptions about the nature of the model system. For me it only matters that a model system is used to *attribute properties to its target* in such a way that we can start thinking about a model partially latching onto reality by virtue of correctly attributing to the target some properties, but not others, if taken as a fully faithful representation of reality.[3] This focus on properties is profitable, I will argue, because in analyzing the way that models approximate, idealize, or abstract, we can mobilize conceptual resources associated with properties and especially the way that properties are related in various ways.

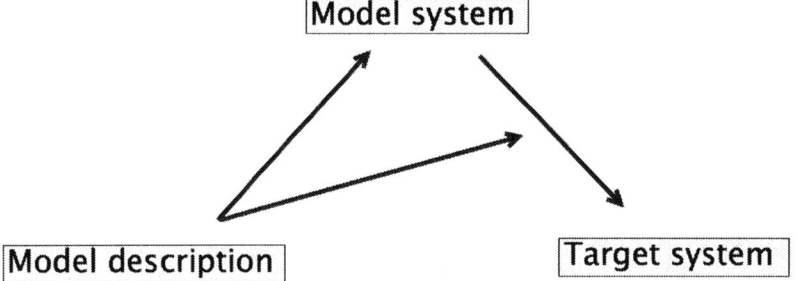

Figure 12.1 A schematic representation of a model, comprising a model description and a model system (modified from Giere 1988). The two arrows from the model description are (typically) linguistic specifications of a model system, on the one hand, and of the non-linguistic representation relation between the model system and the target, on the other.

Some properties are related by virtue of laws of nature, while other properties are related by some stronger modal connection. For example, we have *determinable* properties that logically necessitate the instantiation of a corresponding *determinate* property and vice versa. We have properties that are *multiply realizable* by virtue of some law-like connection and so on. I will argue below that we can make headway in analyzing at least some idealized models by focusing on non-causal relations that hold between various properties that models attribute to their target systems. A little more precisely, I will be arguing that at least some idealized models can be understood in terms of those *models being 'inferentially veridical' regarding some relevant properties, despite purposefully misrepresenting—for some pragmatic gain—some unnecessarily specific properties of the target system*. I will spell this out by explaining what 'inferential veridicality' means; what 'relevant properties' are; what I mean by 'unnecessarily specific properties'; and what sorts of 'pragmatic gains' can be had by thus misrepresenting reality.[4]

The rest of the paper will proceed as follows. After setting out my initial assumptions in the next section, I will further set the context by adopting a particular conception of idealization. Then I will spend some time spelling out the key notion of inferentially veridical representation, with examples, before finally applying it to idealization.

INITIAL ASSUMPTIONS

In this section I present four initial assumptions regarding properties and representation underlying my analysis. But first, here are some prefatory remarks on properties.

Properties are features that particular systems can exemplify. The kinds of properties that science is concerned with are *general*: more than one thing can exemplify (or 'share') them. I use the term 'property' to cover both monadic properties and relations. There are many well-known philosophical questions concerning the nature and ontology of properties. I'm not interested in these questions here. Rather, I just want to help myself to the various well-established *conceptual* distinctions and resources that property talk offers. For example, consider the common idea that psychological facts may be 'fixed' by the physical facts (together with the laws of nature) but not vice versa. The key concepts here, *supervenience* and *multiple realizability*, are typically construed as relationships between pairs of families of properties. Or consider the fact that the electron and the positron are similar in a certain respect by virtue of being charged, unlike neutrons. This similarity is construed as sharing a *determinable* property, *being charged*, that relates to two more specific *determinate* properties, *being positively charged* and *being negatively charged*, respectively. These are examples of typical property talk in analytic philosophy. According to

Idealized Models As Inferentially Veridical Representations 237

(the Giere-Teller version of) the semantic view models represent by specifying properties exemplified by their target systems, and I want to investigate what use can be made of such conceptual resources in analyzing models that are less than fully veridical representations.

Now, my four basic assumptions:

1. Given the way that properties can be (non-causally) related, *when one explicitly represents a system as exemplifying some particular property, one ipso facto implies (implicitly, perhaps) that the system exemplifies some other properties.* For example, by representing a system comprising 10 electrons and 10 positrons, one ipso facto implies that the system comprises 20 charged particles. One also implies that the system exemplifies the property *having a composite number of particles*, which is a property also shared by a collection of 21 neutrons but not by a set of 19 neutrons (given the primality of the number 19), and so on, for countless other properties. The implication regarding such properties is usually only implicit, of course.

   In general, by explicitly representing the system as exemplifying a *more specific* property one ipso facto implies that the system exemplifies some *less specific* properties. By '*more specific*' property, I mean just this: exemplifying a more specific property is *one way (out of many possible ways) of exemplifying* a less specific property. For instance, one way of comprising 20 charged particles is to comprise 10 positive and 10 negative particles. Another way is to comprise 15 positive and 5 negative and so on. Familiar examples are afforded by other determinate-determinable pairs: being red is one way of being colored; weighing 10 kg is one way of weighing between 9 and 11 kg; being an equilateral triangle is one way of being a triangle; being neuro-toxic is one way of being poisonous, etc.[5]

2. *Misrepresentation regarding a more specific property is compatible with veridicality regarding some implied less specific property.* This is obvious, as it just follows from the fact that for any less specific property there are more than one possible ways of exemplifying it. It is trivial to come up with examples with gerrymandered, disjunctive properties. To give a more interesting example, consider a detailed model that represents a gaseous system of particles in a box, specifying the momentum of each particle. In as far as the temperature of this system is determined by the mean kinetic energy of its constituent particles, our model may be fully veridical with respect to the property *temperature*, while widely misrepresenting the system regarding some (or even all) particular momenta.

3. *There is a sense in which a model's predictive success may be due to the model being veridical with respect to some appropriate less specific properties that it implies.* This happens if *any* model, such that it explicitly attributes to the target *some* way of exemplifying these less

specific properties, entails commensurate predictive accuracy (when conjoined with the same laws of nature). For example, consider the model of gas above, assuming that we are predicting the system's temperature: any specific distribution of particular momenta will do, as long as the average kinetic energy is right.

Here's a less obvious example. It has been recently argued that some properties of soft condensed matter (polymers, e.g., DNA in solution) are independent of the specific interactions between the individual molecules and of the exact geometrical properties of the macromolecule but rather depend only on their (less specific) *topological* properties. A model of the inter-molecular interactions and the molecule geometry that gets the topological property right can make right predictions by virtue of this fact. After all, different more specific modeling assumptions regarding the macromolecule's geometry would entail the same predictions, since these are just different ways for the macromolecule to have the appropriate topological feature on which the relevant behavior exclusively depends.

Viewing this from a realist perspective—in term of how predictive success and truth are arguably typically connected—we can put it as follows: *the more specific properties and the less specific properties need not be on a par in explaining a model's predictive success.* The less specific properties can be solely responsible for predictive success—or *'success-inducing'*—in the following sense: if the prediction hangs on getting a less specific property right, then *any* more specific property that is just a way of exemplifying this success-inducing less specific property, is equally good in the model (as far as predictive accuracy is concerned). A more specific property, on the other hand, can be play various less essential roles in bringing about success, or be *'success-catalyzing.'* For example, we might be able to explain how scientists were able to cognitively access some success-inducing properties by dealing with more concrete, more specific success-catalyzing properties that turned out to misrepresent reality.

4. *A model that is veridical vis-à-vis some success-inducing properties, but misrepresents some more specific properties, can be an idealized model.* This follows from the intuition that (many) idealized models are models that make 'simplifying' assumptions that are not too detrimental for their (predictive) purpose. Consider a model that is idealized by virtue of providing a simplified representation of some aspects of its target. If those simplifying assumptions concern purely some more specific properties that are suitably related to the success-inducing less specific properties, then the misrepresentation that issues from the idealization should not be detrimental for predictive success since the model is veridical regarding the success-inducing properties themselves.

How can misrepresentation at the level of more specific properties yield an *idealization*? That is, in what sense can a misrepresentation vis-à-vis more specific properties *simplify* a model? The idea here is that an idealizing assumption that misrepresents the target plays a purely pragmatic role in giving us cognitive or mathematical handle on the success-inducing properties. In other words, idealizing assumptions are success-catalyzing by virtue of allowing us—with our limited cognitive and mathematical powers—to build models which are veridical in the respects that are required for bringing about an accurate prediction. We could try modeling directly in terms of the success-inducing properties, but this can be cumbersome or impossible both cognitively and mathematically, because success-inducing properties can be highly abstract and unspecific. This is why there is often *pragmatic need* to introduce more concrete and more specific assumptions into a model, even if those assumptions per se are not really success-inducing.

In principle we could try modeling a system with any of the more specific properties that are ways of exemplifying the success-inducing less specific properties, since these less specific properties are really responsible for the model's success, and these properties underdetermine the more specific properties. But it may happen that some of the more specific properties are privileged in terms of yielding a model that is the simplest (for us) to operate with, and those privileged more specific properties need not be the ones that the target actually exemplifies. When it comes to pragmatic considerations, it is better to build an idealized model that lies about the more specific properties, but is easier to work with, than a model that is veridical with respect to the more specific properties, but leads to mathematical complications. I will look at an example of this later.

These are—in very general terms—the initial assumptions underlying the conceptual framework that I wish to erect. No doubt these will have to be refined and clarified in due course. I will spend the rest of the paper developing these initial ideas in some detail and situating my position in wider context, beginning with a review of the clearest extant analysis of idealization. This paper can be viewed as an attempt to clarify and extend some aspects of that analysis.

JONES ON IDEALIZATION

We should follow Jones (2005) in partially defining 'idealization' and 'abstraction' as follows. *It is a necessary condition for a model to contain an idealization (viz. to be an idealized model) that the model misrepresents at least some properties of its target system. Models incorporate an abstraction in those respects in which they completely omit certain properties, i.e,. say absolutely nothing about them.* There are further conditions that we will want to impose on idealizations, but stipulating this much is already clarifying given the various conflicting uses of these terms in the literature. I maintain that this is the best starting point for thinking about idealization.

A couple of remarks about Jones's framework before proceeding further: (i) It should be seen as a regulative framework for philosophers of science to define their terms. The literature is replete with contradicting uses of these terms, so this is much desired. (ii) These definitions are not meant to capture the various ways in which scientists use terms 'idealization' and 'abstraction.' (iii) A model that only contains abstractions is simply veridical regarding the target system but limited in its content. There is no good sense in which abstractions are non-veridical. (iv) When we say that abstractions omit properties by saying nothing about them, we really mean that. For example, if we have a model of a particle that is an abstraction with respect to the property of being electrically charged, it doesn't mean that the charge is zero, so that if the particle was placed in an electromagnetic field, the field would have no effect on it. For this would be an idealization, falsely representing the charge as vanishing.

Saying this much only partially defines idealization as opposed to abstraction, and Jones (2005) recognizes that there must be some further factors that are often present when we speak of an idealization, simply because not all misrepresentations are idealizations. Jones takes the following further factors to be 'typical':

- TRUTH APPROXIMATION: what an idealized model says about its target is often 'approximately true.'
- SIMPLIFICATION: the function of an idealization is to simplify by misrepresenting some features of the target.
- GOAL RELEVANCE: somehow what gets misrepresented are the properties relevant for our modeling goals.

There's broad agreement in the literature that all of these are typical aspects of idealization. But so far this only amounts to little more than sign posting, and it would be good if we could sharpen the key notions 'approximate truth,' 'simplicity,' and 'relevant features.' Below I will try to elaborate on Jones's framework by focusing on the kinds of conceptual resources already alluded to.

INFERENTIALLY VERIDICAL REPRESENTATIONS

Let's begin with approximate truth. First of all, I propose that we save the term 'truth' (approximate or otherwise) for the logico-semantic notion

Idealized Models As Inferentially Veridical Representations

that relates to linguistic representation and propositions. Instead, let's use 'veridical representation' in connection with models that can be construed as incorporating non-linguistic representations: a model is a (fully) veridical representation if all the properties the model attributes to the target are actually exemplified by the target. For most models full veridicality is a chimera, of course, and we need a notion qualified somehow. I propose two qualifications below, to define *'partially veridical'* and *'inferentially veridical'* representations. After outlining these concepts, I will illustrate them with some simple examples. (My ultimate realist agenda is worth keeping in mind.) In the next section, I will return to idealization.

Having a *partially veridical representation* can be initially construed simply as a matter of having a representation that is correct about some of the properties it attributes to the target. This notion as such, taken at face value, is rather trivial and uninteresting, for just about any model in science is a partially veridical representation in *some* respects. Teller (2001) suggests a good way to proceed in order to avoid triviality. What really interests us is not just the partial veridicality of a representation, as defined above, but partial veridicality *with respect to those properties that are appropriate for the modeling purpose*. From here on I will use 'partially veridical representation' in this *contextualized* sense that takes into account the use to which a model is put. Hence, from now on, a model is a partially veridical representation (relative to some context of use) if and only if it correctly attributes to the target some properties that are appropriate for the purpose of our modeling. This is a wholly contextual notion, since the partial veridicality of a model depends on what we are trying to do with it.

Defining partial veridicality in this way as a representation that is correct about some relevant properties attributed to the target is ambiguous: do we mean the properties that are *explicitly* attributed to the target, or do we consider the properties that are implied *implicitly*? I propose that we disambiguate by reserving 'partial veridicality' for explicitly attributed properties and introduce the term *'inferentially veridical representation'* to cover both explicitly attributed and implicitly implied properties that the target exemplifies. For example, a detailed particle model of a gaseous system may not explicitly represent its target as having a temperature, even though it implies it (assuming that temperature is identified with mean kinetic energy). The model may misrepresent all individual particle momenta, yet be partially veridical with respect to particle number, for example, and only inferentially veridical with respect to the property *temperature*. The notion of inferential veridicality is still used in the contextualized sense that takes into account the use of a model. Otherwise the notion threatens to become trivial by virtue of every model being inferentially veridical in countless ways, due to the fathomless gerrymandered properties implied by a model.

Inferential veridicality, thus sketched, is serviceable for analyzing how non-veridical models can latch onto reality in ways that make those models successful. But it raises issues about property individuation. How do we identify a model's *success-inducing* properties to see whether or not (or to what extent) the model is an inferentially veridical representation? At this point it is paramount to recognize the importance of various (non-causal) relations between properties, as we need to be able to consider and contrast the role of less specific versus more specific properties in producing a given success. We need to be able to consider the possibility that a model is an inferentially veridical representation due to some relevant less specific properties being success-inducing, while more specific properties are misrepresented. This only requires that the properties critical for our modeling interest are actually less specific than the properties the model falsely attributes to the system. We can then go on to consider the role played by the more specific properties in the model. If not success-inducing, they may nevertheless be 'success-catalyzing' in some sense, e.g., by making the model simpler, or by making a heuristic contribution.

I will give illustrative examples of inferentially veridical models before connecting inferential veridicality with Jones's understanding of idealization in the next section.

Example 1

Here's an intuitive toy-example, adapted from Strevens (2004), to begin with. Assume that you are interested in a system comprising a white billiard ball, with momentum p, hitting a glass window pane. In your model a black billiard ball travels at the same velocity but has momentum p' (with $p'>p$ due to larger mass), hitting and breaking the glass. The model is inferentially veridical with respect to predicting the glass breaking. We can ignore the properties of the system it misrepresents: the color of the ball (obviously), and also the exact mass and momentum of the ball. The model's predictive success is due to it correctly attributing the following key property to the target: the ball's momentum is higher than some critical p_c (with $p_c<p$) required for breaking the class. This property is less specific than the exact momentum the model attributes to the ball. That is, the model correctly attributes (implicitly) to the system a property it exemplifies—*the ball having momentum bigger than p_c*—and latching onto this property is appropriate vis-à-vis your predictive use of the model. (The model is also veridical with respect to the exact speed of the ball, but this in itself is not success-inducing property.) This is an apt model for its purpose, and its aptness is down to it being an inferentially veridical representation in this way.

Example 2

Assume that you have a model that purportedly represents a particular type of knot puzzle. By using the model you are able to predict that most people in the test group cannot systematically figure out how to untie this type of knot when first presented with it. Your prediction is based on a topological property of the model knot, combined with cognitive science knowledge of humans' limited ability to mentally manipulate spatial figures of certain similar topological types. It might happen that the actual knot used in testing human subjects is different from the model knot but not too different: it exemplifies the same topological property. (The model knot is right-handed, while the actual knot is left-handed.) How do we explain your predictive success in the face of this representational discrepancy? We should attribute this predictive success to the model being veridical with respect to the topological property of the knot. After all, the regularity discovered by cognitive science (we assume, for the sake of the argument) concerns specifically *topological* properties of a spatial figure—not this or that more specific geometrical property that is a way of realizing the topological property. Our model, though strictly speaking non-veridical, is an inferentially veridical representation by virtue of correctly attributing the more abstract topological property that is appropriate for this purpose of modeling.

Example 3

The toy examples above are meant to give an intuitive handle on the kind of inferential veridicality of models that I have in mind. Here's a more realistic example from the realist literature.

Augustin Fresnel's ether theory of light has been much discussed as a historical case that demonstrates how a radically false theory can yield novel predictive successes. One of these successes—Fresnel's derivation of the so-called Fresnel equations for the intensity of reflected and refracted polarized light—has been particularly prominent in the literature. Should we accept the anti-realist contention that Fresnel's predictive success relied on false theoretical assumptions? Or should we explain Fresnel's success perhaps in terms of whatever 'structural' assumptions he made regarding light? (Worrall 1989)

I have argued that neither response is appropriate in this case. (Saatsi 2005, 2008) A closer look at Fresnel's derivation indicates that underlying the derivation there was a set of modeling assumptions according to which the undulating elastic solid ether realizes some rather abstract spatial continuity properties. The mathematical equations that Fresnel used to derive his end result capture these more abstract properties. With respect to these properties, there is a perfect correspondence between Fresnel's theorizing, on the one hand, and the contemporary derivation of Fresnel's equations from Maxwell's equations, on the other. Hence, the realist can put her finger on this kind of continuity in the less specific properties in spelling out her realist commitments. Fresnel's theorizing was latching onto reality, and the model underlying his derivation was inferentially veridical.

Example 4

Assume that you wish to model, with good predictive accuracy, a tennis ball rolling down an inclined asphalt plane. You use a simple model found in introductory physics textbooks, with good results. The model misrepresents the target in various respects but very little in each respect: the gravitational field is not *absolutely* homogeneous; the ball is not *absolutely* smooth, hard, and spherical, etc. But intuitively speaking, the model latches onto reality in a way that ensures its predictive success.[6] How should we conceptually precisify this intuition about the model latching onto reality? I prefer to think in terms of the properties that the model explicitly attributes to the system—the ball is perfectly spherical; there is absolutely no slippage, and so on—and see how they relate to the various less specific properties that the model in a sense must get right in order to be accurate.

For the model to be considered successful, there has to be a bit of leeway between the actual motion of the ball and what the model says: the model's predictions are good but not perfect. Once we (contextually) decide how much leeway is accepted, we can start thinking about the various ways in which the model could be modified without too big an effect on predictive accuracy. For example, we can replace the homogeneous gravitational field by an inhomogeneous one without too big an effect, as long as the inhomogeneities are very small. The perfect sphericity of the ball and the perfect smoothness of the surface can be replaced by properties that are, intuitive speaking, "close enough".[7] There is an enormous number of ways in which a model's assumptions can be thus modified without affecting the model's predictive accuracy too much. Most of such modifications lead to mathematical complexities in practice, of course, and one has to resort to various approximation schemes or computer simulations in order to get predictions out of most of the alternative models.

But ignoring such usability issues altogether, and focusing purely on the connection between a model's predictive accuracy and the properties it attributes to the target, we see that there is clearly a sense in which the original model is *unnecessarily specific* in representing the system as having the exact properties. For there is an enormous range of alternative models—each more or less on a par in "approximating" the actual system—that are equally good in terms of predictive success (relative to the amount of leeway accepted). What really matters for predictive success, then, is that each exact property that a model attributes to the system belongs to some appropriate *range* of specific properties. Or, in other words, that each exact property attributed by a model is a way of exemplifying some appropriate less specific property. Any predictively successful model has to imply that the ball exemplifies the less specific property *being approximately spherical*, which is a shorthand for a disjunction over the set of those more specific properties that entail an outcome within the limits of agreed accuracy.[8] (This set includes the exact shape

Idealized Models As Inferentially Veridical Representations 245

actually exemplified by the ball.) In the same sense, the surface has to be *approximately smooth*, and the gravitational field has to be *approximately homogeneous* and so on. These are the properties that any predictively successful model will correctly attribute to the system, and any model that is not successful in this way will fail to attribute at least one of these properties to the system. The intuition that the original model latches onto reality gets thus analyzed in terms of the model being inferentially veridical by virtue of correctly implying, implicitly, that the target exemplifies these critical less specific properties.

All the above examples indicate that inferential veridicality of a model can be due to a representation latching onto reality via a false property attribution that is *unnecessarily specific*.[9] In each case one could try to present a more abstract model that only attributes a less specific property, correctly, to the target system. For example, one could have represented the white billiard ball as an object of some appropriate size hitting the glass with *some* momentum larger than p_c (without specifying any particular momentum). Similarly, one could have represented the knot mathematically as having a particular topological property, without saying more precisely whether it is this or that specific knot that realizes this particular topological character. Likewise, Fresnel could have (in principle, even if not in practice) abstracted away from the unnecessarily specific properties that he mistakenly attributed to light, focusing purely on the less specific spatial continuity properties required for his derivation. Also, one could try presenting a less specific model for the rolling tennis ball, only attributing to the system those less specific properties that the original model correctly implies together with all the different models that are predictively successful (where predictive success is determined by some accuracy requirement).

But even if there exists a corresponding abstraction that is fully veridical, in many cases it is quite natural not to present these less specific properties directly but instead give a model that is more specific in such a way that the appropriate less specific properties are ipso facto implied. There are various reasons why our models often involve such unnecessarily specific properties, related to various features of our modeling practices. I will next finally argue that some idealized models can be analyzed in terms of such unnecessarily specific properties playing a pragmatic, simplifying role.

ELABORATING ON JONES'S FRAMEWORK

The ubiquity of idealization in science presents the scientific realist with some difficult questions. What good can it be for a model to *lie* about the system represented by the model? This general question can be narrowed down by asking it with respect to different uses of models; we can ask the question, for example, with respect to explanatory or predictive worth of idealizations. In this paper I have focused on the predictive aspect, which I consider to be more pressing for the realist.[10]

The realist intuition is to claim that we misrepresent in order to simplify our model, because misrepresentation in some respects does no (significant) harm for the predictive ability of the model, whereas not misrepresenting in those respects would lead to enormous complexities or to the lack of a usable model altogether. Conceptually precisifying that intuition is a great challenge: the realist needs to account for the indispensability of idealizations, on the one hand, and at the same time defend the realist claim that the predictive success of idealized models is in a substantial sense down to those models latching onto reality and not down to their lying about the world. My contention that some idealized models may be construed as inferentially veridical representations goes some way toward answering that challenge, by providing a conceptual framework for analyzing a model's 'latching onto' its target in a way that can be explanatory of its predictive success.[11] Finally, I will now explain how misrepresenting in some respects can *simplify* a model and thus be indispensable to our modeling practices.

Recall the initial assumption (4): an idealized model can be a model that (i) is inferentially veridical by virtue of correctly implying some less specific success-inducing properties exemplified by its target but (ii) purposefully lies about some more specific properties, because doing so does no harm, but rather provides some pragmatic benefits. This is a way of elaborating on Jones's idea that idealizations do not misrepresent any old property but typically properties that are *somehow* relevant for the purpose of the model. The properties being misrepresented are those properties that are very intimately connected to the properties that are success-inducing. Namely, the misrepresented properties are possible ways of the target system having the less specific, success-inducing properties.[12] All that is required for such misrepresentation to count as an idealization is that the misrepresented more specific property is somehow pragmatically useful.

Some of the examples in the section above indicate how it may be natural not to present the success-inducing less specific properties directly, but instead give a model that is more specific in such a way that the relevant less specific properties are ipso facto implied. For example, any *material* model of the billiard ball hitting the glass is necessarily going to attribute this or that specific momentum to the target, even though the success-inducing property is *having a momentum greater than p_c*. Regarding the knot example, we can just note that abstract topological properties or equivalence classes cannot be simply pictured, unlike the particular knots realizing those properties. These are not clear cases of idealization, however, since there is nothing to make it pragmatically useful to adopt any particular way of implying the success-inducing property via some non-veridical more specific property.[13] Also, the model that Fresnel employed is not idealized, for it is not the case that due to some pragmatic constraints Fresnel could only latch onto the more abstract success-inducing properties by way of making knowingly false, more specific assumptions regarding the ether. Rather, it was natural for Fresnel to make such more specific assumptions, and they were heuristically indispensable for him, given the metaphysical and explanatory presuppositions of the ether paradigm.

The final example in the section above is different, however. The inclined plane model of the tennis ball is clearly idealized by virtue of the fact that the various false assumptions it incorporates yield the mathematically simplest—it is safe to assume—model among all the alternative models of acceptable predictive accuracy. So when it comes to demonstrating or predicting the speed and position of the tennis ball over time (within given error bounds), we can do it with the least amount work with this model. There are other models that are equally good if judged purely with respect to our goal of predictive accuracy, each specifying a particular property to the ball, to the inclined surface, to the gravitational field, etc. Not all these models are on a par regarding their mathematical simplicity, however, and the mathematical description of the actual system may also be overly complicated and unworkable. We might try and operate with the less specific, success-inducing properties directly, without specifying this or that particular way of being *approximately* spherical, *approximately* homogeneous, and so on. But doing that would be extremely cumbersome or impossible mathematically. For example, it is so much easier to write down and deal with an equation for a model that attributes a precise force acting on the ball at each moment of time, than it is to deal with the much more abstract model that *only* says that there is *some* force, within some specified bounds, acting on the ball at each moment.

Idealization like this is rife in science, but admittedly there are different kinds of cases that may push the conceptual framework that I have developed here over its limits. For example, it is not obvious how to characterize the success-inducing properties specified by infinite population models in biology or by a model of waves in an infinitely deep pond of water, and I don't claim that the framework presented here is applicable across the board.[14] But the realist's conceptual framework needs to cover these cases, too, for how can she take these models to 'latch onto' any real system, unless we can somehow discount those infinite numerical differences between the model and the actual system as kinds of differences that are ultimately immaterial to explaining the success of the model in terms of its truth-latching aspects?

NOTES

1. I wish to thank two anonymous referees for their valuable comments.
2. The second step in the realist argument would be to argue that the predictive success of some idealized models can indeed be *explained*, in a realist sense, in terms of those models' partial veridicality. The third and final step would be to argue that the second step *justifies* realism regarding these kinds of idealized models. These further steps are beyond the scope of this paper.

Here I just focus on elucidating a kind of partial veridicality, which is of interest independent of the realism debate. Nevertheless, I will be open about my ultimate realist agenda throughout.

3. I do not take the semantic view to be wedded to structuralism of any sort, either. A structuralist rendering of the semantic view allows one to analyze approximation and partial truth in terms of formal notions such as partial structures and partial isomorphisms, introducing a degree of welcome precision (e.g., da Costa and French 2003). At the same time, focusing on such formal notions doesn't itself bring out certain conceptual resources that allow us to understand how certain idealizations naturally arise and function in science.

4. The notion of 'inferentially veridical representation' developed here is not a philosophical theory of representation per se, but rather an account of how false scientific representations can latch onto reality. Hence, it is quite distinct from the inferential conception of scientific representation developed by Suárez (2004).

5. See Funkhouser (2006), for example, on the determination relation and how it differs from other kinds of the specification relations, such as multiple realization. My distinction between 'more specific' and 'less specific' is meant to cover all of these relations.

6. I'm not *arguing* for realism about such a model here. Rather, I'm illustrating how the notion of inferentially veridical representation can be employed to articulate the sense in which the model is arguably latching onto reality in a way that is responsible for its predictive accuracy.

7. Models represent by specifying properties exemplified by their targets, and we can contemplate various changes to the model *directly* in terms of these properties, regardless of whether or not these changes can be nicely captured through the equations that partly define the original model.

8. The disjunction is fixed by a contextual error bound together with the other properties attributed to the system.

9. Unnecessarily specific in the sense that some of the specific properties the model attributes to the system are not really responsible for predictive success but at best play some lesser, 'success-catalyzing' role.

10. For the explanatory dimension of this question, see Strevens (2008) Chapter 8.

11. The next step in the realist argument would be to defend the idea that the explanations furnished by my conceptual framework really are *realist* explanations. I believe (but will not argue) that this step can be taken on a case-by-case basis for many idealized models.

12. This dovetails with the intuition that idealized models represent *counterfactual* systems, which demonstrate, explain, or predict the relevant phenomenon in fiction or in other possible world. Much more work is needed in order to make precise the sense of modality involved, however. It's true that some idealizations seem to represent systems that are nomologically possible in the actual world, but most idealizations are not nomologically possible. So we need to relax at least some laws of nature to view such idealized system as a possible system. But at the same time, idealized models are faithful to some actual laws of nature.

13. A mathematical model of the window-breaking billiard ball might be idealized assuming that some relevant calculations are simplified somehow by attributing a round number 10 kgms ($>p_c$) as the momentum. Thanks to an anonymous referee for this suggestion.

14. Cf. Batterman (2001) for examples of models with infinities.

REFERENCES

Batterman, Robert W. (2001). *The Devil in the Details: Asymptotic Reasoning in Explanation, Reduction, and Emergence.* Oxford: Oxford University Press.
Cartwright, Nancy (1983). *How the Laws of Physics Lie.* Oxford: Oxford University Press.
DaCosta, Newton, and Steven French (2003). *Science and Partial Truth.* Oxford: Oxford University Press.
Funkhouser, Eric (2006). "The Determinable-Determinate Relation." *Noûs* 40: 548–569.
Giere, Ronald (1988). *Explaining Science: A Cognitive Approach.* Chicago: University of Chicago Press.
Jones, Martin R. (2005). "Idealization and Abstraction: A Framework." In *Idealization XII: Correcting the Model. Idealization and Abstraction in the Sciences*, Vol. 86 of *Poznan Studies in the Philosophy of the Sciences and the Humanities*, eds. Martin R. Jones and Nancy Cartwright, 173–217. Amsterdam: Rodopi.
Morrison, Margaret (2000). *Unifying Scientific Theories.* Cambridge: Cambridge University Press.
Saatsi, Juha (2005). "Reconsidering the Fresnel-Maxwell Case Study." *Studies in History and Philosophy of Science* 36: 509–538.
Saatsi, Juha (2008). "Eclectic Realism—The Proof of the Pudding: A Reply to Busch." *Studies in History and Philosophy of Science* 39: 273–276.
Strevens, Michael (2004). "The Causal and Unification Approaches to Explanation Unified—Causally." *Noûs* 38: 154–176.
Strevens, Michael (2008). *Depth: An Account of Scientific Explanation.* Harvard: Harvard University Press.
Suárez, Mauricio (2004). "An Inferential Conception of Scientific Representation." *Philosophy of Science* 71: 767–779.
Teller, Paul (2001). "Twilight of the Perfect Model Model." *Erkenntnis* 55: 393–415.
Worrall, John (1989). "Structural Realism: The Best of Both Worlds?" *Dialectica* 43: 99–124. (Reprinted in *The Philosophy of Science*, ed. D. Papineau, 139–165. Oxford: Oxford University Press).

13 Formats of Representation in Scientific Theorizing

Marion Vorms[1]

INTRODUCTION

Scientists use models as representations of systems or types of systems in order to gain knowledge about them: the representational relationship between features of the model and features of the system enables scientists to draw inferences concerning the system by reasoning with the model. Theoretical models, such as the simple pendulum, as well as data models, can be displayed in different formats: the equation of the simple pendulum can be written in Cartesian coordinates as well as in polar coordinates; temperature values in function of time can be displayed in a list as well as in a graph or a table. Lists and graphs can contain the same information though they convey it in different formats. This paper intends to show that the particular format in which a model is presented matters to the reasoning processes of its users. Moreover, it aims at clarifying the very notion of "format."

The importance of formats for the problem-solving and decision-making performances of agents is a well-studied phenomenon in the area of cognitive science and artificial intelligence (Larkin and Simon 1987; Larkin 1989; Kleinmuntz and Schkade 1993; Zhang 1997). However, its relevance to the study of theorizing is insufficiently acknowledged and analyzed. Philosophers of science interested in the representational function of models generally concentrate on the questions raised by the semantic relationship *between the model and its target, namely the physical system it stands for*: for instance, how does this relationship enable scientists to infer successfully from properties of the model to properties of the target? (see, e.g., Suárez 1999, 2003, 2004; Frigg 2002, 2006, 2010; French 2003) But the relationship *between the model and its user* is generally neglected. The question is: How do models enable agents to obtain information *within* them by reasoning with them, independently of the true or approximately true statements that can be inferred about their target? In order to address such a question, I shall concentrate on *concrete* representations couched in a particular format: I assume that agents do not reason with models such as, for instance, the simple pendulum, *in abstracto*, but rather with a particular concrete presentation of it, namely a set of marks, such as an equation, a graph, or a schematic drawing.[2] And, for the sake of simplicity, I will restrict myself to *external* representations, as opposed to mental representations.

Some philosophers of science, following historians and sociologists (see, e.g., Lynch 1988; Lynch and Woolgar 1990), have emphasized the importance of non-linguistic representations in scientific practice (see, e.g., Griesemer and Wimsatt 1989; Wimsatt 1990; Baigrie 1996; Perini 2004, 2005). These studies generally aim at showing that diagrammatic and pictorial representations play a crucial role in explanation and discovery in some disciplines and cannot be conceived of as mere substitutes for linguistic representations. Such an emphasis on diagrammatic and pictorial representations relies most often on a sharp distinction between these representations and linguistic ones. I, however, aim at integrating in my analysis both the use of non-linguistic representations and of linguistic representations in various formats: I am interested in the possibility of presenting the same information under various formats and in the consequences of a difference in format for the agents' reasoning processes.

Following Paul Humphreys (2004), I shall focus on the "concrete pieces of syntax" that are constructed and manipulated in scientific practice. My perspective, though, is substantially different from Humphreys'. His proposal is meant to do justice to the extension of our computing abilities due to the increasing use of computers in scientific practice, and he therefore concentrates on the importance of the syntax of linguistic representations. However, I am interested in the way agents with limited cognitive abilities reason with representations of various kinds.

I will start ("Formats Matter: Some Examples") by considering a handful of examples that highlight the fact that representations in different formats can contain the same information though they facilitate different inferential processes, and that such phenomena have non-trivial consequences on scientific practice.

"Formats as Symbol Systems?" aims at clarifying the intuitive definition of a format used in "Formats Matter: Some Examples" by means of the notion of a symbol system. A symbol system is defined as a set of syntactic and semantic rules governing the construction and interpretation of representations.

It will finally become apparent ("Relativity to Agents. Towards a Definition of Formats") that defining formats in terms of rules of construction and interpretation is insufficient to account for the complexity of the cognitive interaction between agents and the representations they construct and manipulate. I shall argue that the format of a representation ineliminably involves the epistemic interests and skills of the users of the representation.

252 Marion Vorms

FORMATS MATTER: SOME EXAMPLES

Let me start with a toy example, drawn from (Kulvicki 2010). The results of a temperature survey can be displayed in different ways (Figures 13.1–13.3).

1, 1, 06	3, 1, 33
1, 2, 22	3, 2, 39
1, 3, 27	3, 3, 45
1, 4, 33	3, 4, 54
2, 1, 27	4, 1, 37
2, 2, 24	4, 2, 46
2, 3, 35	4, 3, 45
2, 4, 49	4, 4, 65

Figure 13.1 By John Kulvicki, "Knowing with Images: Medium and Message," *Philosophy of Science* 77(2): 303 (2010), University of Chicago Press. Copyright 2010 by the Philosophy of Science Association. All rights reserved.

23	49	54	65
27	35	45	45
22	24	39	46
06	27	33	37

Figure 13.2 By John Kulvicki, "Knowing with Images: Medium and Message," *Philosophy of Science* 77(2): 304 (2010), University of Chicago Press. Copyright 2010 by the Philosophy of Science Association. All rights reserved.

23	49	54	65
27	35	45	45
22	24	39	46
06	27	33	37

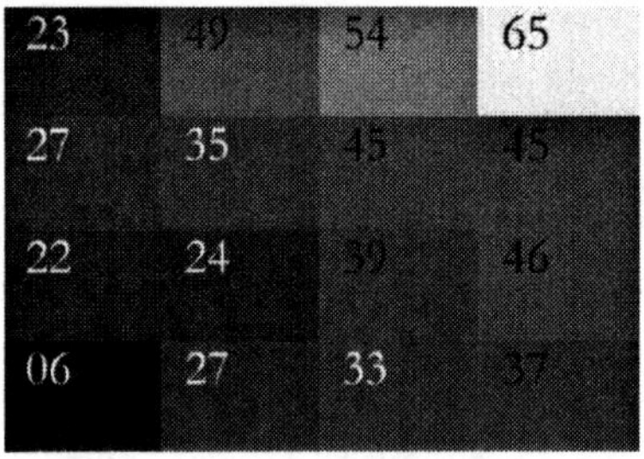

Figure 13.3 By John Kulvicki, "Knowing with Images: Medium and Message," *Philosophy of Science* 77(2): 305 (2010), University of Chicago Press. Copyright 2010 by the Philosophy of Science Association. All rights reserved.

One can present them, as on the left-hand figure, as a list of triples of numerals, the first two standing for the coordinates of the different places where measurements were taken and the third one for the corresponding temperature values. The very same data can be presented in a two-dimensional diagram, where the locations of the numerals standing for the temperature values respect the relative distance between the places where measurements were taken: the tables in Figure 13.2 display spatially the data from the list. Shades of grey corresponding to ranges of temperatures could also be added, as in the right-hand table.

The data displayed in these three figures are identical. The tables, though, enable one to draw some information more quickly and easily. Without any precise knowledge of how inferential processes are implemented in agents' brains, it is a difficult task to define precisely the rapidity or the length of an inferential process. Nevertheless, it seems *prima facie* possible to assess it by measuring the amount of time involved in performing this task, or by counting the number of (at least conscious) steps involved in it.[3] For an average user with normal cognitive abilities,[4] it seems quite uncontroversial that the process of comparing between the values of the temperature in two distinct areas is much quicker– according to either criterion– than this operation would be using only the list.

Indeed, in virtue of their format, tables explicitly preserve the topological relations between the areas represented. This means that, in virtue of the spatial relationships between the numerals standing for temperature values *within* them, the tables exempt one from drawing some inferences, which are indispensable when one uses the list. If one wants to know about the relative location of the temperature values, one does not need to infer it from the numerals standing for the coordinates: the very format of the tables, so to speak, itself "computes" this information on behalf of the agent.[5] The table with shades of grey, moreover, increases the ease and quickness of some operations: the relative intensity of the different squares makes immediately perceptible the relative values of temperature. A quick glance suffices to conclude that temperature is higher in the northeast than in the southwest of the area represented, without the need of a comparison of the numerical values. Reaching this conclusion with the middle table would require a longer inferential process, although not a difficult one for an agent with normal cognitive abilities. On the other hand, reaching it by using the list would be quite laborious: one would have to compare and memorize various numerical values displayed in different lines. A natural way of proceeding would be to draw a table; that is, to transform the list into another representation containing the same information and displaying it in a format more appropriate to the task.

This first example highlights a well-known phenomenon: according to the format in which data are presented, an agent's performances in solving a problem can be increased or reduced. As Larkin and Simon (1987) put it, two representations can be *informationally equivalent* and *computationally different*: though containing the same information (since the representations can be obtained by transformation from each other), they do not make the various pieces of information they contain equally accessible to the agents. I shall clarify this distinction in "Informational Content *versus* Cognitively Accessible Content". Larkin and Simon (1987) suggest that different kinds of representations typically display, in an explicit form, different kinds of information: diagrams preserve topological relations, outlines preserve hierarchical relations, and languages are well fitted to display logical or temporal relations.[6] Therefore, these representations do not require the same inferential processes from the agents: entering data into them as well as computing information does not rely on the same operations. Depending on the kind of problem to be solved, and on the information one has at one's disposal, a certain format may be more appropriate.

Scientific activity, however, does not consist solely in manipulating data. As we shall now see, changes in format also happen in the case of general representations such as theoretical models[7]. Let me first consider classical mechanics. Solving a problem in classical mechanics, say, predicting and explaining the dynamical evolution of a system, typically consists in finding the functions that describe the temporal evolution of the position and velocity of the system under study. First, one writes down the differential equations governing the dynamics of the system with the help of the information one has about it, and then, one solves these equations.

The equations of mechanics can be formulated in different ways, according to the kind of coordinate system used to describe the motion of a physical system. One distinguishes generally between the Newtonian formulation and the analytical ones (Lagrangian and Hamiltonian). According to the problem at hand, using one or the other formulation can dramatically facilitate both the process of writing the equations and of solving them. The Newtonian formulation relies on a description of the configuration of systems by means of Cartesian (and sometimes polar) coordinates, which represent the position and velocity of each point of the system at some instant t. Newtonian equations of motion, which govern the dynamical evolution of a system so represented, have the form of Newton's Second Law ($F=ma$, where F, the force, and a, the acceleration, are vectorial quantities). The first step in solving a problem in this framework consists in specifying the forces exerted on the various points of the system in order to write down the corresponding equations. In other words, the Newtonian format requires that one *enter data concerning the forces*, since the value of forces is explicitly displayed in the Newtonian equations.

In the case of constrained systems, namely systems with internal forces maintaining constraints between different points (thus preventing them from moving independently from each other) the identification and specification of each force is practically impossible. In such cases, the Lagrangian formulation is more appropriate: it relies on a description of the configuration of systems by means of so-called "generalized" coordinates, which correspond to the degrees of freedom of the system. Transforming the description of the system from Cartesian coordinates into generalized ones (q_i) enables one to write down the Lagrangian equations of the system, without needing to know the forces maintaining the constraints. The Lagrangian equations have the form

$$\frac{d}{dt}\left(\frac{\partial L}{\partial \dot{q}_i}\right) - \frac{\partial L}{\partial q_i} = 0$$

Forces do not appear in them; the dynamical evolution of the system is entirely governed by a scalar quantity, the Lagrangian L, which typically corresponds to the difference between the kinetic and potential energies of the system. These equations *implicitly* contain the forces maintaining the constraints, since they are expressed in coordinates taking these constraints into account. Therefore, it is possible to retrieve information concerning the forces; nevertheless, the constraints need not appear explicitly. In terms of the analysis of the previous example, what may be called the "Lagrangian format" consists in presenting, in an explicit form, the information concerning the energy of the system, rather than the information concerning the forces. Despite their equivalence to the Newtonian equations, the Lagrangian ones are therefore much more appropriate to cases where some forces are unknown.

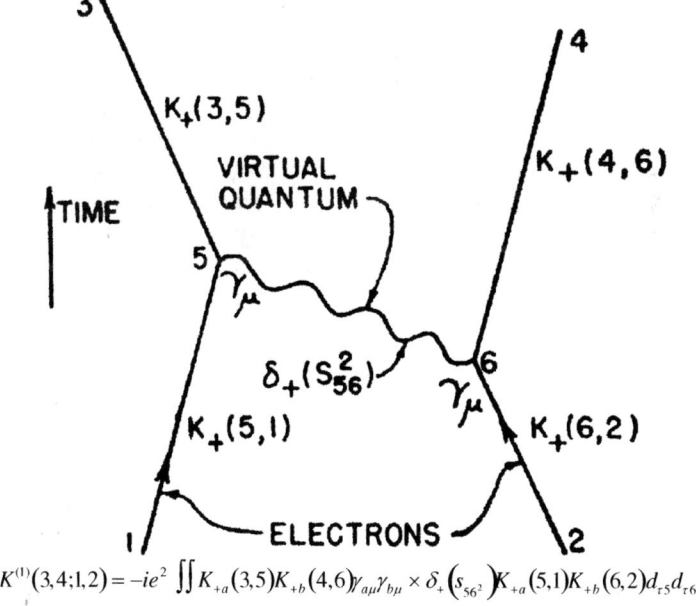

Figure 13.4 Diagram and its corresponding equation. Reprinted with permission from Richard P. Feynman, *Physical Review* 76(6): 772, 1949. Copyright 1949 by the American Physical Society.

In some cases, though, Lagrangian equations are only partially integrable; it is thus impossible to achieve the second step of the problem-solving process, namely finding the analytical solutions of the equations. The Hamiltonian formulation, which uses a different kind of generalized coordinates[8], enables one to change an intractable Lagrangian equation into two corresponding first order equations, by means of a mathematical transformation called "Legendre transformations." Such first order equations are integrable:

$$\frac{\partial H}{\partial p_i} = \dot{q}_i \qquad \frac{\partial H}{\partial q_i} = \dot{p}_i$$

The Hamiltonian H typically equals the total energy of the system.

One can easily show the inter-deducibility of these three kinds of equations. Nevertheless, as we have seen, changing from one to the other can considerably enhance our problem-solving performances: changing from Cartesian to generalized coordinates sometimes facilitates the process of writing the equations, which is otherwise practically impossible; changing from a Lagrangian representation to a Hamiltonian one transforms one intractable equation into two tractable ones. Therefore, despite their equivalence *in principle*, these various equations are not equally usable *in practice* (Humphreys 2004).

Moreover, this variety of formulations has consequences on the development of the theory itself. The Lagrangian formulation is suitable for the expression of the theory of relativity, and the Hamiltonian formalism is used in Quantum Mechanics. As Feynman said concerning the various ways of expressing the law of gravitation (via Newton's law, field theory, or minimum principles), these formulations are "equivalent scientifically. [. . .] But psychologically, they are very different" (Feynman 1965, 53).

Finally, let me consider the case of Feynman's diagrams.[9] Feynman first introduced his famous diagrams (see Figure 13.4) at the Pocono meeting of April 1968 (see Kaiser 2005, 27–59), as a mean to help physicists get rid of the infinities of quantum electrodynamics (QED), which prevented them from giving predictions about complex interactions of atomic particles. At that time, he presented them as "mnemonic devices" (Kaiser 2000, 52) to complete complex higher order calculations without confusing or omitting terms, that task being practically impossible by means only of the mathematical formulae.

A year later, Dyson (1949a, 1949b) demonstrated the equivalence of Feynman's diagrams with the mathematical derivations given at the same time by Schwinger (1948a, 1948b) as a workable calculational scheme for QED. Moreover, Dyson made Feynman's methods "available to the public,"[10] by codifying the rules for constructing the diagrams, stating the one-to-one correspondence of features of the diagrams to particular mathematical expressions.

As is well known, this was the beginning of an amazingly successful career for these diagrams, which, as Kaiser (2005) describes in detail, were eventually used in almost every field of theoretical physics. The diagrams were initially intended to relieve physicists' memory and help them in performing difficult calculations; they finally became genuine theoretical tools, which went beyond the theoretical frame within which they were first designed. They indeed played a crucial role in the research in high-energy post-war physics, and are still taught and used today. Beyond the equivalence of mathematical formulae and diagrams in principle, the latter acquire a genuine independence. As Kaiser suggests, Dyson demonstrated "the mathematical [. . .] equivalence between Schwinger's and Feynman's formalisms," but "by no means" their "conceptual equivalence"[11] (Kaiser 2005, 75).

In all these cases, representations whose equivalence can be mathematically proven happen to have different consequences in problem solving and theory development. According to the intuitive understanding of the notion of "format" suggested by the toy example I gave first, it makes sense to claim that all these are *differences in format*. Newtonian, Lagrangian, and Hamiltonian equations are in different formats; so are Schwinger's formulae and Feynman's diagrams. Although they contain, at least partially, the same information, these representations do not convey this information the same way. The different pieces of information they contain are not equally accessible to agents. Constructing and using them do not consist in the same cognitive operations: identifying and specifying the forces exerted on a system by manipulating vectors does not amount to the same processes as identifying the Lagrangian of the system; drawing a diagram enabling one to visualize the different quantities to be remembered in a calculation does not consist in the same operation as performing this calculation by means of a mathematical formula. *These different representations do not play the same role in the agent's inferential[12] processes.*

Consequently, a first intuitive definition of the notion of format would be: The format of a representation is the "way" this representation conveys to its users the information it contain. A change in format implies a change in the *inferential function* a representation fulfills for its users. In the next section, I shall try to give a more precise definition of the notion of format.

FORMATS AS SYMBOL SYSTEMS?

The intuitive notion of format, as it is used in the above analysis, appears to rely on an analogy with computer science. A format serves to specify a procedure for both encoding data and for retrieving them. Depending on what information is needed and on what kind of "processor" is available[13], some particular procedure for encoding the data may be the most appropriate. Likewise, the differences in format given in the examples imply differences in the inferential processes of agents, which can be described as the application of transformation rules. Depending on the information one is seeking and on one's cognitive abilities, some formats are more appropriate than others. Following this analogy, the format of a representation could be defined as a set of rules of construction and interpretation.

However, the notion as it is used in computer science is not sufficient to account for the cases I am interested in. The idea of a set of transformation rules to process data is certainly a suggestive analogy to describe the cognitive processes of humans, but it is doubtful that such algorithms could be stated for cases as complex and diverse as those offered by the use of external representations in science, which are infinitely more sophisticated than sequences of binary numerical symbols.

The Goodmanian notion of a "symbol system" (Goodman 1976), and the various analyses of representations in structural terms that arose from Goodman's works (see, e.g., Kulvicki 2006), seem likely to offer tools to clarify our notion of format. I will briefly recall in what the notion of symbol system consists, before trying to see whether and how it enables us to capture the phenomena described in "Formats Matter: Some Examples".

The Notion of a Symbol System

According to Goodman, a set of marks (visual or auditory) is a representation of some feature of the world (its *target*)[14] in virtue of a symbol system. Like a language, a symbol system is characterized by a syntax and a semantics. Its syntax defines the set of relevant perceptual properties and their rules of arrangement and transformation (this set of perceptual properties together with their syntactic rules constitute the "symbolic scheme" of the system). The semantics governs the way these perceptual properties so arranged denote different elements of the domain of reference of the system. Unlike the notion of format as it is used in computer science, the syntactic and semantic rules of a system are not supposed to be explicitly stated. The mastering of a symbol system is akin to the mastering of a natural language. The two share various properties, such as generativity, which means that one's competence in using a language or a symbol system is characterized by one's ability to generate and identify novel contents. To sum up, a symbol system determines the content of a representation, and it allows for the production of an infinite number of representations with a novel content.

The notion of a symbol system seems to capture an important aspect of formats. Indeed, one of the goals of Goodman's definition of symbol systems is to characterize different types of systems by referring only to their syntactic and semantic properties. If this enterprise is successful[15], it enables one to distinguish between linguistic, diagrammatic, and pictorial systems in terms of the structural properties of representations functioning in such systems. For instance, whether a "black wiggly line on white backgrounds" is a "momentary electrocardiogram" representing

Formats of Representation in Scientific Theorizing 259

heartbeats or "a drawing of Mt. Fujiyama" (Goodman 1976, 229–230) depends on the system in which it functions. Moreover, some philosophers of science, who claim to be Goodman's followers, have offered analyses of the epistemic virtues associated with different types of systems often used in scientific practice. Laura Perini (2005), for instance, shows that diagrammatic systems are particularly appropriate to the expression of some biological explanations.

However, by concentrating on *differences between types of systems* (linguistic, diagrammatic, pictorial), such analyses do not account for more fine-grained phenomena such as a change in format within the same type of system, as in the case of classical mechanics. Neither do they account for the possibility of transforming a representation that is intractable or difficult to use for some task into a more tractable one, albeit with the same content. Generally, the representations taken as examples in this literature intended to contrast different types of symbol systems do not have the same content; the question of how the same content can be expressed under different formats is not addressed. In consequence, the Goodmanian notion of symbol system provides a way to analyze the *relationship between the structure of a representation and its content,* but it is not aimed at accounting for the *relationship between the presentation of its content and the operations agents* have to perform to access it.[16] As such, the notion of symbol system is therefore insufficient to account for differences in inferential function between equivalent representations in different formats.

Let me not rule out, yet, the notion of a symbol system. It needs to be refined, but nothing excludes, at this stage, the possibility that one could account for the inferential differences between representations in terms of a set of syntactic and semantic rules relating a representation with its content. The rest of the current section investigates the possibility of this.

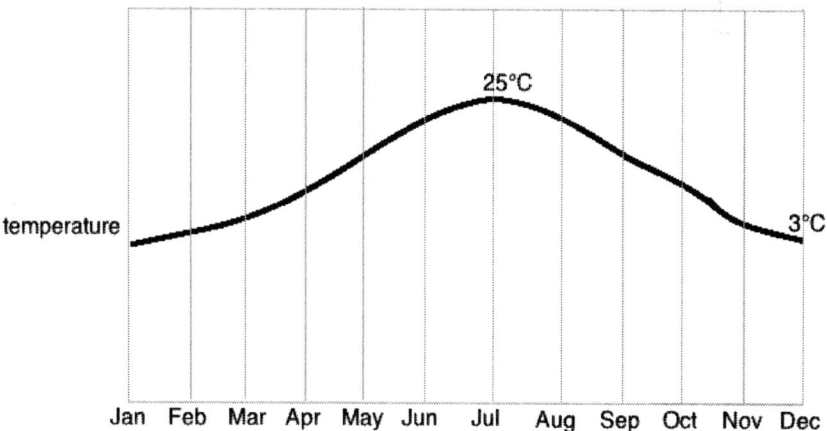

Figure 13.5   Evolution of temperature in Paris.

Informational Content *versus* Cognitively Accessible Content

Let me go back to our intuitive notion of format and to the phenomena that have to be accounted for: two representations in different formats can be informationally equivalent although inferentially different. That means that they can *contain the same information* without *making it equally accessible* to agents with limited cognitive abilities. I will now try to clarify what is meant by "informational content" and by "cognitively accessible content".

What is the Informational Content of a Representation?

There are various theories of information[17], but it is unnecessary, for my purpose, to enter into any detail: let me simply state that a piece of information is a proposition that can be object of belief. One can start by defining the informational content of a representation as the set of all the pieces of information that an agent mastering its symbol system could in principle extract from it. This definition requires several refinements.

The epistemic enterprise of using a representation to gain knowledge concerning its target consists in two steps, which are often simultaneous, but need to be distinguished. For such an enterprise to be successful, one needs to be aware of the source and the precision of the information one can extract from the representation. For instance, one has to know what approximations and idealizations were made in collecting the data. Consider the graph in Figure 13.5. If one believes the values shown in it to represent the average temperature over one month, whereas measurements were taken at 8 a.m. every 10th of the month, one's epistemic enterprise fails. Likewise if one takes the graph to be a representation of the variations of temperature in Paris, whereas measurements were in fact taken in Madrid. This step consists in interpreting the information carried by the graph *as information about* some particular target.

But, *before* inferring—soundly or not—from the features of the graph to the features of its target, one has to *know how to read the graph itself* in order to extract information from it. For a given graph, the system that defines it determines which of its features are syntactically relevant, and how they are to be interpreted, *within the graph*. Let me insist: reading off the information from the graph—even before interpreting it *as about* some target—requires the knowledge of the system's semantics (although not necessarily of its actual referent), and not just its syntax. When a teacher draws a graph on a blackboard in order to show how one has to read it, he does not intend the graph to represent the evolution of any real quantity. Nevertheless, there is a sense in which the graph "says" to its readers that the quantity represented increases or decreases over time. It does contain such information, whether or not it is true of any real place.

Whether one can genuinely speak of representation and informational content when there is no actual referent is a difficult issue, one I shall not tackle here. Since I will not consider issues concerning successful representations or misrepresentations, I shall not use the term "information" as a success term; rather, I use it to refer to any propositional content that an agent who knows how to read the graph can extract from it, whether or not this agent is mistaken concerning the target, and whether or not there is any such target.

Of course, this definition of the informational content of a representation involves some further difficulties.[18] First, as I have said above, some information is presented in an explicit way, while other information requires some inferential process to be extracted. Performing such inferences relies on the mastering of the rules of the system. But it is far from clear, for a given representation, what rules are to be included in the system's syntax and semantics. For instance, the graph in Figure 13.3 "says" in an explicit way that temperature in July is 25°C and that temperature in December is 3°C. One would also like to say that it contains– though implicitly –the information that temperature in December is 22°C less than in July; therefore, it seems reasonable to consider that the basic arithmetic rules, which enable one to make the subtraction, are part of the system's rules. Now, should we consider that the transformation rules from Celsius to Fahrenheit are part of the system's rules? If yes, then the graph also says –implicitly –that temperature in July is 77°F. If not, the graph does not contain this information. Here, one has to acknowledge that whether these transformation rules are part of the system depends on the context in which the graph is used.

Moreover, the informational content of a representation seems to depend on the context for (at least) one more reason. As Haugeland (1991) suggests, one should distinguish between what he calls the "bare-boned" content of a representation and its "fleshed-out" content. The bare-boned content of a representation needs no further assumption to be extracted. On the other hand, the fleshed-out content is obtained *via* a deduction which implies some background knowledge. As Haugeland notes, this distinction does not correspond to the implicit/explicit one. Some implicit information can be extracted from the graph without any further factual knowledge: inferring that temperature in December is 22°C less than in July only requires the mastering of arithmetic rules. However, inferring, for instance, that temperature in Paris in July is 15°C less than in Madrid implies possessing the knowledge that temperature in Madrid in July is 40°C. In some contexts– and particularly in scientific ones –a considerable amount of background knowledge is indispensable to draw important information from a representation. Here again, what knowledge is supposed to be possessed by the user of a representation depends on the context; therefore, the informational content of a representation is context dependent.

However, acknowledging this fact does not dangerously challenge the characterization of the informational content of a representation in terms of its symbol system. In any case, the very system in which a representation has to be read is settled by the context: whether a "black wiggly line" has to be read as an electrocardiogram or as a drawing of Mt Fujiyama obviously depends on the context (for instance, on the caption). One can therefore state that, once what is part of the system is settled by the context, the informational content of a representation is the set of all pieces of information an agent mastering the system can in principle extract from it.

As such, the informational content— despite its context-dependence —does not depend on the relative order in which these pieces of information can be accessed; nor does it depend on the inferential processes agents have actually to perform in order to access them. It is also independent of how difficult it might be to extract different pieces of information. Therefore, the informational content of a representation depends on the representation's perceptual properties and on the system under which it functions; but it does not depend on the rules agents *actually* have to follow, given their cognitive abilities, in order to extract different pieces of this information. It certainly depends on the existence of some agent, since nothing is a representation unless someone uses it as such, but this agent is an *ideal* one, who perfectly masters the system, and whose cognitive abilities are not limited. Differences in format do not have any consequences regarding the information such an ideal agent can draw from a representation and regarding the relative difficulty of the inferences he has to draw: the various representations in Figure 13.1, for instance, have strictly the same informational content.

Cognitively Accessible Content: Cognitive Cost and Types of Inferences

All the pieces of information contained in a representation are not, in practice, equally easy to access. Some are practically inaccessible without the aid of external devices (e.g., computers). As I have already noted, the relative easiness of the inferential process needed to access a piece of information within a representation seems to depend *both* on the symbol system under which the representation has to be read *and* on the cognitive abilities of its user. As Larkin and Simon note "when we compare two representations for computational equivalence [as opposed to informational equivalence], we need to compare both data and operators. The respective value of sentences and diagrams depends on how these are organized into data structures and on the nature of the processes that operate on them"[19] (Larkin and Simon 1987, 67). Therefore, two identical representations, i.e., two sets of data having the same informational content and structured in the same way— following the same syntactic and semantic rules —may have different cognitively accessible contents for two different users. The notion of a symbol system enables us to capture the informational content of a representation; the notion of a format should enable us to capture the cognitively accessible content of a representation. Therefore, format has to be agent dependent.

In the following, I try to clarify what I call the "cognitively accessible content of a representation"; I shall first assume an average user, and try to define what is meant by "accessibility" in terms of syntactic and semantic properties. We shall meanwhile see how agents enter into the picture.

Since accessibility is a matter of degree, let me first clarify what "immediate accessibility" is. John Kulvicki (2010) has given a definition of this notion in terms of syntax and semantics. According to him, a piece of information is immediately accessible when it is (i) displayed in an extractable form, (ii) syntactically salient, and (iii) semantically salient. A piece of information is extractable if there is a "non-semantic feature of the representation in virtue of possessing which it carries the piece of information in question and *no other, more specific piece of information*" (Kulvicki 2010, 299).[20] It is syntactically salient if this feature is perceptually salient. For instance, a map on which red and green hues stand for ranges of temperature makes some information perceptually salient for an agent with normal perceptual abilities (but not for a color-blind agent). According to the color contrast, this information is more or less syntactically salient. Semantic salience depends on the relationship between the perceptual properties of the representation and the data to which they correspond: a map of temperature where shades of red stand for cold areas and shades of blue for warm areas would make the information less semantically salient, given our habits, than the reverse; so would a diagram representing the succession of historical events on a time's arrow starting from the right, rather than starting from the left.[21] Syntactic and semantic saliences depend on the user's cognitive abilities, and they are a matter of degree. Finally, the relative accessibility of a piece of information for an agent depends, first, on the relative syntactic and semantic salience of the perceptual features of the representation responsible for the desired piece of information. These features either display it in an extractable form or facilitate the inferential processes needed to make it so (either mentally or there in black and white). Second, the relative accessibility of the piece of information depends on the relative length of this latter process.

One's ability to extract a piece of information from a representation depends on one's mastering the syntactic and semantic rules of the system and on the perceptual properties of the representation. I will show in the next section that the very notion of "mastering the rules" is far from clear. For now, let me just state that in order to obtain a piece of information, one has to be able to find and recognize the perceptual properties of the representation responsible for its carrying this piece of information; one has also to be able to manipulate and interpret them correctly. The various pieces of information contained in a representation are not equally costly to extract. The *cognitive cost* an agent has to pay in order to extract a piece of information from a representation is inversely proportional to the relative accessibility of this piece of information: the more accessible the piece of information, the less costly the process. Consider again a temperature graph. Reading off the value of the temperature at time t, as well as assessing its global evolution, are easy (not costly) tasks for an average user; these pieces of information are immediately available. However, even if it is not very costly, giving the exact value of the difference between the temperatures at two different moments nevertheless requires some cognitive operations. In some cases, it might even require the use of pencil and paper to write down the values and make the subtraction (e.g., if the values are very precise, with many decimals). The tables in Figure 13.1 make information concerning the relative temperature of different areas less costly to access than the list does. Shades of grey on the right-hand table make this information syntactically and semantically more salient.

A change in format does not only modify the cost of the information-extracting process. As demonstrated in the examples in "Formats Matter: Some Examples", it also modifies the *types of inferences* and more generally the cognitive operations agents have to perform. Reading and comparing numerals on a list is not the same operation as identifying colors and comparing their topological relations in the right-hand table. Likewise, calculating the value of a variable x at time t from the corresponding function is not the same operation as drawing the graph and visually identifying the value of x at t. It is certainly difficult to give a definition of these types of inferences. Nevertheless, without knowing what "happens inside the head" of an agent (which is the subject of cognitive science), and without being able to state algorithms that would model these inferential processes, we have an intuitive grasp of what a type of inference is, by analogy with the notion of algorithm. This intuitive grasp enables us to say that representations in different formats facilitate different types of inferences. There is a sense in which the process that leads one to conclude from the table in Figure 13.1 that the northeast area is warmer than the southwest area, in addition to being quicker, is not the same kind of cognitive operation.

Where are we now? As we have seen, a *difference in format* between two informationally equivalent representations is characterized by a *difference in the cost and in the type of the inferential processes* needed to extract the different pieces of information contained in the representation –that is, a difference in the *inferential function*[22] of the representation for agents with limited cognitive abilities. Although it is practically impossible (because of our poor knowledge of the neural processes of agents) to state explicit rules agents have to follow in order to construct, manipulate, and interpret the representations, one can still define the *format of a representation as a set of procedures agents have to perform in order to access the various pieces of information contained in this representation*. I will now show that such a definition of format is untenable; as will become apparent, it is not only impossible in practice to state such rules, but rather the very idea of a set of rules *to be followed* is a rough idealization, which does not do justice to the complexity of the cognitive interaction of agents with external representations.

RELATIVITY TO AGENTS. TOWARDS A DEFINITION OF FORMATS

We have seen that the inferential function of a representation, defined by the cognitive cost and the type of inferences agents have to perform in order to extract such and such a piece of information, depends on the cognitive abilities of agents. Until now, I have assumed a standard agent, namely a human adult with average cognitive abilities; moreover, this agent is presumed to "master the system" in which the representation has to be interpreted. I will now show that such an assumption relies on an illegitimate idealization.

I have mentioned that, in order to extract a piece of information from a graph, one has to know how to read the graph, that is, one *has to master the rules of construction, transformation, and interpretation* of such a representation. Consider now the representations that are used in scientific practice; mastering their rules requires a certain degree of expertise. Someone who does not know how to solve a differential equation, even when it is integrable, would not gain anything if provided with the Legendre transformations, in addition to a non-integrable Lagrangian equation. That, however, does not challenge the definition of the format of a representation as I have sketched it in the last section. This definition, indeed, is a *normative* one: the format is a set of procedures an agent *has to* perform in order to extract such and such a piece of information. The existence of incompetent agents is not an objection to this definition: such agents do not master the symbol system; moreover, expertise in some field could be defined as the mastering of the interpretation rules of some types of representations. The definition given in the last section is certainly not relative to an ideal agent with no cognitive limitations, but it is not relative to a layman either. The format of a representation is characterized by the inferential function of the representation for its *standard user*: an agent with limited abilities, but who masters the rules of this kind of representation. I will now show that assuming such a standard user is misleading, and quite problematic for a study of the use of models in theorizing, in particular if one wants to analyze their role in theory development as well as in scientific learning.

Consider again the case of Feynman's diagrams. The interest of this case is not exhausted in the comparison between diagrams and mathematical formulae. As their letters and personal papers show, Feynman and Dyson explicitly disagreed on the legitimate use and status of the diagrams within QED. Dyson conceives of them as secondary, psychological aids to the performance of mathematical calculations. On various occasions, he claimed that their use would be illegitimate if they had not been proven rigorously derivable from mathematical formulae: "Until the rules were codified and made mathematically precise, I could not call [Feynman's method] a theory" (Dyson 1979, 62). For him, diagrams were means to "visuali[ze] the formulae which [he derived] rigorously from field theory" (Dyson 1951,

129–130, quoted in Kaiser 2005, 190); they had a meaning only *within* QED, to which they added nothing except cognitive tractability.

On the other hand, Kaiser reports that Feynman never felt the need to show how to derive diagrams from mathematical expressions, and that he expressed clearly on various occasions his theoretical preference for diagrams over mathematical formulae: "All the mathematical proofs were later discoveries that I don't thoroughly understand but the physical ideas I think are very simple" (Feynman, "Letter to Ted Welton, 16 Nov. 1949," quoted in Kaiser 2005, 178).[23] Hence, unlike Dyson, he thought of diagrams as primary and more important than any mathematical derivation that might be given. In addition to being mnemonic devices, they provided an intuitive dimension to the theory, and Feynman took them as "intuitive pictures" (Kaiser 2005, 176). As Dyson notes, Feynman "regard[ed] the graph as a picture of an actual process which is occurring physically in space-time."[24] Rather than visualizations of the formulae, they were primary visualizations of the physical processes themselves. Despite their agreement on the in-principle equivalence of the diagrams and the formulae, Feynman and Dyson did not construct and use diagrams in the same way, and in the final analysis did not "see" the same thing in them.

As Kaiser suggests, this difference in use by the two physicists can be explained by referring to their own theoretical commitments and preferences. Unlike Dyson, who demonstrated how to cast both Feynman's diagrams and Schwinger's equations within a consistent field-theoretic framework[25], Feynman's renormalization approach, from which the diagrammatic method arose, was based on particles, rather than on fields. More generally, as Kaiser notes, Feynman had a preference for a semi-classical approach, and worked almost entirely in terms of particles, trying to remove fields from theoretical descriptions altogether. Such theoretical commitments and interests, together with individual preferences for some kinds of reasoning (Feynman expressed on various occasions his favoring "visualization" over abstract calculation) must have contributed to giving the diagrams different inferential functions for Feynman and Dyson. Dyson deduces them from mathematical formulae, whereas Feynman draws them intuitively: each one relates them in a different way to other representations, and, finally, to the physical world.

In addition to Feynman and Dyson's using the diagrams along different rules, Feynman himself, as well as other physicists, continuously modified their rules of use. Kaiser (2005) gives an impressive analysis of the "plasticity" of diagrams throughout their "spreading" in theoretical practices in modern physics. He shows their varying uses and interpretations in different contexts and "schools" (Oxford and Cambridge, Japan, Soviet Union). Despite Dyson's efforts, the rules of construction and interpretation of the diagrams have not been strictly followed. Moreover, this is the reason why they were so successful: rather than mere calculation tools, they were genuine discovery tools that contributed to important theoretical developments in modern physics, by being used and applied in new fields.

Who is the "standard user" of these diagrams? Feynman? Dyson? Others? Who is the expert who masters the system defining the format of the diagrams? The inferential function of the diagrams is obviously different for Feynman and for Dyson. Moreover, this function constantly changes for Feynman himself. Reducing the function of the diagrams to a set of rules their user must master and conform to would amount to missing some essential aspects of their role in theorizing.

I suggest that, far from being an exception, this is an exemplary case of the way representations are used in theorizing. Let me come back to the example of classical mechanics. In the case of the Newtonian, Lagrangian, and Hamiltonian equations, there seem to exist fixed sets of rules that any expert masters. Let's suppose that this is so and that every physicist today uses them by following the same inferential processes. Reducing them to representations whose inferential function is strictly fixed would nevertheless prevent us from noticing essential aspects of scientific invention and discovery. Consider Hamilton's use of the Legendre transformations: this innovation does not rely on any empirical novelty but rather consists in the *introduction of new transformation rules* within mechanics, which results in a modification of the inferential functions of the various equations. I suggest that theory development often consists in such a process of modifying the rules of use of the representations. As a chess player who, knowing the rules, invents a new gambit, the theoretician modifies the inferential processes he performs in order to solve problems and sometimes develops novel connections between different equations. As this analogy with chess suggests, the process of inventing new inferential paths is potentially infinite. In the case of Classical Mechanics, as well as in the case of Feynman's diagrams, defining the format of the different equations as a set of rules agents *have to* follow in order to extract some pieces of information relies on an illegitimate idealization: *the* right algorithm does not exist.

These considerations imply that we should reassess the idea of a normative definition of formats. The definition of what we intuitively meant by "format" is too context and agent dependent to be settled in terms of a fixed set of procedures *to be applied*. An agent's use of a representation depends on the particular *situation* in which he is involved and on his particular *goals*. As the case of Feynman's diagrams shows, the inferential function of a representation for the same user– as well as for the community –changes over time: the format of diagrams is not the same for Feynman in 1948 and 10 years later. Likewise, the format of the Lagrangian equations was modified by Hamilton's adding to mechanics new rules of transformation, which changed the inferential function of the various equations of the theory. Therefore, the notion of format is fundamentally dynamical in character; the format of a representation has to be defined (beside the perceptual properties of the representation and a minimal set of construction and interpretation rules) in reference to a particular situation, involving a particular agent, with particular skills, theoretical commitments, preferences, reasoning habits, as well as interests and intentions in the particular inquiry in which he is involved.

By this, I do not mean that the definition of a format has to be strictly descriptive.[26] *Once the highly agent-dependent and situation-dependent nature of the use of representations has been acknowledged*, it is certainly worth trying to find the theoretically interesting regularities in the use of representations in different contexts. Such an enterprise might shed light on various theoretical activities, which are usually not treated under the same heading. Indeed, the development of a theory can be (partially) described as a manipulation and modification of the inferential function– the format –of some representations; likewise, learning consists in the progressive modification of the inferential function of some representations for the novice. Becoming an expert in some field, as suggested by Andrea Woody (2004), consists in acquiring "articulated awareness" of the representations used in this field. Learning how to use a certain type of equations, and becoming more and more skillful in it, consists in modifying their format by learning new transformation rules, which modify their inferential function.[27] Deepening one's understanding of a theory consists in developing new inferential paths between representations and reducing the inferential cost that one would typically pay to draw one from the other. This is what physics students do. This is what Hamilton did, with this (important) difference: he did it first and made his modification publicly available.

There is certainly room for normative claims in the analysis of all these activities. For instance, acknowledging that inferential functions of representation are relative to the agents' background knowledge and skills could help us analyze the role and virtue of various kinds of representations in scientific teaching and popularization. In these activities, as well as in theory development by experts, there are definitely successful as well as failing strategies. My main conclusion, in this paper, is that the idealized– positivist-like– image of a fixed set of rules to be followed does not enable us to capture the complex processes that are in play in theorizing. Certainly, according to the object of philosophical inquiry, one has to make some idealizations: if one is interested in popularization, one should pay attention to various inter-individual differences and to the cognitive abilities of the laypersons[28]; however, if one is interested in the inferential differences between Lagrangian and Hamiltonian equations, one should definitely assume that any expert uses differential equations the same way. As the example of Feynman's diagrams shows, in some cases, inter-individual differences between experts (or groups of experts[29]) are also worth taking into account.

Let me clearly state, before concluding, that my point does not amount to saying that scientists' reasoning does not obey any rule and that there exists no difference between a sound inference and a wrong one. Of course, solving a differential equation implies that one conforms oneself to a whole set of calculation rules; if one obeys them, one cannot deduce contradictory results from the different types of equations of classical mechanics. There is a sense in which these equations are equivalent; one cannot draw just anything from them. Just like a chess game[30], the inferential function of a representation partially depends on a set of rules that are objective, in the sense that they do not depend on the users and on the situation. My point is to claim that these rules are highly insufficient to determine the way representations are used; reducing our analysis of theoretical representations to these objective rules results in a narrow conception of theorizing, which misses essential aspects of discovery and of agents' understanding of theories. Theoretical representations such as the equations of mechanics or Feynman's diagrams are highly sophisticated tools of calculation and inquiry, whose rules of use are potentially infinite. Theorizing partially consists in developing these rules to explore the content of representations.

CONCLUSION

Theoretical representations have most of the time been approached from what Paul Humphreys (2004, 77) calls a "no-ownership perspective." From that perspective, theories and models are taken as abstract entities, whose representational relationship with the phenomena has to be elucidated. My analysis aims at showing the importance of the computational dimension of theorizing: theory learning, development, and application relies on the use of "concrete pieces of syntax" (Humphreys, 2004, 99–100), whose rules of construction and interpretation are not fixed. Once the importance of the format of representations in scientific practice is acknowledged, it becomes clear that a definition of format in terms of rules agents have to master and follow is untenable. Once agents and their limited cognitive abilities get into the picture, it becomes impossible to draw a clear-cut frontier between epistemic differences, which would count for all humans, and purely psychological differences. This is particularly true in the case of scientific representations, which are complex and sophisticated tools, rather than mere displaying of data. Taking into account particular agents, with particular skills, involved in particular situations, finally enables us to enlighten some essential aspects of theorizing, which are often neglected. From this perspective, theorizing partially consists in constructing and manipulating representations, whose role in the agents' reasoning processes –whose format –changes with agents' abilities and interests.

NOTES

1. I would like to thank Anouk Barberousse, Paul Humphreys, Philippe Huneman, Cyrille Imbert, Vincent Israel-Jost, John Kulvicki, and two anonymous referees for comments on earlier drafts and/of for helpful discussions.
2. In other words, I shall focus on model *descriptions*, rather than on models themselves (thanks to an anonymous referee for urging me to clarify this point).

3. Thanks to an anonymous referee for this suggestion and for urging me to add these precisions.
4. As an anonymous referee mentioned to me, the relative length of two inferential processes– assessed in either way– certainly depends on the cognitive abilities and skills of the agent with respect to whom they are compared. As Larkin and Simon (1987, 67) note, "'easily" and "quickly" are not precise terms," since the "ease and rapidity of inference depend upon what operators are available for" computing the information. I shall come back later to the problem of the agent-relativity of the notion of format; this problem will be the core of my argument in "Relativity to Agents. Towards a Definition of Formats."
5. It is quite difficult to give a precise definition of what it means for data to be represented in an explicit form. The notion of "immediacy," which I define in "Formats as Symbol Systems?", may capture one of the senses that can be given to this expression. For now, I take it in the sense of (Larkin and Simon 1987): an information is explicitly represented when no inference is needed to access it. For discussions on the explicit/implicit distinction, see, e.g., (Kirsh 1990; Clark 1992).
6. For an analysis of the types of reasoning associated with the use of graphs, see the works of Tufte (1983, 1990); for diagrammatic logic, see (Shin 1994) and (Stenning and Oberlander 1995).
7. As an anonymous referee suggested, a comparison between the importance of format in data models and in theoretical models would be useful; however, this seems beyond the scope of this paper, which merely aims at emphasizing the importance of formats in the case of theoretical models, while it is generally acknowledged in the case of data models.
8. Lagrangian generalized coordinates have the dimension of positions and of velocities, whereas Hamiltonian ones have the dimension of positions and momenta.
9. For historical and technical details, I refer the reader to the works of David Kaiser (2000, 2005).
10. Dyson, "Letter to His Parents, 4 Dec. 1948," quoted by Kaiser (2005, 77).
11. So-called "conceptual role semantics" or "inferential role semantics" (see Harman and Greenberg 2007), which states that the content of a representation consists in (or depends on) its role in the inferential processes of agents could help us give a precise meaning to Kaiser's suggestive remark: diagrams and equations neither facilitate nor require the same inferential processes.
12. Cognitive processes are certainly not reducible to inferential processes. Since their difference has no consequence on my argument, I shall equally speak of "inferential" and "cognitive processes."
13. As I have already noted, the inferential process needed to extract some information from a representation depends on the agents' cognitive abilities. I shall come back to this point in "Relativity to Agents. Towards a Definition of Formats".
14. The target can be a material object, properties of an object, the evolution of the value of some quantity, the relation between various quantities, an event, a pattern, etc.
15. My point is not to propose an internal criticism of this project and of its realizations. Some philosophers have proposed alternate typologies also grounded on (at least partially) structural criteria (see, e.g., Perini 2004; Kulvicki 2006).
16. As an anonymous referee mentioned to me, although Goodman does not pursue a study of the dynamical phenomenon of information processing by humans, some of his analyses could offer tools for dealing with it (see, e.g., Gardner 1974; Goodman 1974;).

17. See (Floridi 2003) for a recent account.
18. Thanks to an anonymous referee to draw my attention on this problem and to urge me to clarify this point.
19. Larkin and Simon refer to Anderson (1984) who argues "that the distinction between representations is not rooted in the notations used to write them, but in the operations used on them" (Larkin and Simon 1987, 68).
20. Kulvicki (2010, 299) gives the following example: "red regions of the Doppler radar image indicate stormy weather of a certain intensity, nothing more. Being red says nothing about the location of such a disturbance: the relative *location* of the red region is responsible for that."
21. See (Tversky 2001) for an analysis of the cross-cultural aspects of the semantics of directionality in diagrams.
22. I maintain a difference between the *format* of a representation and its *inferential function*; since the notion of a format should enable us say that two representations of different targets, and therefore having different informational content, can be in the same format. For instance, one would like to say that a map of France and a map of Spain are in the same format, if they are at the same scale and if they contain the same *type* of information (e.g., relief and nothing else). However, they do not have the same informational content and *a fortiori* not the same cognitively accessible content; therefore, they do not have the same inferential function (they do not enable agents to draw the same inferences). But using them requires the same *type* of inferences. They therefore have the same *type* of inferential function. Accordingly, formats could be defined as types of inferential functions.
23. Feynman also spoke of the "physical plausibility" of the diagrammatic approach (quoted in Kaiser 2005, 177).
24. Dyson 1951, 129 (quoted in Kaiser 2005, 190).
25. Dyson (1965, 23) claims that he contributed to allow "people like Pauli who believed in field theory to draw Feynman diagrams without abandoning their principles."
26. Thanks to an anonymous referee for his/her very valuable comments on this point.
27. As noted by Kuhn (1974), learning consists in acquiring skills (know-how) and not only in learning explicit rules.
28. Although one can assume *types* of laypersons.
29. In most cases, one can certainly define *types* of agents according to their cognitive abilities, skills, habits, and research interests. This enables us to characterize scientific communities by referring to the way their members use some types of representations. Such an approach would account for what Kuhn (1974) had already noted: that different groups of agents, according to the scientific field they work in, will use, e.g., Schrödinger's equation in some consistent way (that is, agreeing on some deductive rules), without applying it to the same cases and without giving it the same interpretation. Kaiser's study of the uses and interpretations of the diagrams by different schools is an example of the fruitfulness of this approach.
30. Of course, the example of chess is a mere analogy: chess playing does not consist in drawing inferences.

REFERENCES

Anderson, John R. (1984). "Representational Types: A Tricode Proposal." Technical Report ONR–82–1, Office of Naval Research, Washington, D.C.

Baigrie, Brian, ed. (1996). *Picturing Knowledge. Historical and Philosophical Problems Concerning the User of Art in Science*. Toronto: University of Toronto Press.

Clark, Andy (1992). "The Presence of a Symbol." *Connection Science* 4: 193–205.

Dyson, Freeman J. (1949a). "The Radiation Theories of Tomonaga, Schwinger, and Feynman." *Physical Review* 75: 486–502.

Dyson, Freeman J. (1949b). "The S Matrix in Quantum Electrodynamics." *Physical Review* 75: 1736–1755.

Dyson, Freeman J. (1951). "Advanced Quantum Mechanics." Mimeographed notes (by Stan Cohen, Don Edwards, and Carl Greifinger) from lectures delivered at Cornell (during autumn) 1951. Copy available in the Physics Department Library, University of California, Berkeley.

Dyson, Freeman J. (1965). "Old and New Fashions in Field Theory." *Physics Today* 18: 21–24.

Dyson, Freeman J. (1979). *Disturbing the Universe*. New York: Basic Books.

Feynman, Richard (1949). "Space-Time Approach to Quantum Electrodynamics." *Physical Review* 76(6): 769–789.

Feynman, Richard (1965). *The Character of Physical Law*. MIT Press.

Floridi, Luciano (2003). "Information." In *The Blackwell Guide to the Philosophy of Computing and Information*, ed. Luciano Floridi, 40–61. Oxford –New York: Blackwell.

French, Steven (2003). "A Model-Theoretic Account of Representation (Or, I Don't Know Much about Art . . . but I Know It Involves Isomorphism)." *Philosophy of Science* 70(5): 1472–1483. Proceedings of the 2002 Biennial Meetings of the Philosophy of Science Association.

Frigg, Roman (2002). "Models and Representation: Why Structures Are Not Enough." In *Measurement in Physics and Economics Project Discussion Paper Series*, DP MEAS 25/02, London School of Economics.

Frigg, Roman (2006). "Scientific Representation and the Semantic View of Theories." *Theoria* 55: 49–65.

Frigg, Roman (2010). "Models and Fictions." *Synthese* 172(2): 251–268.

Gardner, Howard (1974). "A Psychological Investigation of Nelson Goodman's Theory of Symbols." *The Monist* 58(2): Symposium on skills and symbols in the arts 319–326 (Suppl.).

Goodman, Nelson (1974). "On Reconceiving Cognition." *The Monist* 58(2): Symposium on skills and symbols in the arts 339–342 (Suppl.).

Griesemer, James R., and Wimsatt, William C. (1989). "Picturing Weismannism: A Case Study of Conceptual Evolution" In *What the Philosophy of Biology Is. Essays for David Hull*, ed. Michael Ruse, 75–137. Dordrecht: Kluwer.

Goodman, Nelson (1976). *Languages of Art. An Approach to a Theory of Symbols*. Hackett Publishing Company.

Harman, Gilbert, and Greenberg, Mark (2007). "Conceptual Role Semantics," In *Oxford Handbook of Philosophy of Language*, eds. Ernest LePore and Barry Smith, 295–322. Oxford University Press.

Haugeland, John (1991). "Representational Genera." In *Philosophy and Connectionist Theory*, eds. William Ramsey, Stephen Stich, and David Rumelhart, 61–89. Hillsdale, NJ: Lawrence Erlbaum.

Humphreys, Paul (2004). *Extending Ourselves. Computational Science, Empiricism, and Scientific Method*. Oxford University Press.

Kaiser, David (2000). "Stick-Figure Realism: Co Hillsdale, NJ nventions, Reification, and the Persistence of Feynman Diagrams 1948–1964." *Representations* 70: 49–86.

Kaiser, David (2005). *Drawing Theories Apart: The Dispersion of Feynman Diagrams in Postwar Physics*. University of Chicago Press.

Kirsh, David (1990). "When Is Information Explicitly Represented?" In, *Information, Language and Cognition*, ed. Philip Hanson, 340–365. Vancouver: University of British Columbia Press.
Kleinmuntz, Don N., and Schkade, David A. (1993). "Information Displays and Decision Processes." *Psychological Science* 4(4): 221–227.
Kuhn, Thomas S. (1974). "Second Thoughts on Paradigms" In *The Structure of Scientific Theories*, ed. Frederick Suppe, 459–482. Urbana: University of Illinois Press.
Kulvicki, John (2006). "Pictorial Representation." *Philosophy Compass* 1/6: 535–546.
Kulvicki, John (2010). "Knowing with Images: Medium and Message." *Philosophy of Science*, 77(2): 295–313.
Larkin, Jill H. (1989). "Display-Based Problem Solving" In *Complex Information Processing: The Impact of Herbert A. Simon*, eds. David Klahr and Kenneth Kotovsky, 319–341. Hillsdale, NJ: Erlbaum.
Larkin, Jill H. and Simon, Herbert A. (1987). "Why a Diagram Is (Sometimes) Worth Ten Thousand Words." *Cognitive Science* 11: 65–99.
Lynch, Michael (1988). "The Externalized Retina: Selection and Mathematization in the Visual Documentation of Objects in the Life Sciences." *Human Studies* 11: 201–234.
Lynch, Michael, and Woolgar, Steve, eds. (1990). *Representation in Scientific Practice*. Cambridge, MIT Press.
Perini, Laura (2004). "Convention, Resemblance and Isomorphism: Understanding Visual Representations." In *Studies in Multidisciplinarity*, Vol. 2, ed. Grant Malcolm, 37–47.
Perini, Laura (2005). "Explanation in Two Dimensions: Diagrams and Biological Explanation." *Biology and Philosophy* 20: 257–269.
Schwinger, Julian (1948a). "On Quantum-Electrodynamics and the Magnetic Moment of the Electron." *Physical Review* 73: 416–417.
Schwinger, Julian (1948b). "Quantum Electrodynamics." *Physical Review* 74: 439–461.
Shin, Sun-joo (1994). *The Logical Status of Diagrams*. Cambridge University Press.
Stenning, Keith, and Oberlander, Jon (1995). "A Cognitive Theory of Graphical and Linguistic Reasoning: Logic and Implementation." *Cognitive Science* 19(1): 97–140.
Suárez, Mauricio (1999). "Theories, Models, and Representations." In *Model-Based Reasoning in Scientific Discovery*, eds. Lorenzo Magnani, Nancy Nersessian, and Paul Thagard, 77–83. Dordrecht: Kluwer.
Suárez, Mauricio (2003). "Scientific Representation: Against Similarity and Isomorphism." *International Studies in the Philosophy of Science* 17(3): 225–244.
Suárez, Mauricio (2004). "An Inferential Conception of Scientific Representation." *Philosophy of Science* 71(5): 767–779. Proceedings of the 2002 Biennial Meetings of the Philosophy of Science Association,.
Tufte, Edward (1983). *The Visual Display of Quantitative Information*. Cheshire, CT: Graphics press.
Tufte, Edward (1990). *Envisioning Information*. Cheshire, CT: Graphics Press.
Tversky, Barbara (2001). "Spatial Schemas in Depictions." In *Spatial Schemas and Abstract Thought*, ed. Merideth Gattis, 79–111. Cambridge: MIT Press.
Wimsatt, William (1990). "Taming the Dimensions-Visualizations in Science." In Arthur Fine, Mickey Forbes, and Linda Wessels (eds), *Philosophy of Science Association (PSA): Proceedings of the Biennial Meeting of the Philosophy of Science Association*, Vol. 2, 111–135.
Woody, Andrea. (2004). "More Telltale Signs: What Attention to Representation Reveals about Scientific Explanation." *Philosophy of Science* 71(5): 780–793. Proceedings of the 2002 Biennial Meetings of the Philosophy of Science Association.
Zhang, Jiajie (1997). "The Nature of External Representations in Problem Solving." *Cognitive Science* 21(2): 179–217.

Contributors

Mark A. Bedau is Professor of Philosophy and Humanities at Reed College, Editor-in-Chief of the journal *Artificial Life*, and a regular Visiting Professor in the Foundations of Life Sciences and their Ethical Consequences (FOLSATEC) program in the European School of Molecular Medicine in Milan. His research interests include emergence, evolution and adaptation, the nature of life, using machine learning to optimize complex systems, and the social and ethical implications of creating new forms of life in the laboratory.

Brian Epstein is an Assistant Professor at Tufts. He works in the philosophy of social science, metaphysics, and the philosophy of language, particularly focusing on issues in the metaphysics of the social world and the semantics of social kind terms. He is currently at work on a book on the shortcomings of individualism in the social sciences.

Axel Gelfert received his PhD from the Department of History and Philosophy of Science, University of Cambridge, in 2005 and is now an Assistant Professor of Philosophy at the National University of Singapore. He has published on issues in the philosophy of science and technology, as well as social epistemology and its history.

Ronald N. Giere is Professor of Philosophy Emeritus at the University of Minnesota. He is the author of *Understanding Scientific Reasoning* (5th ed., 2006), *Explaining Science: A Cognitive Approach* (1988), *Science without Laws* (1999), and *Scientific Perspectivism* (2006). Professor Giere is a former director of the Minnesota Center for Philosophy of Science and past president of the Philosophy of Science Association.

Paul Humphreys is Professor of Philosophy at the University of Virginia. He is author of *Extending Ourselves: Computational Science, Empiricism, and Scientific Method* (2004) and editor with Mark Bedau of the collection *Emergence* (2008).

Contributors

Cyrille Imbert holds a research position at Centre National de la Recherche Scientifique (CNRS; Archives Poincaré, Université Nancy 2). He has previously studied philosophy of science at University Paris 1—Panthéon Sorbonne, where he received his PhD, and physics at University Pierre et Marie Curie (Paris 6). His primary areas of research are general philosophy of science, especially scientific explanation and the epistemology of computational science, and issues related to the connections between physics and computer science, in particular complexity theory. He has coedited (with Roman Frigg and Stephan Hartmann 2009, 2011) two special issues of *Synthese* about models and simulations.

Tarja Knuuttila is a senior researcher in philosophy at the University of Helsinki. The main themes of her work have been modeling and scientific representation and methodology of economics. She has published in *Philosophy of Science*; *Erkenntnis*; *Science, Technology and Human Values*; *Science Studies;* and in numerous edited books.

Jaakko Kuorikoski is a post-doctoral researcher in the Trends and Tensions in Intellectual Integration project at the University of Helsinki, where he also did his PhD work on model-based explanations in the social sciences. He has published on models, simulations, mechanisms and explanation, especially in relation to economics and other social sciences.

Andrea Loettgers is a historian of science with training in physics. She is currently located at the California Institute of Technology where she is working on a laboratory study in synthetic biology, focusing on the application and change of engineering concepts and metaphors and in the study of organizational principles in biological systems.

Tracy Lupher is an Assistant Professor in the Philosophy and Religion Department at James Madison University. His primary areas of research are the history and philosophy of physics, especially quantum field theory, and broader issues in the philosophy of science, metaphysics, and logic.

Uskali Mäki is Academy Professor at the Academy of Finland and former professor at Erasmus Institute for Philosophy and Economics. He is a former editor of the *Journal of Economic Methodology* and the editor of *The Handbook of the Philosophy of Economics* (Elsevier) and of *The Methodology of Positive Economics: Reflections on the Milton Friedman Legacy* (Cambridge University Press). Much of his current work is on models, realism, and interdisciplinarity.

Isabelle Peschard received her PhD in Philosophy in 2004 (University of Paris I: Panthéon-Sorbonne), after a PhD in Fluid Mechanics in 1995

(University of Provence, Aix-Marseille I). She is Assistant Professor at San Francisco State University since 2008 and is currently engaged in a research project funded by the National Science Foundation (NSF), investigating the experimental dimension of scientific modeling.

Juha Saatsi is Lecturer in Philosophy at the University of Leeds, United Kingdom (UK). He is the co-editor of *The Structural Foundations of Quantum Gravity* (Oxford University Press (OUP), 2006), *The Continuum Companion to the Philosophy of Science* (Continuum Press, 2011), and has published in many of the leading journals in the field.

Michael Stöltzner is Associate Professor of Philosophy at the University of South Carolina. He has studied physics and philosophy at Tübingen, Trieste, Vienna, and Bielefeld and held positions at the Universities of Salzburg, Bielefeld, and Wuppertal. His main areas of research are philosophy and history of physics and applied mathematics, core principles of mathematical physics, history of logical empiricism, the development of formal teleology, and the philosophy of applied science, in particular the role of models and *ceteris paribus* laws.

Patrick Suppes is the Lucie Stern Professor Emeritus of Philosophy at Stanford University. He has published widely in philosophy and the social sciences, especially psychology. He is doing research on the brain, with current emphasis on language and emotion. He is a member of the United States (U.S.) National Academy of Sciences. His last book appeared in 2002, *Representation and Invariance of Scientific Structures*.

Marion Vorms is a postdoctoral fellow at Institut d'Histoire et de Philosophie des Sciences et des Techniques (CNRS, Paris). She holds a PhD in philosophy of science (2009).

Index

'n' refers to a single note and 'nn' refers to multiple notes

A
abstract
  abstract algebra: 25, 27–28, 30–34, 36–38, 39 (n3)
  abstract entities: 209, 269
  abstract model: 91, 213, 245, 247
  abstract representation: 5 , 176
  abstract state: 30–33, 36, 38, 39 (n14), 40 (n21), 48
abstraction: 4, 7, 11, 65, 76, 160, 168, 180, 201, 239–240, 245, 249
accessibility/inaccesssibility: 151, 263
agent-based: viii, 3, 61, 112, 115–119, 121, 123, 125–126, 129, 131–135, 137, 139, 141–144, 176, 178, 180, 184–185, 215, 279, 286
  agent based computational economics: 144, 180
  agent-based modeling/models: viii, 3, 112 (n20), 115–119, 126, 132–135, 139 (n6), 144 196
algebra/algebraic: 25–34, 36–38, 39 (nn2–9, 14), 40 (nn15–16), 41, 147
algebraic quantum field theory: vii, 25, 41
analogy: 3, 5–6, 8, 13, 15–18, 20, 22, 64–65, 67–68, 72, 81, 86, 111, 123, 147–148, 150, 162, 164, 183, 257–258, 264, 267
approximate truth: *see* truth
artifact: xiv, 5, 44–47, 49–54, 56, 58–59, 60 (n4), 78, 160
artificial life: xv, 91–92, 99–100, 107, 112–114, 275
associative learning/association: 188, 201–202
assumption
  applicability assumption: 221, 225–226, 232
  heuristic assumption: 221, 228
  idealizing assumption: 216–218, 220–223, 225–229, 231, 239
  negligibility assumption: 221–224, 226, 229
  early-step assumption: 226–230
  tractability assumption: 168, 228–229, 232
astrophysics: xiii, 145, 151
atom: 15, 64, 71, 73–74, 76, 78, 84, 86, 124, 188, 213, 256
audience: 21, 122, 176, 183, 217–219, 222–225, 227, 230, 232
audition/ auditory: 191, 205, 258

B
Bailer-Jones, Daniela: 17, 22, 69–70, 72–73, 84–85
behavioral: 188–189, 200, 203–204, 206
brain: viii–ix, xi, xvi, 171, 176, 188–189, 191–195, 197–206, 253, 277
brain computer interaction (BCI): 189, 191, 193

C
calibration: 156–157, 161–162, 174, 184 (n10)
Cartwright, Nancy: ii, 11, 22, 55, 60, 140, 142, 214–215, 234, 249
category: 64, 77–80, 86
causal inaccessibility: see accessibility
cellular automata: 97, 16, 107, 111 (n10), 112 (n20), 115, 117, 124, 139 (n5), 141 (n34), 175–176, 180–181, 184 (n8)
classical mechanics: xiv, xvi, 10, 64, 77, 254, 259, 267–268

classical statistical of significance: 192
classroom: 66, 69–72, 76, 86–87, 209, 228
cognition: ii, 13, 181–182, 192, 197, 272–273
cognitive agent: 181, 184
cognitive science: ii, 62, 76, 85–86, 114, 171, 185–186, 201, 243, 250, 264, 273
Collins, Harry: 145–147, 150, 157, 166–167
complex systems: 22 (n6), 91, 97–99, 101, 106–107, 110, 111 (n1), 112, 114, 144, 151, 175, 184, 275
complexity: xv, 6, 8–10, 12, 22 (n6), 94, 95, 99, 108–109, 111, 114, 118, 143, 169, 180, 185–186, 205, 226, 251, 254, 276
  explanatory complexity: 94, 111 (n7)
  incompressible complexity: 99, 109
compressible/incompressible: xv, 93–99, 106–110, 111 (nn9–10), 112 (n11), 120
computational: xv, xvi, 3–6, 8–12, 15–20, 21 (n1), 22–23, 70, 84, 105, 107, 110, 112 (n20), 115, 119, 128, 30, 133–135, 141 (n34), 142, 144, 147–149, 151, 153–155, 159, 161, 165 (n5), 166, 171, 180–181, 185, 191–192, 201–202, 209, 212–213, 214 (nn2, 14), 229, 253, 262, 269, 272, 275–276
  computational device: 3, 155, 161
  computational economics: 144, 180, 185
  computational embodiment: 106–107
  computational method: 3, 20, 153
  computational model: xvi, 4, 23, 70, 84, 119, 185, 192, 209, 212–213, 214 (nn2, 14)
  computational physics/physicist: 142, 147–148, 151, 153
  computational power: 168, 201, 229
  computational procedure: 159
  computational science: xv, 166, 181, 185, 215, 272, 275, 276
  computational system: 105–106, 112 (n20)
  computational template: 3–6, 8–12, 15–21 (n1), 22 (n6), 149, 154–155, 159, 165(n5)

computational tractability: 134, 229. *See also* tractability.
computational template: 3–6, 8–12, 15–21 (n1), 22 (n6), 149, 154–155, 159, 165
computer simulation studies: 145, 147, 149–150, 153, 158–15
conceptual: ii, viii, xiv-xv, 5, 20, 25–26, 37–38, 41, 42, 45, 56–57, 184 (n3), 193, 212, 234–237, 239–240, 244, 246–247, 248 (nn3, 11), 257, 270 (n11), 272
condensed matter physics: 161, 238
connection: 80, 117, 188, 219, 272
consistency/inconsistency: 66, 76, 79, 81–83, 92–93, 96, 98–99, 110–111, 112 (n14), 156, 174, 234, 266, 271 (n29)
content: xvi, 25–27, 30, 32–33, 35–38, 112, 147–148, 168, 177, 183, 201, 205, 220, 240, 254, 258–263, 269, 270 (n11), 271 (n22)
  accessible content: 254, 260, 262–263, 271 (n22)
  informational content 254, 260, 261–262, 271 (n22)
  physical content: 26, 36
  propositional content: 177, 261
  representational content: xvi
  theoretical content: 25–27, 30, 35–38, 254
Contessa Gabriela: 214 (n5)
controversy: 42–43, 45, 47–48, 51–52, 91, 150–151, 156, 165 (n4), 166
coupled/coupling: xvi, 12, 60, 195, 200, 203–204
credit default swaps: 130–131
cross-disciplinarity: 3–6, 21, 175, 214
Culp, Sylvia: 145, 157, 166

D

data: ix, xi, xix, xvi-xvii, 7–8, 16–17, 21, 42, 46–47, 52–53, 61, 63, 68, 70, 73–76, 107, 142 (n37), 145–147, 149–154, 156–158, 160–166, 176, 180–181, 191–193, 195, 197–200, 204, 212, 223, 250, 253–254, 257–258, 260, 262–263, 269, 270 (nn5, 7). *See also* phoneme, signaling.
  brain data: ix, xi, 195, 197–200
  data analysis: 21 (n1), 52
  data-technique circles: 145–146, 156–157, 160–161, 164–166

Index

empirical data: 7, 17, 63, 68, 70, 150–152, 154, 162, 223
experimental data: 74, 147, 152, 161–162, 192
model of data: 61
sparsity of empirical data: 150, 152
De Chandarevian, S.: 214 (n3), 215
DeepView: 212–213
dependence: ix, 14, 92–93, 97, 110, 120–127, 129–133, 135–137, 139, 140 (n12), 141 (nn21, 26, 31), 142 (n35), 152, 157, 158, 160, 162–163, 177, 218, 222, 257, 262
   cross-level dependence: 126, 129, 133
   ontological dependence: ix, 120, 122 (fi6.4), 123, 126, 136, 141 (nn21, 26)
   simple dependence: 127, 129, 131, 141 (n31)
deoxyribonucleic acid (DNA): 157, 211–212, 238
detection/detectors: 109, 141 (n34), 146–147, 150, 158
determinable vs. determinate properties: 236–237, 249
   psychology: 62, 75–76, 85–87, 186, 188, 192, 206, 277
      developmental psychology: 62, 269
      experimental psychology: 192
diagram: 76, 126, 177, 251
discontinuity: 42–43, 45–56, 58, 76, 253–259, 262–263, 265–269, 270
disorder: 98–100, 102–103, 105, 108–110, 112 (n17)
DNA: see deoxyribonucleic acid
domain-specific 65, 84 (n3), 85
Duhem-Quine thesis: 146, 271
Dyson, Freeman: 256–257, 265–267, 270 (n11), 271 (n25), 272

E

education/educational: vii; xiv, 62, 65, 66, 68–71, 73–74, 76, 84 (n2), 85–87, 231
   educational science: 62
electroencephalography (EEG): ix, 189–192, 197, 206
Elgin, Catherine: 171, 173, 185, 216, 224, 232
emergence: ii, vii, xv, 2, 62, 91–99, 101, 105–111 (nn2, 5, 7–9), 112 (nn13, 14, 17), 113–114, 150, 175, 186, 249, 275
   weak emergence vii, xv, 91–99, 101, 105–111 (nn2, 7–9), 112 (nn13, 17), 113
   degrees of emergence: 95
   robustness of emergence: *see* robustness
empirical: see data, empirical
empirical adequacy: 70
epistemic opacity: 169, 177
epistemological anthropocentrism: 182
evolution: ii, 13–15, 17, 26, 29, 43–47, 49–52, 54, 56, 100–103, 105, 109, 113–114, 117, 127, 135, 157, 185–186, 231, 254, 259–260, 264, 270 (n14), 272, 275. *See also* evolutionary.
   evolution of evolvability: 101–102, 105, 109, 119
   dynamical evolution: 231, 254–255
evolutionary: 23, 99–103, 105, 109, 113, 185
existence claim: 150–153
experiment/experiment-al,-ing,-ation: vii–viii, xiv–xv, 42, 44, 46, 48 51, 53, 59, 69 145–147, 149, 153, 155–158, 164, 166–167, 177, 179, 186, 188–189, 191–195, 197, 200, 202, 205, 277. *See also* brain computer interaction (BCI), electroencephalography, experimentation, functional magnetic resonance imaging, magnetic encephalography, numerical, positron emission tomography.
Amyotrophic Lateral Sclerosis (ALS): 191
behavorial experiment: 188. *See also* behavioral.
brain experiment: viii, xi, 188, 197, 200
experiment in physics: 166
experimental activity: 42, 44, 46, 51, 59, 158
experimental set-up: 48, 53
experimenter's regress: viii, xv, 145–147, 149, 153, 155–158, 164, 166
experimentation: xiv, 42, 44–45, 49, 51–61, 64, 69, 71, 74–75, 83, 101–102, 107, 113–114, 145–165, 165 (nn2, 8, 9), 166–167, 177, 179

282 Index

expert / novice: xiv, 64–65, 67, 69, 70–71, 75, 77, 84–85, 87, 107, 176, 265, 267–268
expertise: 265
explanation: xv–xvi, 8, 16, 22, 49–50, 57–58, 65, 78, 83, 86, 91, 93–99, 103, 106–113, 140–144, 163, 170, 172–173, 177, 183, 184 (n3), 185–186, 201, 248 (n11), 249, 251, 259, 273, 276
 compressible explanation: 95, 98–99, 107–108
 incompressible explanation: xv, 95, 99, 107–108, 110, 111 (nn9–10), 112 (n11)
 generative explanation: 94–97
explanatory relevance: 172, 179–180
exploratory: 44–45, 51, 59, 159
extended cognition: 182

F

falsehood: 119, 168, 216–218, 220, 224
 apparent: 216, 218, 220
Feynman, Richard: 255–257, 265–269, 271 (nn23, 25), 272
Feynman's diagrams: 256–257, 265–269
format: xvi, 250–251, 253–255, 257–260, 262, 264–265, 267–269, 270 (nn4, 7), 271 (n22) frequency: 42–52, 54–59, 60 (nn4), 61, 202–204
Fresnel, Augustin: 243, 245–247, 249
Friedman, Milton: 173, 185, 216, 231, 232–233, 276
functional magnetic resonance imaging (fMRI): 189

G

Game of Life: 111 (n10), 106
Gaster, Michael: 49–50, 57, 60
general systems theory: 13–14, 17
Giere, Ronald: viii–ix, xvi, 52, 55, 60 (n1), 61, 62–65, 68–69, 71–72, 77–78, 84 (n5), 85, 182, 185, 209, 212 (nn4, 6), 215, 234–235, 237, 249, 275
Godfrey-Smith, Peter: 214 (n8), 215
Goodman, Nelson: 258–259, 270 (n16), 272
goodmanian: 258–259
graph dynamical systems: 117, 127, 139 (n6)
gravitational waves: 146–147, 150, 158, 215

H

Hamilton, William Rowan: 267–268
Hamiltonian (mechanics, equations, coordinates, etc.): xvi, 254, 256–257, 267–268, 270 (n3)
harmonic oscillator: see oscillator
Haugeland, John: 261, 272
Hempel, Carl: 91, 98, 113, 170, 185
Hindriks, Frank: 216, 228–229, 232
history of science: 69, 71, 80, 83, 172
Hopwood, Nick: 214 (n3), 215
Humphreys, Paul: ii–iii, xvii, 3–4, 10, 16, 18, 21, 23, 91, 112–114, 149, 154, 161, 163, 165 (n5), 166, 168, 175, 177, 181, 185, 212, 215, 251, 256, 269, 272, 275

I

idealization: viii, xiii, xvi, 4, 7–8, 25, 30–32, 38, 41, 63, 70, 132, 141, 144, 160, 168, 215, 216–223, 225–226, 228–229, 231–233, 234–236, 238–242, 245–247, 248 (n3, 12), 249, 260, 264–265, 267–268
 Galilean: 141, 228–229
image: ix–x, 106, 182, 191–193, 202, 212–213,214 (n14), 252, 268, 271 (n20), 273
imagination/imaginary/imagined: 13, 20, 44, 48, 54, 152, 189, 192–193, 209–211, 214, 217–219, 222
in practice/in principle: 93–98, 112 (n15), 154, 162, 168, 174–175, 178, 184 (n7), 213, 239, 245, 256, 257, 260, 262, 264, 266
independence: 152, 157–158, 160, 162–163, 257
inference: 66, 85–86, 148, 166, 172–173, 175, 177, 179–181, 184 (n4), 217, 219, 250, 253, 261–262, 264–265, 268, 270 (nn4–5), 271 (nn22, 30)
information: 26–28, 32, 34, 39 (n10), 40 (n8), 45, 52, 63, 65, 76, 81, 94, 96, 103, 106–107, 109, 111 (n11), 122,131, 142, 148, 167, 168–169, 171, 173, 180, 191, 205, 206, 212, 216, 220–221, 250–251, 253–255, 257–258, 260–265, 267, 270 (nn4–5, 13, 16), 271 (n22), 272–273

intentional: xiii, xvi, 124, 170, 210, 212–214
interdisciplinary: 3, 18, 20
intrinsic: 16, 42, 48, 52–53, 56, 58, 108, 111 (n9), 114, 117, 120, 130, 135–136, 140 (nn18, 20), 156, 183 (n1)
isolation: 21, 153, 218, 220, 232, 233
isomorphism: 34, 40 (n16), 63, 194–195, 201, 210, 215, 248 (n3), 272–273
   partial isomorphism: 248 (n3)
   structural isomorphism: 194–195, 201

J

Jones, Martin: 215, 239–240, 242, 245–246, 249

K

Kaiser, David: 256–257, 266, 270 (nn9–11), 271 (nn23–24, 29), 272
Keil, Frank: 174, 185–186
Keynes, J.M.: 225, 232
knowledge: 7, 65–67, 71–72, 77–78, 81–82, 85–87, 98, 114, 137, 159–160, 165 (n8), 169–173, 175–179, 182–183, 186, 243, 250, 253, 260–261, 264, 268–269, 270 (n7), 272
   teachers' knowledge: 86–87
Kuhn, Thomas: 3, 23, 68, 81, 85 (n11), 86, 271 (nn27, 29), 273
Kuhnian gestalt switches: 62
Kulvicki, John: ix, x, 252, 258, 263, 269, 270 (n15), 271 (n20), 273

L

Lagrangian: xvi, 25, 41, 254–257, 265, 267–268, 270 (n8)
Landau (model): 42, 46, 51–52, 59, 61
language: xiii, 7, 11, 63, 86, 188–189, 191, 193–194, 204–206, 210, 214 (n6), 215, 235, 254, 258, 272, 273, 275, 277
   brain language: 193
Larkin, Jill: 250, 253–254, 262, 270 (nn4–5), 271 (n19), 273
learner: 69, 78, 84
learning: ix, xiv, 64–67, 69, 71, 74–76, 79–82, 84–85, 87, 103, 159, 174, 181, 192, 200–204, 206, 227, 230, 265, 268–269, 271 (n27), 275

Lenhard, Johannes: 55, 61, 168–169, 173, 175, 177, 186
level
   phenomenological level: 64
   cognitive level: 66
   cross-level: 125–126, 129, 133, 135, 116
   macro-level: 99–101, 104, 107, 109, 116
   micro-level: 94, 96–98, 100–101, 105–106, 108–109, 116
   system level: 175
locality: 120–121, 123, 126, 133, 139, 143. See also nonlocality.
logical positivism: 73
Lotka, Alfred: 6–7, 13- 19–23
Lotka-Volterra model: xiv, 3–4, 6, 8, 13, 18–21 (n2)

M

macro-property: 94–95, 97, 108, 110, 111 (n9), 115–116, 123, 126, 132–135, 142 (nn35–37)
magnetic encephalography (189)
maker's knowledge:169, 178–179
many-body models: 151–152, 161–162, 166
Mayr, Ernst: 109
measurement: 22, 44–47, 49, 51–52, 56–60 (n4), 61, 70, 113, 142, 145–146, 158, 163,166, 196, 214 (n14), 253, 260, 272
mechanical: 8–9, 23, 46, 54, 56, 78, 185
mechanics:
   classical mechanics: xiv, xvi, 10, 64, 77, 254, 259, 267–268
   fluid mechanics 60–61, 276
   quantum mechanics: xiv, 25, 32, 35–36, 41, 64, 74, 91, 256, 272
   statistical mechanics: 14–15, 17, 39 (n7), 41, 85 (n9), 167
mechanism: vii, xiv, xvi, 3, 5–6, 8, 9, 11, 15–17, 19–21, 21 (n2), 22 (n6), 23, 77, 100–101, 105, 114, 151, 172, 174, 177–178, 181, 183, 186, 202–203, 216, 218, 222, 233, 276
metamodel: 169, 180, 184 (n10), 185
method (of)
   axiomatic analysis: 200, 254
   choice of method: 192
   classical statistical method: 192
   component analysis: 192
   conditioning: 188, 200–202

formal clarity: 200
linear discrimination classification (LDC): 192
measurement theory: 196
method of hypothesis: 21 (n2)
reinforcement: 181, 200–201
stimulus: 188, 191–193, 200–204, 206
stimulus sampling: 200, 206
statistical: 192
Surface-Laplacian (SL): 192
methodology: 61, 74, 84 (nn7, 8), 86, 93, 115, 119, 143, 166, 170, 185–186, 232–233, 276
microfoundations: 115, 133, 143
micro-property: 142 (n37)
misconception: 78–83, 115
models
  abstract model: 91, 213, 245, 247
  audience of a model: *see* audience
  autonomy of models: 65
  "ball and stick" model: 213
  biological model: 19
  boundary layer model: 63–65
  classroom models: see classroom
  computational model: xvi, 4, 23, 70, 84 (n9), 119, 185, 192, 209, 212–213, 214 (nn2, 14)
  concept of model: 63
  conceptual categories and models: *see* category
  consensus model: 66–67, 69, 75
  consistency of different models: *see* consistency
  dynamic model: 34
  enrichment of models versus revision: 81
  equivalent model: 37
  functional decomposition approach about modeling: 217, 231, 233
  hybrid model: 74–75
  intermediate model: 67–69, 73–76, 78, 81–83
  isomorphism between a model and x: *see* isomorphism
  Landau model: see Landau
  learning model: 201. *See also* association, learning.
  Lotka-Volterra model: *see* Lotka-Volterra
  mental: ix, 65–67, 80–83, 84 (nn5–6), 85–87, 171, 183 (n2)
  minimal model: 33, 37, 39, 85, 165, 223, 226
  model and learning: *see* learning
  model and reality: 75
  models and theory: 68–69, 77
  models as measurement devices: 70
  model commentary: 217–221, 225, 230
  model construction: 16, 67–68, 120–121, 141 (n30)
  model dynamics: vii, 73, 75, 80, 83–84, 100
  models in the history of science: *see* history of science
  models of the atomic nucleus: 64
  numerical model: 152
  phase oscillator: xvi, 200, 203, 206
  physical model: viii, xvi, 65, 70, 92, 202, 209–214, 214 (n3)
  purpose of a model: 217–219, 222–225, 227, 229–230
  revision of models: *see* revision
  stimulus-response model: *see* stimulus response
  student ownership of models: 71, 77. *See also* student.
  substitute system: 219
  surrogate system: 219, 233
  synthetic model: 23, 83
  target model (of instruction): 66–67, 69, 75, 154
  theoretical model: xv, xvi, 4, 42–44, 51–52, 55, 63, 76, 78, 151, 159, 209–211, 213, 216, 222–223, 250, 254, 270 (n7)
  truth and model: *see* truth
Models as Mediators: 22, 63–64, 69, 71, 83, 86, 215
model builders: 175
  constructivist attitude of model builders: 71, 73
Monte Carlo simulation: 120, 147, 154, 160
Morgan, Mary: 22, 63–64, 69–70, 72, 76, 78, 84, 86, 214–215
Morrison, Margaret: 22, 45–46, 61, 63–64, 69–70, 72, 76, 78, 84, 86 148, 159, 166, 214–215, 234, 249
multiple realization: 97, 248 (n5)
Musgrave, Alan: 86, 221, 228, 231–233
mutation rate: ix, 101–105, 108–110, 112 (n18), 113

N

Navier-Stokes equations: 46, 64, 73

network: 21: 94, 106, 113, 117, 128, 139 (n6), 142, 185
neural: 16, 21, 115, 135, 171, 191, 201, 205
  neural oscillator: 203–204, 206
  neural network: 21 (n1), 205
  neural process: 171, 264
neuron/neuronal: 65, 188–189, 203–206
neuroscience: xvi, 22, 186, 188, 201
non-linearity / non-linear: 12, 18–19, 21, 49, 51, 73, 106, 176
nonlocality: 121, 123
normative: xiv, 46–47, 53, 55–56, 59, 184, 265, 267–268
novice: *see* expert
null hypothesis: 192
numerical: 42, 43, 46, 53–54, 85, 113, 147–155, 161–162, 176, 196, 204, 247, 253, 258
  numerical experiment: 148–150
  numerical model: 152
  numerical simulation: 42, 46, 53
  numerical procedure/software/technique: 113, 153, 155

O

ontological commitment: 67, 69, 81
ontological individualism: 119, 140 (nn12, 13), 143
ontology: 4, 80, 115, 119–120, 127–135, 141 (n34), 143, 144
  base ontology: 128–132, 133–135, 141 (n29)
  causal ontology: 126–131, 133–135, 141
  modeled ontology: 127–129, 132–134, 141
  target ontology: 127, 131, 132, 135
Oppenheim, Paul: 91, 98, 113, 124, 141, 143
Oppenheim-Putnam: 124, 134–135
order
  partial order: ix, 197–198, 206
  semiorder: ix, 194, 196–200
oscillation/oscillator/oscillatory: xvi, 8–10, 12–13, 15, 19–21, 23, 176, 200, 202–206, 209–210. *See also* coupling.
  (un)-damped oscillator: 12–13, 15, 23
  harmonic oscillator: 12, 209–210
  neural oscillator: 203–204, 206

P

Packard's Bugs (model): 92, 95, 98–99, 101, 104, 105–106, 108, 110

Index 285

Packard, Norman: 95, 100, 102, 112 (n18), 113–114
paradigm: 18, 21, 65, 68, 81, 86, 95, 99, 209, 247, 273
parameter: xiv, 14, 29, 34, 42, 44, 46–47, 51–54, 56–60, 63, 116, 133, 142 (n37), 149, 152, 155–156, 160–161, 177, 181, 197, 202, 205
  relevant (parameter): xiv, 42, 44, 46–47, 52–54, 56–60
phase transitions: 102, 151–152, 160, 162, 164
phoneme: 188, 194–195, 197, 200
positron emission tomography (PET): 190
practice: see also in principle/in practice.
  experimental practice: 149, 164, 202, 223,
  modeling practice: xv, 4–6, 16–17, 20, 223, 231, 245–246
pragmatic: xvi, 125, 149, 170, 173–174, 177, 181, 218–219, 222–224, 229, 231, 236, 239, 245–246
pragmatics: 218, 223–224
predator-prey
  predator-prey dynamics: 6, 9, 15–16, 115
  predator-prey system: 7, 8, 12, 18, 20
probability: xi, 4, 178, 192, 195–197, 201, 203. *See also* null hypothesis.
  conditional probability: xi, 195–197
problem of relevance: see relevance
property: *see also* see micro-property, macro-property.
  local/nonlocal property: 112 (n20), 121–123, 135–136, 139, 140 (n20)
  intrinsic property: 117, 135–136, 140 (n20)
  social property: xv, 119–126, 129, 133–139, 140 (nn12, 14, 20), 141 (n26)
  topological property: 238, 243, 245, 246
  typical property: 236
protein: 106, 212–213
psychological: xv, 62, 65, 78, 81, 83, 106, 124–125, 130, 169–171, 173–174, 182–183 (n6), 189, 194, 196, 206, 236, 256, 265, 269, 272, 273

psychology: 62, 75–76, 85–87, 186, 188, 192, 206, 277
developmental psychology: 62, 269
experimental psychology: 192
psychology of learning: 11, 75, 76

Q
quantum field theory: vii, 25, 40–41

R
rational reconstruction: 66, 73, 75–76, 84
realism: 41, 70, 73, 234, 247–248 (n6), 249, 272, 276
realist: 73, 172, 231, 234, 238, 241, 243, 245–247, 248 (nn2, 11)
realistic (assumption, model, simulation, etc.): 4, 6, 8, 16, 73, 101, 134, 154, 216, 218–219, 226–227, 231–232, 243
reasoning: xiv, 15, 60, 85–87, 124, 171, 177, 185, 249–251
recognition: 13, 192, 197
recording: ix, 189–191, 197
reduction: 50, 57, 94–97, 109, 112 (n14), 113–114, 133, 249
  in principle reduction: see in practice/in principle
  in practice reduction: see in practice/in principle
reductionism: 5, 19, 23, 41, 109, 114, 140 (n12), 186
redundancy: 115, 120, 132–135
reinforcement: see method of reinforcement
relevance: 45, 58–59, 62, 119, 170, 172, 179–180, 240, 250. See also explanatory relevance.
reliability: 46, 49, 154, 167, 177
replicability: 46, 49, 145–146, 148–150, 153–156
representation:
  (in)-equivalent representations: 25–26, 30, 33, 35–38, 41, 259, 264
  inferentially veridical representation: viii, xvi, 236, 240–243, 246, 248 (nn4, 6)
  intentional conception of representation: 212
  (non)-linguistic representation: 211, 234–235, 241, 251
  mental representation: 183, 251
  misrepresentation: 237–240, 246, 261
  non-linguistic representation: 234–235, 241, 251
  partially veridical representation: 234, 241
  pictorial representation: 251, 273
  theoretical representation: xvii, 269
  visual representation: xiv, 175, 273
research programs: xiv, 20, 62, 73–74, 76, 81, 84, 86
response: 70, 82, 104, 189, 191, 193, 195, 197, 200–204, 206
response to an image: 193
revision: 68, 75, 78, 81, 149, 216
revolution: 68, 78, 80, 86, 107
rigorous results: 145–146, 161–167
robust: 24, 66, 79, 97–98, 102, 104, 108, 110, 114, 145, 156–157, 182, 200, 203, 205
robustness: 145–146, 156–158, 161, 163–164, 167, 184 (n10), 205, 229
Roschian: 65, 77
Roshko: 47–49, 61

S
sample/sampling/sampled: ix, 70, 163, 195, 197, 199–204, 206. See also stimulus sampling.
Schelling: 122, 144
Schwinger, Julian: 30, 256–257, 266, 272–273
scientific revolution: see revolution
semantic view: 63, 69, 78, 83, 235, 237, 248 (n3), 272
signaling (of neurons): 188–189, 191
similarity: xvi, 63–64, 143, 148, 194–197, 200, 206, 210–213, 215, 235–236, 273
  degrees of similarity: 210
  selective similarity: 210, 213–213
Simon, Herbert: 13, 19, 23, 147, 167, 184 (n11), 186, 250, 253–254, 262, 270 (nn4–5), 271 (n19), 273
simulation: see also Monte-Carlo, numerical.
  agent-based simulations: 143, 144, 185
simulationist's regress: xv, 145–146, 149–150, 152–153, 156, 158, 162, 164
statistical mechanics: see mechanics
Steinle, Friedrich: 44–45, 51, 59, 61
stimulus: 188, 191–193, 200–204, 206. See also response, sampling.
stimulus response model/theory: 200, 206

stimulus sampling: 200–206
Strevens, Michael: 141 (n32), 144, 242, 248 (n10), 249
student: xiv, 66–67, 69, 71, 77–83, 85, 176, 209, 219, 230, 268,
Suárez, Mauricio: ii, 210, 214 (n9), 215, 233, 248 (n4), 249–250, 273
Sugarscape model: 115–116
supervenience: xv, 114, 119, 135–139, 140 (n13), 141 (nn26, 31), 142–143, 236
systems and synthetic biology: 21, 276
Swarm (modeling program): 118–119, 139, 143
SWISS-MODEL: 212
symbol system: 251, 257–260, 262, 265, 270 (n5)

T
teacher: 66, 70–73, 75, 86–87, 260
teachers' knowledge: 86, 87
Teller, Paul: 60 (nn1, 3), 140 (n17), 144, 168, 186, 235, 237, 241, 249
template
  computational template: 3–6, 8–12, 15–21
  theoretical template: 4, 20, 154
theory-testing: 44–45
theorizing: viii, 6, 120, 149, 235, 243, 250, 265, 267–268, 269
thermodynamics: 14
Thompson-Jones, Martin: see Jones
thought experiment: 54, 69, 101
Tomasello, Michael: 214–215
tractable/tractability: xv, 4–6, 12, 18, 20–21, 38, 114, 117, 128, 130–131, 133–135, 154–156, 168, 172, 180, 226, 228–232, 256, 259, 266
  computationally tractable: 128, 154, 229
  tractable model: 128, 180
  tractable simulation: 172
  analytic tractability: 168
traffic jam: 91–93, 96, 98–99, 106–108, 110, 111 (n4)
Tritton, David J.: 48–50, 52
Trout, J.D.: 170–171, 184, 186
truth: viii, xvi, 63, 95, 156, 210, 216–233
  approximate truth: xvi, 234, 240
  partial truth: xvi, 248 (n3), 249
  relevant truth: 217, 219–222, 227, 231–232

true enough: 224, 232
truth re-nomination: 219–220, 224, 231
  by paraphrase: 220–222, 225–226, 228, 231–232
  by meta-claim: 220–221, 226, 228, 230–232

U
understanding
  illusion of understanding: xv, 169, 173, 175–176, 178
  pragmatic conception of understanding: 173, 177
  psychology of sense of understanding: xv, 169, 182–183
  sense of understanding: xv, 169–171, 173–176, 179, 182–183, 183 (n2), 184 (n5), 186
  theoretical understanding: 43, 73
  understanding, as inferential ability: 172–174, 180
unitarily inequivalent representations: 25–26, 30, 33, 35–38, 41
universal register machine: 201
universal Turing machine: 201

V
van Fraassen, Bastien. C.: 214
verisimilitude: 252
vesicle: 92, 93, 96, 98–99, 110
visual/visualization: xiii–xv, 43, 48, 57, 63, 98, 107, 169, 175–177, 183, 191, 213, 257–258, 264–266, 273
Volterra, Vito: xiv, 6, 7–13, 15–20, 21 (n2), 22 (n4), 23–24, 139, 144
von Bertalanffy, Ludwig: 13–14, 16, 24

W
wake: ix, 42–61
Weisberg, Michael: 21, 24, 141, 144, 177, 184, 186, 223, 228–229, 233
Williamson, Charles H. K.: 44, 48, 52, 56–58, 60–61
Winsberg, Eric: 150, 159, 161, 167, 175, 186
Wittgenstein, Ludwig: 169–171, 184, 186, 215
Wolfram, Stephen: 106–107, 112 (n11), 114
Woodward, James: 46, 60, 172–173, 186